CONTEMPORARY ISSUES IN ROAD USER BEHAVIOR AND TRAFFIC SAFETY

Contemporary Issues in Road User Behavior and Traffic Safety

Dwight A. Hennessy
David L. Wiesenthal
Editors

Nova Science Publishers, Inc.
New York

For permission to use material from this book please contact us:
Telephone 631-231-7269; Fax 631-231-8175
Web Site: http://www.novapublishers.com

NOTICE TO THE READER

Library of Congress Cataloging-in-Publication Data
Hennessy, Dwight A.
Contemporary issues in road user behavior and traffic safety / Dwight A. Hennessy and David L. Wiesenthal.
p. cm.
Includes index.
ISBN 1-59454-268-6 (hardcover)
1. Motor vehicle drivers--Psychology. 2. Traffic safety. I. Wiesenthal, David L. II. Title.
TL152.35.H45 2005
629.28'304--dc22 2005000186

Published by Nova Science Publishers, Inc. ✦ New York

DEDICATIONS

This book is dedicated to the memory of Professor Jim Rotton, late of Florida International University, whose sudden death occurred shortly after sending us his two chapters. We both first corresponded with Jim over his organizing a paper session on driver research for the American Psychological Association's conference in Toronto last year. Despite the possible cancellation of the meeting because of the SARS outbreak, terrorism induced travel difficulties, along with tight budgets, Jim was unflappable and his energy provided sufficient force for the session to be proceed. A long, leisurely lunch followed our session where we had the opportunity to chat. One thing led to another, and before long, Jim had agreed to start writing and organize the work he'd been doing with his graduate students, Paul Gregory and David Van Rooy, which we are delighted to feature in this collection.

Jim was a productive researcher whose many publications in social and environmental psychology are testimony to both his productivity and curiosity. His interest in the relationship between temperature and crime, the behavioral effects of air pollution, stress following natural disasters, the soundness of alleged lunar influences on interpersonal behavior, and various aspects of scholarly publication stands in testimony to his lively mind. His publications reflect the broad nature of his interests and were usually co-authored with his students. His mentorship of students over a career spanning three decades will not be forgotten.

Thanks, Jim, this one's for you.

Dwight A. Hennessy and David L. Wiesenthal

Most specifically I would like to dedicate this to Sandra for tolerating my random rantings on the psychology of driving and to AJ^2 for reminding me about the pure joy of learning. I am always grateful to my co-editor David Wiesenthal for his years of mentoring and collaboration. Finally, to all the authors that contributed their ideas, research, and time to this project – I firmly believe we have assembled some of the most influential minds in traffic safety research.

Dwight A. Hennessy

I would like to acknowledge and thank my wife Sandy for the steadfast love, encouragement, cheer, and humor given to me over the many years we've been together. Her support and unfailing tolerance made this book (and so much more in our lives!) possible. I would also like to acknowledge the considerable infusion of ideas, energy and enthusiasm that Dr. Dwight Hennessy brings to all his undertakings—without them, this volume would not have been possible. I am truly fortunate to have such a talented collaborator. I would also like to acknowledge the considerable skills of Ms. Judy Manners, who always greeted my last minute word processing problems with a smile and prompt remedy. Finally, a special thanks to all the students I have been privileged to work with both at York and abroad.

David L. Wiesenthal

CONTENTS

Dedications v

Foreword xi

Introduction xiii

PART 1 THEORETICAL PERSPECTIVES/MODELS **1**

Chapter 1 On the Homeostasis of Risk **3**
Gerald J. S. Wilde

Chapter 2 The Influence of the Actor-Observer Bias on Attributions of Other Drivers **13**
Dwight A. Hennessy, Robert Jakubowski and Alyson J. Benedetti

PART 2 ANGER/AGGRESSION **21**

Chapter 3 Motorists' Perceptions of Aggressive Driving: A Comparative Analysis of Ontario and California Drivers **23**
Christine M. Wickens, David L. Wiesenthal and Kathy Rippey

Chapter 4 Behind the Wheel: Construct Validity of Aggressive Driving Scales **37**
James Rotton, Paul J. Gregory and David L. Van Rooy

Chapter 5 Understanding and Treating the Aggressive Driver **47**
Tara E. Galovski and Edward B. Blanchard

Chapter 6 On the Road: Situational Determinants of Aggressive Driving **61**
David L. Van Rooy, James Rotton and Paul J. Gregory

Chapter 7 Field Methodologies for the Study of Driver Aggression **71**
Andrew R. McGarva

PART 3 DRIVING VIOLATIONS AND COLLISIONS **79**

Chapter 8 Observing Motorway Driving Violations **81**
A. Ian Glendon and Danielle C. Sutton

Chapter 9 Traffic Safety in Hong Kong: Current Status and Research Directions **101**
J. P. Maxwell

Chapter 10 Speeding Behavior and Collision Involvement in Scottish Car Drivers **113**
Stephen G. Stradling

Chapter 11 The Use and Misuse of Visual Information for "Go/No-Go" Decisions in Driving **125**
Rob Gray

Chapter 12 Road Safety Impact of the Extended Drinking Hours Policy in Ontario **135**
Evelyn Vingilis, Jane Seeley, A. Ian McLeod, Robert E. Mann, Doug Beirness and Charles Compton

PART 4 ALCOHOL AND DRIVING **151**

Chapter 13 Characteristics of Persistent Drinking Drivers: Comparisons of First, Second, and Multiple Offenders **153**
William F. Wieczorek and Thomas H. Nochajski

Chapter 14 Personal Drinking and Driving Intervention: A Gritty Performance **167**
J. Peter Rothe

PART 5 TREATMENT/DRIVER CHARACTERISTICS **183**

Chapter 15 The Effects of Multiple Variable Prompt Messages on Stopping and Signalling Behaviors in Motorists **185**
David L. Wiesenthal and Dwight A. Hennessy

Chapter 16 Early Indicators and Interventions for Traumatic Stress Disorders Secondary to Motor Vehicle Accidents **199**
Connie Veazey and Edward B. Blanchard

Chapter 17 Supplemental Speed Reduction Treatments for Rural Work Zones **215**
Eric D. Hildebrand, Frank R. Wilson and James J. Copeland

PART 6 ENGINEERING/HUMAN FACTORS **225**

Chapter 18 Is it Safe to Use a Cellular Telephone While Driving? **227**
David L. Wiesenthal and Deanna Singhal

Chapter 19 Cognitive Distraction: Its Effect on Drivers at Intersections **245**
Joanne L. Harbluk and Patricia Trbovich

Chapter 20 The Use of Event Data Recorders in the Analysis of Real-World Crashes: Tales from the Silent Witness **251**
Kevin McClafferty, Paul Tiessen and Alan German

Chapter 21 The Effectiveness of Airbags for the Elderly **263**
Eric D. Hildebrand and Erica B. Griffin

Chapter 22 The Role of Control Data in Crash Investigations: Haddon Revisited **275**
Mary L. Chipman

Index **285**

FOREWORD

This book contains 22 contributions from authors working in several countries in various disciplines of road safety including vehicle and road engineering, collision investigation, epidemiology, ergonomics and psychology. It represents the state of the art in theory and research on driver behavior such as impaired driving, aggressive driving, speeding, as well as the interface between the vehicle and the driver. Several contributions also advance the methodology used to conduct road safety research (e.g. event data recorders) and others address the effects of technology such as air bags and cell phones on road safety. This book will serve as an excellent text for courses in road safety and applied psychology.

Currently, about 2800 Canadians are killed and another 17,000 are seriously injured annually on our roads. It is estimated that this carnage costs our economy as much as $25 billion dollars each year. In addition, there is the enormous pain and suffering that these collisions cause for the victims and their loved ones.

Canada's Road Safety Vision 2010 is aimed at Canada's roads being the safest in the world by 2010. This vision is being pursued by the federal, provincial and territorial governments in partnership with a variety of non-governmental partners. In 2002, Canada ranked eighth among member countries of the Organization for Economic Cooperation and Development behind countries like the United Kingdom, Sweden and the Netherlands and slightly ahead of the United States, based on the number of fatalities/100 million kilometers traveled. In order to advance toward our vision, Canada must affect a 30% reduction in fatalities and serious injuries. Hence, it is important to know what has been effective in other OECD countries in reducing collisions and casualties and this volume assists in the diffusion of these best practices.

The World Health Organization recently published a report about the state of road safety internationally noting that by 2020, road injuries are predicted to rank as the third leading cause of disease or injury surpassing HIV, cerebral vascular disease and respiratory infections. The situation within developing countries is becoming critical as these countries motorize, resulting in skyrocketing levels of injuries and fatalities, particularly among the vulnerable road users such as pedestrians and cyclists who must interact with the growing vehicular traffic. April 7, 2004 was designated by the United Nations as World Health Day, a day devoted to road safety, the theme of which was "Road Safety is no accident". Many countries have national road safety plans with targets and strategies and look to best practices in other countries to see what works and what doesn't. This book will also assist those policy makers in these countries decide how best to address their road safety needs.

Brian Jonah, Ph.D.
Director, Road Safety Programs
Transport Canada

INTRODUCTION

This volume presents the work of researchers from around the world and from a variety of disciplines who are actively searching for ways to make our roadways a safer and more pleasant place to be. Although behavioral scientists have long been interested in learning about what drivers do (Allport, 1934) the study of driving behavior has only recently attracted the dedicated interest of psychologists and other researchers. Roadways are now increasingly recognized as an excellent naturalistic setting to study a variety of behaviors that were previously constrained to laboratories. Streets and roads are ubiquitous, constituting an integral part of most people's everyday environment or life space. As with other environmental features, emotional meanings are attached to our subjective perceptions of roadways which ultimately influence immediate and long term thoughts, feelings, and actions (Proshansky, Ittelson, & Rivlin,1970).

The very high traffic volumes that have characterized our urban environments facilitates the study of infrequently displayed behaviors, while the driving context provides a wealth of opportunity to understand the factors that instigate such actions. Hence, with millions of daily commuting trips, rarely exhibited behaviors may be manifested in sufficient quantities to allow for their study. For example, while acts of violence are rarely displayed by the average individual, given the millions of daily highway commuters, the frequency of such acts should allow its scientific examination. Similarly, the desire for revenge can lead to aggression under a variety of conditions that are common to the traffic environment, but difficult to recreate in an ecologically valid manner under controlled laboratory settings (Hennessy & Wiesenthal, 2001a; Hennessy, Wiesenthal & Piccione, 2001; Wiesenthal, Gibson, & Hennessy, 2000). Another important contextual factor is deindividuation, which refers to the behavioral consequences of having an individual's identifiability reduced (Zimbardo, 1969). Tintend windows and night time driving (Wiesenthal & Janovjak, 1992) may facilitate anonymity and increase perceptions of deindividuation, which, coupled with the low probablility of future encounters with other drivers during commuting trips, may increase the possibility of dangerous, antisocial and illegal behavior (Novaco, 1991). The automobile is also both a weapon and a means of escape, making it a power "equalizer" across all drivers. Women's aggression (but not extremely violent responses) have been found to be equal to the levels seen in males (Hennessy & Wiesenthal, 2001b; Hennessy, Wiesenthal, Wickens & Lustman, 2004), suggesting that the vehicle may engender a perception of ability and opportunity to be aggressive, reducing the inhibitions (including gender roles) against proscribed actions.

The importance of understanding the effects of driving are compounded by the fact that it is intricately entwined with other life contexts, such as the workplace and home environments. Daily life hassles have been found to increase the experience of driver stress (Hennessy, Wiesenthal, & Kohn, 2000), which can subsequently increase psychological and physical illness (Gulian, Matthews, Glendon, Davies, & Debney, 1989; Hennessy & Wiesenthal, 1997; Matthews, Dorn, & Glondon, 1991; Novaco, Stokols, Campbell, & Stokols, 1979) and dangerous driving behaviors (Hennessy & Wiesenthal, 1999; Selzer & Vinokur, 1974; Westerman & Haigney, 2000). Further, a stressful commute can spill over into the workplace and increase the probability of verbal, symbolic, and passive forms of aggression (Hennessy, 2003). According to Evans and Phillips (2002), even rail commuters are susceptible to increase stress response, particularly under unpredictable conditions. The perils of the traffic environment, however, are no more discernable and socially relevant than in the risks of physical injury and death from traffic collisions. The World Health Organization (2004) has estimated that more than 20 million people are killed or injured on roadways worldwide, costing over $500 billion annually. Ultimately this is the reason so many researchers have dedicated their time and resources to traffic research – to make the traffic environment safer for all.

This volume describes the growing body of research on driver behavior and traffic safety, including the nature, measurement and treatment of roadway aggression, types of traffic violations in diverse parts of the world, the pervasive concern with the alcohol and driving, attempts to modify problematic driver behaviors, engineering and human factors concerns such as cell phone operation by drivers, the use of vehicle "black box" recorders, and the safety of airbags. We also present some examples of theoretical models and their usefulness in stimulating research and providing an overall explanatory model for a diverse range of driving behaviors.

The chapters in this book explore many of these issues with driver behaviors being investigated by psychologists, sociologists, engineers and others. These authors have contributed the unique perspective of their disciplines, but it will be clear to the reader that these disciplinary perspectives lead to a healthy cross-fertilization of ideas rather than acting as blinders.

References

Allport, F. H. (1934). The J-curve hypothesis of conforming behaviour. *Journal of Social Psychology, 5*, 141-183.

Evans, G. W., & Phillips, D. (2002). The morning rush hour: Predictability and commuter stress. *Environment and Behavior, 34*, 521-532.

Gulian, E., Matthews, G., Glendon, A. I., Davies, D. R., & Debney, L. M. (1989). Dimensions of driver stress. *Ergonomics, 32*, 585-602.

Hennessy, D. A. (2003). *From driver stress to workplace aggression.* Symposium presented at Annual American Psychological Association Convention. Toronto ON, Canada: August 7 – 10.

_____ . (1999). The influence of driving vengeance on aggression and violence. *Proceedings of the Canadian Multidisciplinary Road Safety Conference XI*. Daltech Vehicle Safety Institute: Dalhousie University, Halifax, Canada.

Hennessy, D. A., & Wiesenthal, D. L. (1997). The relationship between traffic congestion, driver stress, and direct versus indirect coping behaviours. *Ergonomics, 40*, 348-361.

_____ . (2001a). Further validation of the Driving Vengeance Questionnaire. *Violence and Victims, 16,* 565-573.

_____ . (2001b). Gender, driver aggression, and driver violence: An applied evaluation. *Sex Roles, 44*, 661-676.

Hennessy, D. A., Wiesenthal, D. L., & Kohn, P. M. (2000). The influence of traffic congestion, daily hassles, and trait stress susceptibility on state driver stress: An interactive perspective. *Journal of Applied Biobehavioral Research, 5,* 162-179.

Hennessy, D. A., Wiesenthal, D. L., & Piccione, G. (2001). Aggressive driving, stress, and vengeance in Canada and Italy. *Proceedings of the Canadian Multidisciplinary Road Safety Conference XII.* University of Western Ontario: London, Ontario, Canada.

Hennessy, D. A., Wiesenthal, D. L., Wickens, C., & Lustman, M. (2004). The impact of gender and stress on traffic aggression: Are we really that different? In S. P. Shohov (Ed.), Advances in psychology research. *Nova Science Publishers.*

Matthews, G., Dorn, L., & Glendon, A. I. (1991). Personality correlates of driver stress. *Personality and Individual Differences, 12*, 535-549.

Novaco, R. W. (1991). Aggression on roadways. In R. Baenninger (Ed.), *Targets of violence and aggression* (pp. 253-326). North-Holland: Elsevier Science Publisher.

Novaco, R. W., Stokols, D., Campbell, J., & Stokols, J. (1979). Transportation, stress, and community psychology. *American Journal of Community Psychology, 7*, 361-380.

Proshansky, H. M., Ittelson, W. H., & Rivlin, L. G. (1970). *Environmental psychology*. New York: Holt, Rinehart and Winston.

Selzer, M. L., & Vinokur, A. (1974). Life events, subjective stress and traffic accidents. *American Journal of Psychiatry, 131*, 903-906.

Westerman, S. J., & Haigney, D. (2000). Individual differences in driver stress, error and violation. *Personality and Individual Differences, 29*, 981-998.

Wiesenthal, D. L., Hennessy, D. A., & Gibson, P. M. (2000). The Driving Vengeance Questionnaire (DVQ): The Development of a scale to measure deviant drivers' attitudes. *Violence and Victims, 15,* 115-136.

Wiesenthal, D. L., & Janovjak, D. P. (1992). Deindividuation and automobile driving behaviour. *The LaMarsh Research Programme Report Series, No. 46*, May 1992, LaMarsh Research Programme on Violence and Conflict Resolution, York University, North York, Ontario, Canada.

World Health Organization (2004). Road traffic hazards: Hidden epidemics. *World Health Report. Chapter 6: Neglected Global Epidemics*. Retrieved online on August 25 from *http://www.who.int/whr/2003/chapter6/en/index3.html*.

Zimbardo, P. G. (1969). The human choice: Individuation, reason, and order versus deindividuation, impulse, and chaos. In W. J. Arnold, & D. Levine (Eds.), *Nebraska symposium on motivation (Vol. 17)*. Lincoln: University of Nebraska Press.

PART 1
THEORETICAL PERSPECTIVES/MODELS

In: Contemporary Issues in Road User Behavior and Traffic Safety ISBN:1-59454-268-6
Editors: D. Hennessy and D. Wiesenthal, pp. 3-11

Chapter 1

ON THE HOMEOSTASIS OF RISK

Gerald J. S. Wilde
Queen's University

INTRODUCTION

As there is arguably no human action with total certainty of outcome, all behavior may be viewed as risk-taking behavior. It is, therefore, of interest to identify the factors and mechanisms that determine people's perception of risk, their acceptance of risk and the actions they take to keep risk in check. No accident countermeasure can be expected to have much effect unless it is based on a valid theory of how people behave in the face of danger.

Consider for a moment the strange coexistence between two current, but opposite safety and health policies that may, surprisingly, well be implemented by one and the same accident prevention agency. The first seeks to improve safety by alleviating the consequences of risky behavior and may take the form of seatbelt installation and wearing, airbags, crashworthy vehicle design or forgiving roads (collapsible lampposts and barriers), safety shoes, low tar/nicotine cigarettes, low alcohol beer and the like. This policy offers forgiveness for a moment of inattention or carelessness and an opportunity for moderation in the consumption of hazardous substances.

The opposite policy, however, seeks to improve safety by making the consequences of risky behavior more severe, comprising speed bumps, narrow street passages and road throttles (a.k.a. choke or pinch points). By making speeding more dangerous to the driver, the authorities expect that the hazard of speed bumps will lead drivers to slow down. However, the authorities do not seem to expect drivers to speed up when the physical environment is made less hazardous. Does that make sense?

While the coexistence of these two policies is puzzling, the first being the more common approach, and the second a kind of "safety engineering in reverse," neither is likely to reduce the injury rate, if it is true that people adapt their behavior to changes in environmental conditions to offset any changes in safety that would have occurred if people had not altered their behavior. We will present both theory and data indicating that such behavioral adaptation does indeed occur and, thus, that safety and lifestyle-dependent health is unlikely to improve unless the level of health and safety risk that people are willing to take is reduced. This can be achieved by interventions that reduce what is called the target level of risk, which is the level of physical risk that people aim to maintain.

THE ACCIDENT RATE PER DISTANCE DRIVEN VERSUS THE ACCIDENT RATE PER CAPITA

In any discussion about injury prevention the first requirement is that the criterion of success be clearly specified, or else confusion abounds (Wilde, 1984, 1988). What exactly is it that we want to achieve: fewer accidents per unit distance driven or fewer accidents per head of population?

Sometimes the choice of denominator is obvious. We wish to reduce the number of suicides per head of population, not per handgun, not per km of available rope, nor per cubic meter of cooking gas. Success in promoting electrical safety is not measured in terms of fewer cases of electrocution per kWh consumed. Rather than aiming a reduction in the number of deaths per cigarette smoked, health officials are interested in reducing the smoking-related death rate in the nation.

In the domain of traffic, is the purpose of accident prevention to offer more mobility per death or injury, or do we want fewer cases of death and injury? That these two measures of success are not interchangeable is clearly demonstrated by data from many countries. Between 1923 and 1996, for instance, the death rate per km driven in the US fell by a factor of about 12, but the death rate per 100,000 residents showed no clear change, neither upward nor downward; in fact, the death rate per capita showed no difference between 1923 and 1996!

In the years between 1955 and 1972, a period of steady economic growth, the province of Ontario saw a major reduction in the traffic death rate per unit distance driven, but an increase in the traffic death rate per capita (Wilde, 1984). Improvement of the death rate per unit distance driven may be viewed as a (very worthwhile) economic gain: more mobility per death, but interventions that bring this about may be more appropriately labeled 'mobility promotion measures' than safety measures. From a public health point of view, the injury rate per head of population would seem the most relevant criterion of progress, and that is why that denominator will be used here.

MACRO-ECONOMIC EFFECTS ON THE YEAR-TO-YEAR VARIATIONS IN THE DEATH RATE PER CAPITA

Traffic injury rates show major fluctuations from one time period to another even in the absence of a lasting and continuous trend, and it has been established that these fluctuations cannot be attributed to particular "safety measures," but go hand in hand with the business cycle. In eight countries studied, years of high unemployment and low industrial production are marked by a low per capita traffic death rate (Wilde, 2001).

The increased traffic death rate in periods of relative economic prosperity may be attributed to an increased inclination of people to expose themselves to risk on the road. This is because in these periods more money is to be made by more and faster mobility while the costs (relative to disposable income) of gasoline, car repairs and insurance surcharges are reduced. There is less of an urge to avoid accidents. In depressed economic times, the death rate on the road is reduced because there is less to be gained from speed and mobility and more to be lost.

THE ACCEPTED LEVEL OF INJURY RISK

Besides macro-economic influences, there are other factors that influence the level of accepted risk; these are of a cultural, social or psychological kind. In general, the amount of risk that people are willing (in fact, prefer) to take can be said to depend on four utility factors and will be greater to the extent that Factors 1 and 4 are stronger, and Factors 2 and 3 are weaker:

1. The expected *benefits of risky behavior* alternatives (examples: time gained by speeding, the experience of excitement, fighting boredom, increased mobility, showing bravado to others).
2. The *expected costs of risky behavior* alternatives (examples: speeding tickets, car repairs, insurance surcharges, accident likelihood, social disapproval).
3. The expected *benefits of safe behavior* alternatives (examples: insurance discounts for accident-free periods, enhancement of reputation of responsibility, longevity).
4. The expected *costs of safe behavior* alternatives (examples: using an uncomfortable seatbelt, being called a coward by one's peers, time loss).

The level of risk at which the net benefit is expected to maximize is called *the target level of risk* in recognition of the realization that people do not try to minimize risk - the risk of dying in traffic could, of course, be easily eliminated, but at the price of zero mobility -, but instead attempt to optimize it (Wilde, 2001). Risk homeostasis theory posits that people at any moment of time compare the amount of risk they perceive with their target level of risk and will adjust their behavior in an attempt to eliminate any discrepancies between the two. Each action carries a certain level of injury likelihood such that the sum total of all actions taken by people over one year determines the accident rate for that year. This rate, in turn, in combination with the attendant personal experiences on the road and communications in the mass media, has an effect on the level of risk that people perceive and thus upon their subsequent decisions, and so forth.

This homeostatic mechanism is depicted in Figure 1 and constitutes a case of circular causality: a change in the degree of caution displayed in behavior brings about a change in the injury rate, while a change the injury rate also leads to a change in behavior. The phenomenon is similar to a thermostat: this instrument controls the actions of the heating/cooling unit, which determines the temperature, and the temperature in turn controls the actions of the thermostat. There will be fluctuations in the room temperature, but averaged over time, the temperature will remain stable, until the thermostat is set to a new target (set-point) level.

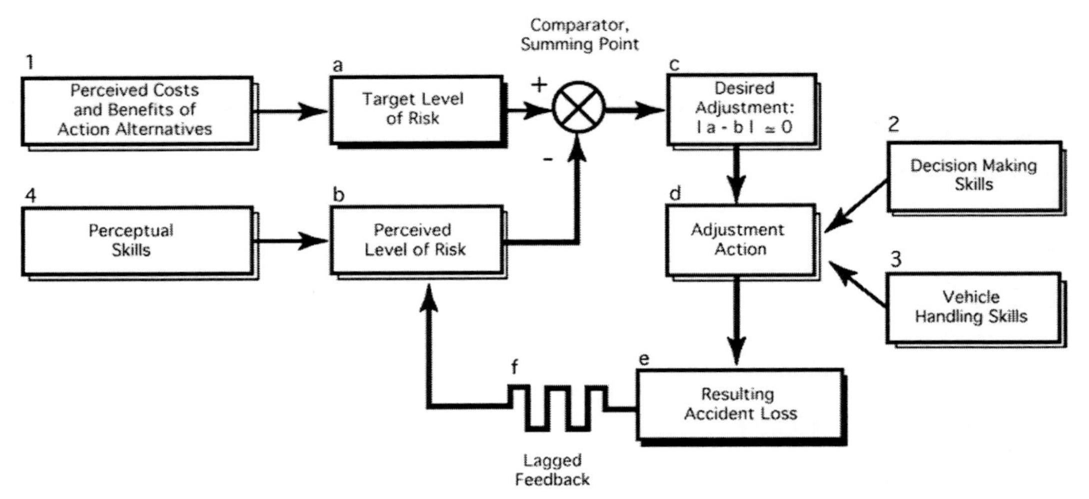

Figure 1. Homeostatic model relating the accident rate per head of population in a jurisdiction to the level of caution in road-user behavior and vice versa, with the average target level of risk as the controlling variable (after Wilde, 2001, p. 33).

Similarly, the target level of risk is seen as the controlling variable in the causation dynamic of the injury rate. It follows that the basic strategy of injury prevention should be to reduce the level of risk that people are willing to accept. Because the nation's accident rate reflects the target level of risk in the population, the art of traffic safety management is synonymous with the art of reducing that target level.

Because of the closed-loop nature of the control process, all other variables, such as variations in skill or environmental conditions can only produce minor and/or short-term fluctuations, and even these are often reduced or virtually eliminated through anticipatory adaptation ("feed-forward control;" Wilde, 2001).

Evidence of Risk Homeostasis

In the fall of 1967, Sweden changed over from left-hand to right-hand traffic. This change-over was followed by a marked reduction in the traffic fatality rate. About a year and a half later, the accident rate returned to the pre-change-over trend. In terms of Figure 1, what happened was a sudden surge in box b as a result of the change-over coming into effect. Perceived risk (box b) was suddenly significantly higher than the target level of risk (box a). Road users adjusted their behavior by choosing much more prudent behavior alternatives (box d). As a result, the fatal injury rate dropped (box e). After some time (box f), however, people discovered - through the mass media as well as their own experiences - that the roads were not as dangerous as they had thought they were. The level of perceived risk dropped below the target level of risk. Consequently, road users opted for less cautious behavior alternatives and the fatal injury rate rose again. The same pattern was seen in Iceland after it changed the direction of traffic one year later (Wilde, 2001).

Numerous other findings from time-series analyses, field studies, field and laboratory experiments and can be explained by risk homeostasis theory. In road sections where the accident rate per km driven is low, drivers move commensurably faster and thus keep the accident rate per hour of driving at the same level (May, 1959). Mandatory wearing of seatbelts reduces the likelihood of death or injury in case an accident happens, but does not reduce the death rate per capita. An American study published in 2002 reported that car-occupant seatbelt usage, as determined by roadside surveys, has risen from a low of 10% (in 1985 in Indiana) to a high of 87% (in 1996 in California). This study found that seatbelt usage did not have the effect of reducing the traffic fatality rate per head of population (or even per unit distance driven) in 14 years of data (Derrig, Segui-Gomez, Abtahi, & Liu, 2002).

Similarly, "[...] air-bag-equipped cars tend to be driven more aggressively and that aggressiveness appears to offset the effect of the air bag for the driver and increases the risk of death to others." (Peterson, Hoffer & Miller, 1995). In the period from 1994 to 1996, about half of all 50 American states plus the federal district had laws compelling all motorcyclists to wear a helmet, while the other half of jurisdictions did not. Through a comparison of the law states with the no-law states over this period, it was found that the helmet laws failed to have a significant impact on the fatality rate per 10,000 registered motorcycles (Branas & Knudson, 2001). The authors mention "risk compensation" as a possible explanation, and quote several earlier authors on the topic of crash helmet legislation in the US who also identified risk compensation as an explanatory factor.

Cars outfitted with anti-lock brakes are driven faster, more carelessly and closer to the car in front, braked more abruptly and have no lower accident rate per hour of exposure than cars without these devices (Asschenbrenner & Biehl, 1994; Grant & Smiley, 1993; Fosser, Sagberg, & Sætermo, 1997). Similarly, with better road lighting motorists drive faster and pay less attention (Assum, Bjørnskau, Fosser & Sagberg, 1999). Traffic lights have been found to affect the type of collisions at intersections (fewer at 90 degrees, more rear-enders), while overall numbers and severity remain unchanged (Wilde, 2001) Providing better visibility at level railway crossings causes drivers to approach at higher speeds while safety margins are not favorably influenced (Ward & Wilde, 1996).

A rather startling case of technical improvements that have failed to reduce the accident rate is presented by a study on parachute jumping. This case is so startling that outsiders might find it difficult to believe at first. In the past many of the fatalities in this sport occurred when parachutes failed to open. In more recent years, however, the engineering of the ripcord function has been improved and parachute deployment brought more fully under the skydiver's control. Thus, one might have expected a major reduction in skydiving fatalities, but this did not occur. Indeed, the new ripcord did produce a major reduction in accidents where the parachute was closed at the time of landing, but these accidents were replaced by fatalities in which the parachute was open at landing. A graduate student at Western Oregon University wished to test the hypothesis, that he derived from the concept risk homeostasis, that "accident records will show that when fatal skydiving accidents in one category [closed parachute] decline, there will be an increase in fatal skydiving accidents in a different category [open parachute]." This is indeed what he found for the time period in which the new ripcord gained in popularity; the "total annual fatalities have remained the same during this period [because] people adjust their behaviors to maintain arousal at optimal levels." (Napier, 2000). There was no accident reduction, but accident metamorphosis. As people become more confident that the parachute will open, they will open it later, and sometimes too late.

Driver training or a mandatory course of driving on slippery roads does not reduce accident risk (Christensen & Glad, 1996). Such training does indeed improve skill, but it apparently increases confidence even more, with the end effect that driver education graduates show a *higher* accident rate per capita (Brown, Groeger, & Biehl, 1987). A recent Swedish study showed that the more traffic safety education children in kindergarten and primary school had received the higher their traffic injury rate. This was attributed to the greater independence and mobility, including the use of a bicycle, which better trained children were allowed by their parents as they felt that their children had received superior traffic education. (Johansson, 1997).

Outside the world of traffic, the introduction of child-proof medicine vials has failed to limit the number of cases of accidental poisoning. Such deaths, in fact, became more frequent, apparently as the result of parents becoming less careful in the handling and storing of the "safer" bottles. (Viscusi, 1984). An interesting case of risk and protective behavior is what people do in the face of the threat of attack by a computer virus. One study found, consistent with risk homeostasis theory, that as the perceptions of threat of the Michelangelo virus changed over time, so did the protective behaviors with the effect that personally experienced risk remained unchanged (Sawyer, Kernan, Conlon, & Garland, 1999). Give smokers cigarettes to smoke with low nicotine content – even without telling them about the nicotine content - and they will alter their smoking behavior in one or several ways, by smoking more cigarettes or by inhaling more deeply or smoking to shorter butts. Give smokers cigarettes with a high-nicotine content and they alter their behavior in the opposite direction. These behaviors, occur so naturally that they are sometimes not even noticed by the smokers themselves, but others can readily observe them (Chapman, Haddad, & Sindhusake, 1997; McKim, 1997). It is not surprising, then, that there is no evidence that the marketing of low tar/nicotine cigarettes has reduced the smoking-related death and disease rate (Warner & Slade, 1992).

INCENTIVES FOR SAFETY

To counter the erroneous impression received by some that risk homeostasis theory casts a pessimistic perspective on the preventability of accidents and lifestyle-dependent ill health, we will now turn to effective intervention efforts that follow directly and logically from the theory. The bad news is that people will persist in maintaining their target level of risk. And it is also true that the traditional safety measures do not alter that target level of risk; hence these safety measures do not enhance safety. The good news is that it is rather easy to alter a person's target level of risk.

As subsequent data demonstrate, an individual's target level of risk is determined by the good things he or she is able to look forward to in an accident does not occur. These good things function as effective incentives in enhancing actual safe conduct. The incentives may be significant and long-term as staying alive for the birth of a grandchild, or trivial and short-term, as a fifty dollar award for accident-free driving in the next two months.

Incentive systems for accident-free operation have shown to be a very powerful method for the reduction of injury rates. Incentives, i.e., future rewards (for instance in the form of cash or other bonuses) contingent upon fulfilling a future condition, increase the perceived benefits of safe behavior alternatives (utility Factor 3 above). There have been many studies of the effectiveness of incentive schemes, both in industrial settings and traffic and their most productive features have been identified. Injury rate reductions ranging from 10 to 90% have been observed and this at very favorable benefit/cost ratios, usually exceeding 2 to 1, but often very much higher (Fox, Hopkins & Anger, 1987). The only undesirable side effect noted so far is the underreporting of accidents, but this phenomenon is limited to relatively minor injuries and property damage. Among their positive side effects is a more congenial social climate and greater productivity, especially if the program is well-designed in cooperation with those to whom it is addressed (Wilde, 2001).

An incentive program for safe driving in California, which led to a 22% reduction in accidents in the first year of operation and to 33% fewer accidents in the second, was seen especially effective in young drivers (Harano & Hubert, 1974). A Norwegian program that promised substantial insurance rebates for accident-free driving by novice drivers produced a 35% reduction in accident frequency (Vaaje, 1991). An American team-based incentive program aimed at transit bus operators yielded a 25-35% reduction in accident rates as compared to randomly selected controls within the same company. The ratio between program costs and benefits was estimated at almost seven-to-one. After the program was withdrawn, the safety records of the incentive group dropped to a level that was still better than that of the no-treatment employees, but no longer significantly so (Haynes, Pine & Fitch, 1982).

The remarkable effectiveness of incentive programs is arguably due to the fact that these programs enhance people's perceived value of the future. The prospect of future rewards causes people to look forward to the future with positive expectation. Thus, they will realize they have more to lose and thus have a greater desire to be alive and well when that future comes (utility factor 3 above), and be more inclined to take action to protect their health and safety. And indeed, there is evidence accumulating that individuals who are marked by a high valuation of the future relative to the present display fewer unsafe behaviors and unhealthy (Björgvinsson & Wilde, 1996). In a Canadian study of late adolescents and early adults, a significant relation was found between safe driving practices, regular seatbelt use and moderate alcohol consumption, a wholesome diet and regular physical exercise on the one hand, and a high valuation of the future, and more deliberate planning for that future, on the other (Björgvinsson, 1998). An earlier study of Quebec motorists found that individuals characterized by a comparatively high valuation of the future had more favorable attitudes to automobile safety, fewer demerit points, and fewer road accidents (Chebat & Chandon, 1986).

Expectationism

The theory of risk homeostasis (also known as "risk compensation" or “behavioral adaptation”) was primarily developed and validated in the area of road safety. Some of the supporting data, however, come from quite different behavior domains including smoking and settling in flood-prone territories (Lave & Lave, 1991). This is not surprising because the mechanisms that are involved in risk homeostasis are probably universal. Moreover, the accident-prevention strategy that logically follows from risk homeostasis theory has been found effective in many areas. Incentives for safety and health may be viewed as one example of a wider class of "expectationist" interventions (as distinguished from technological interventions). These are the interventions that offer people more positive anticipations regarding their future than is currently the case and thus motivate them to be more cautious with life and limb. If the rate of accidents and lifestyle-dependent poor health essentially depends on the level of risk people are willing to take, then the concept of expectationism would offer a meaningful rationale for effective prevention.

REFERENCES

Aschenbrenner, M., & Biehl, B. (1994). Improved safety through improved technical measures? Empirical studies regarding risk compensation processes in relation to anti-lock braking systems. In R. M. Trimpop and G. J. S. Wilde (Eds.), *Challenges to accident prevention: The issue of risk compensation behavior* (pp. 81-90). Groningen, the Netherlands: Styx Publications.

Assum, T., Bjørnskau, T., Fosser, S., & Sagberg, F. (1999). Risk compensation – the case of road lighting. *Accident analysis and Prevention, 31,* 545-553.

Björgvinsson, T. (1998). Health and safety habits as a function of the perceived value of the future. Unpublished Doctoral Dissertation, Dept. of Psychology, Queen's University, Kingston, Ontario.

Björgvinsson, T., & Wilde, G. J. S. (1996). Risky health and safety habits related to perceived value of the future. *Safety Science, 22*, 27-33.

Branas, C. C., & Knudson, M. M. (2001). Helmet laws and motorcycle rider death rates. *Accident Analysis and Prevention, 33*, 641-648.

Brown, J. D., Groeger, J. A., & Biehl, B. (1987). Is driver training contributing enough towards road safety? In J. A. Rothengatter and R. A. de Bruin (Eds.), *Road users and traffic safety (*pp.135-156). Wolfeboro, New Hampshire: Van Gorcum.

Chapman, S., Haddad, S., & Sindhusake, D. (1997). Do work-place smoking bans cause smokers to smoke "harder"? Results from a naturalistic observational study. *Addiction, 92*, 607-610.

Chebat, J. C., & Chandon, J. L. (1986). Predicting attitudes toward road safety from present and future orientations: An economic approach. *Journal of Economic Psychology*, *7*, 477-499.

Christensen, P., & Glad, A. (1996). Mandatory course of driving on slippery roads does not reduce the accident risk. *Nordic Road & Transport Research, 8*, 22-23.

Derrig, R. A., Segui-Gomez, M., Abtahi, A., & Liu, L. L. (2002). The effect of population safety belt usage rates on motor-vehicle-related fatalities. *Accident Analysis and Prevention, 34*, 101-110.

Fosser, S., Sagberg, F., & Sætermo, A. F. (1997). An investigation of behavioral adaptation to airbags and antilock brakes among taxi drivers. *Accident Analysis and Prevention, 29*, 293-302.

Fox, D. K., Hopkins, B. L., & Anger, W. K. (1987). The long-term effects of a token economy on safety performance in open pit mining. *Journal of Applied Behavior Analysis*, *20*, 215-224.

Grant, B.A., & Smiley, A. (1993). Driver response to antilock brakes: A demonstration of behavioral adaptation. Proceedings of the Canadian Multidisicplinary Road Safety Conference VIII, June 14-16, University of Saskatoon, Saskatchewan, Canada.

Harano, R. M., & Hubert, D. E. (1974). *An evaluation of California's 'good driver' incentive program*. Report No. 6, California Division of Highways, Sacramento.

Haynes, R. S., Pine, A. R. C., & Fitch, H. G. (1982). Reducing accident rates with organizational behavior modification. *Academy of Management Journal, 25*, 407-416.

Johansson, B. S. (1997). Trafiktränade barn löper större olycksrisk. *VTI Aktuellt*, No. 4, June, p. 9.

Lave, T. R., & Lave, L. B. (1991). Public perception of the risks of floods: Implications for communication. *Risk Analysis, 11*, 255-267.

May, A. D. (1959). A friction concept of traffic flow. *Proceedings, 30th Annual Meeting of the Highway Research Board*. Washington, DC, pp. 493-510.

McKim, W. A. (1997). *Drugs and behavior*. Third edition. Upper Saddle River, New Jersey: Prentice Hall, Chapter 8.

Napier, V. (2000). Open canopy fatalities and risk homeostasis: a correlation study. Department of Psychology, Western Oregon University, March 5.. Available online at http://www.noexcusesrigging.com/ArticlesEssays/LayOverview.htm.

Peterson, S., Hoffer, G., & Millner, E. (1995). Are drivers of air-bag-equipped cars more aggressive? A test of the offsetting behavior hypothesis. *Journal of Law and Economics, 38*, 251-264.

Sawyer, J. E., Kernan, M. C., Conlon, D. E., & Garland, H. (1999). Responses to the Michelangelo computer virus threat: The role of information sources and risk homeostasis theory. *Journal of Applied Social Psychology, 29*, 23-51.

Vaaje, T. (1991). Rewarding in insurance: Return of part of premium after a claim-free period. *Proceedings, OECD/ECMT Symposium on enforcement and rewarding: Strategies and effects*. Copenhagen DK, Sep. 19-21, 1990.

Viscusi, W. K. (1984). The lulling effect: The impact of child-resistant packaging on aspirin and analgesic ingestions. *American Economic Review, 74*, 324-327.

Ward, N. J., & Wilde, G. J. S. (1996). Driver approach behavior at railway crossings before and after enhancement of lateral sight distances: An experimental investigation of a risk perception and behavioral adaptation hypothesis. *Safety Science, 22*, 63-75.

Warner, K. E., & Slade, J. (1992). Low tar, high toll. *American Journal of Public Health, 82*, 17-18.

Wilde, G. J. S. (1984). On the choice of denominator for the calculation of accident rates. In S. Yagar (Ed.), *Transport Risk Assessment* (pp. 139-154).. Waterloo, Ontario: University of Waterloo Press, Ontario.

Wilde, G. J. S. (1988). Risk homeostasis theory and traffic accidents: propositions, deductions and discussion of dissension in recent reactions. *Ergonomics, 31*, 441-468.

Wilde, G. J. S. (2001). *Target Risk 2: A new psychology of safety and health* - What works? What doesn't? And why.... Toronto: PDE Publications, pp. 122-125.

In: Contemporary Issues in Road User Behavior and Traffic Safety ISBN:1-59454-268-6
Editors: D. Hennessy and D. Wiesenthal, pp. 13-20

Chapter 2

THE INFLUENCE OF THE ACTOR-OBSERVER BIAS ON ATTRIBUTIONS OF OTHER DRIVERS

Dwight A. Hennessy, Robert Jakubowski and Alyson J. Benedetti
State University of New York College at Buffalo

INTRODUCTION

Attributions are the cognitive process by which we attempt to explain the source or cause of others' behavior. Heider (1958) proposed that at its basic level, attributions tend to focus either on internal causes (personality/dispositions) or external causes (environment/situations). Typically, categorization of other people's actions is a normal process that can help simplify the cognitive resources needed to evaluate the constant barrage of stimuli in the world around us. However, this process can become problematic when the actions and dispositions of others do not correspond. As a result, our attributions of the causes of their behavior become incorrect and oversimplified. The Actor Observer Bias is the stable tendency of individuals to overestimate dispositional and underestimate situational causes of the actions of others, but to conversely attribute more situational and less dispositional causes to their own actions (Jones & Nisbett, 1971; Storms, 1973). According to Jones and Nisbett (1972), this biased evaluation occurs due to 1) differences in visual perspective of an event between actors and observers, and/or 2) due to differences in knowledge of, and experience with, the self versus others. With the former explanation, our focus of attention when viewing the behavior of others is on the individual, so our attributions tend to be based on individual factors. In contrast, when evaluating our own actions, our focus of attention is mainly on the situation, so our self attributions tend to be more situational. With the latter explanation, our in-depth knowledge of the self leads to a greater understanding of situational influences on our behavior, and, conversely, our lack of situational knowledge of the behavior of others leads to greater dispositional attributions.

Storms (1973) conducted a landmark study to demonstrate that attribution biases are due to visual point of view as suggested by Actor Observer Bias. Participants were placed into dyads and given the opportunity to have a brief conversation with their fellow participant.

They were then asked to make attributions of their own and their fellow co-actor's behavior during the conversation. He found that attributions were consistent with Actor Observer Bias in that participants tended to believe their own actions were dictated mostly by the situation, while the actions of their co-actor were dictated mainly by internal dispositions. Participants then watched a videotape of their conversation in which the visual point of view was reoriented to show events from their co-actors perspective, and were again asked to provide attributions of their own and their co-actor's behavior. Storms (1973) found that attributions had reversed, in that participants made fewer situational attributions for their own behavior, while making greater situational and fewer dispositional attributions for their co-actor.

Martin and Huang (1984) have argued that reorientation of point of view may not be sufficient to explain Actor Observer Bias. Rather than use a social task, they had actors conduct a motor coordination task under the scrutiny of an observer, while a video camera recorded the actor and her performance. Their results indicated that *both* actors and observers provided greater situational attributions after watching the videotaped performance, showing minimal evidence of a reorientation effect. Martin and Huang (1984) suggested that the reorientation of attributions found by Storms (1973) may have been a temporal confound, where the time delay between the two rating periods may have led to altered attributions. While this temporal explanation is plausible, there is currently no consistent understanding of the impact of time on attributions. In fact, the same criticism could be made regarding Martin and Huang (1984) in that their method also required a time delay between ratings of the actual and videotaped events. Without knowledge of its exact influence on attributions, time may have led to "consistency" of attributions in their context, especially given the unique and dynamic task performed by the actors. As they intended, the observed task in their study was one that required more directed attention from both the actor and observer, which would likely create greater dispositional and reduced situational attributions at the onset (especially among the actors). Their failure to find a "switch" in attributional focus following the point of view reorientation may have been a function of the task itself. Specifically, the task may have been so engaging or demanding of cognitive resources that it remained so during the viewing of the videotape, leading to similar attributions as with the original action. In sum, the findings of Martin and Huang (1984) do not necessarily negate the reorientation effect or the impact of visual point of view on attributions, but rather suggest it may be more complex than originally proposed and dependent on the nature of the task, the information available to both actors and observers of events, and temporal or situational constraints.

Attributions and attribution biases are important concepts for understanding driver behavior. The driving environment is a common context that can have a significant impact on our daily lives. As in any other social setting, all drivers make evaluations about themselves, as well as the events and people they encounter. However, given the transitory nature of driving interactions – high speeds, combined with visual isolation and anonymity among drivers – the amount of information available when attributions are formed can be very limited and subject to error. Current research into attribution biases among drivers, including attributions of the self, has found that erroneous judgments can lead to a variety of dangerous attitudes and behaviors, such as vengeful aggression (Wiesenthal, Hennessy, & Gibson, 2000), elevated perceptions of personal driving skills (McKenna, Stanier, & Lewis, 1991), downgraded perceptions of other drivers' skill (Walton & Bathurst, 1998), decreased belief in collisions likelihood (Svenson, Fischhoff, & MacGregor, 1985), reduced belief in risk of injury from traffic collisions and increased perception of driving ability while fatigued

(Dalziel & Job, 1997). The present study was designed to investigate the impact of visual point of view on the Actor Observer Bias within the driving context. Specifically, the intent was to examine if visual point of view leads to exaggerated dispositional attributions of other drivers after seeing them engage in negative driving behavior and elevated situational dispositions when viewing the same events from a driver's perspective.

PREDICTIONS

When viewing a driving incident from the perspective of an "offending" driver, participants would attribute more skill and less riskiness (dispositional attributions) to the offending driver and more blame to other drivers (situational attributions).

In contrast, those that viewed the same events as a trailing motorist would attribute less skill and more riskiness to the offending driver (dispositional attributions) and less responsibility to other drivers (situational attributions).

METHODS

Participants

The present study included 87 female and 51 male participants from the student and employee populations of Buffalo State College. Driving experience ranged from 3 months to 33 years, with an average of 5.96 years experience ($M = 4.64$ years for females and $M = 8.23$ years for males). Their ages ranged from 18 – 50 years, with an average of 22.36 years ($M = 21.21$ years for females and $M = 24.32$ years for males). The average driving time ranged from 1 minute to 405 minutes per day, with an average of 45.53 minutes per day ($M = 44.52$ minutes for females and $M = 38.76$ minutes for males).

Measures

The Driving Vengeance Questionnaire (DVQ) (Wiesenthal et al., 2000) was developed to evaluate a general susceptibility toward vengeful driving reactions. Items represent common driving situations in which a participant might be irritated, or feel unjustly treated by another driver. Participants were required to select a likely response from a series of four options involving decreasing levels of aggression. Response alternatives ranged from displays of extreme aggression (e.g., force the other vehicle off the road) to nonaggressive reactions (e.g., do nothing). Scoring consisted of assigning a rank to each item, based on the level of aggression involved in the chosen response option. The first, and most extreme, option was assigned a rank of 4, while subsequent options, which decreased in their level of aggression, were assigned ranks of 3, 2, and 1 respectively. All items also included an open ended response option, to which participants could indicate an alternate response to those provided. All alternate responses were independently rated as to their level of aggressiveness in relation to the options provided for that item. For example, those deemed equivalent in aggression to the most aggressive option for that item were given a rank of 4, while those considered equivalent to the nonaggressive option were given a rank of 1. A vengeance score was calculated as the sum of all individual ratings with higher scores indicating a more vengeful driving attitude. The DVQ has been found to represent a reliable measure of vengeful driving attitudes (alpha = 0.83), and to predict the likelihood of mild driver aggression and violence (Hennessy & Wiesenthal, 2001; Wiesenthal et al., 2000).

The Desirability of Control Scale (DCS) (Burger & Cooper, 1979) includes 20 items designed to tap a motivation to control personal behaviors and life experiences. Items represent statements that indicate a preference or aspiration to gain control (e.g., "I would prefer to be a leader rather than a follower) or give up control (e.g., "I wish I could push many of life's decisions off on someone else") within common life situations. Respondents were asked to rate how much each item applies to them, using a Likert scale from 1-7. An aggregate control score was calculated as the mean response to the positively worded and the reverse key of the negatively worded items. The DCS has been found to have high internal consistency (alpha = .80) and test retest reliability (.75) (Burger & Cooper, 1979).

Procedure

Using a simulated driving program (Need for Speed Porsche Unleashed), a near collision incident was pre-recorded in which an "offending" car partially crossed the center line in front of an oncoming truck (without collision). The offending driver proceeded for approximately one minute prior to the incident and then continued driving for approximately one minute following the incident. The program allows for multiple visual angles and perspectives of the same recorded event. Participants were randomly assigned to either the "inside" or "outside" point of view groups. The "inside" group viewed all recorded events from the perspective of the driver of the offending vehicle, while the "outside" group viewed all events from the perspective of a trailing motorist.

Following the video, all participants completed the DVQ, DCS, and provided ratings and attributions of the parties involved in the simulated incident. Using a scale from 0 – 5, respondents were asked to rate the overall skill of the offending driver, how often they thought the driver engaged in risky action while actually driving, and how responsible the driver of the truck was in the near collision.

Results

Intercorrelations, means, standard deviations, and alpha reliabilities appear in Table 1.

Separate hierarchical entry stepwise multiple regressions were calculated for skill ratings, riskiness ratings, and responsibility of the truck driver using age, gender, vengefulness, desire for control, visual point of view, and all two way cross product interactions as predictors. The procedure for each criterion was to enter gender, age, driving vengeance and desire for control forcibly in the first step, enter point of view in the second step, and add all cross product interactions stepwise in the third step.

Table 1. Intercorrelations and Descriptive Statistics

	1	2	3	4	5	6
1. Skill	—	—	—	—	—	—
2. Riskiness	-.07	—	—	—	—	—
3. Age	.17	-.14	—	—	—	—
4. Gender	-.00	.16	.21	—	—	—
5. Vengeance	-.03	-.10	.31*	-.08	—	—
6. Control	.20	-.18	.04	.05	-.04	—
Mean	2.50	1.23	22.36	—	47.22	4.38
SD	1.33	1.07	7.11	—	7.69	1.33
Minimum	0.00	0.00	18.00	—	28.00	2.55
Maximum	5.00	5.00	50.00	—	60.00	5.80
Cronbach	—	—	—	—	.80	.76

n = 158.
* p<.05.

Skill ratings of the offending driver were predicted by the main effect of point of view and the interaction of age X point of view ($R^2 = 0.18$, $F(6,114) = 4.28$, $p<.01$; see Table 2). Overall, those viewing the event as a trailing motorist provided lower skill ratings to the offending driver, but this was more prevalent with increased rater age (see Figure 1).

Riskiness ratings were predicted by the main effects of point of view and gender ($R^2 = 0.20$, $F(5,115) = 5.89$, $p<.01$; see Table 2). Those viewing the incident from the perspective of a trailing motorist were more likely to see the driver as risky. Similarly, males were more likely to believe the offending driver to be a risky driver in general.

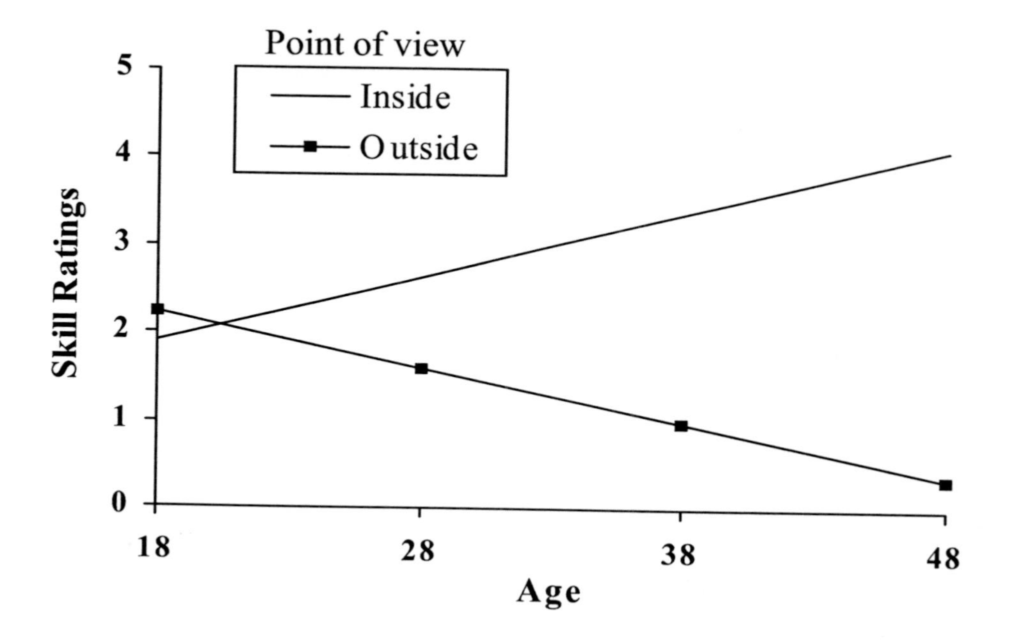

Figure 1. Ratings of "Offending" Driver Skill as a Function of Age X Point of View.

Ratings of responsibility attributed to the truck driver were predicted by the main effect of point of view ($R^2 = 0.14$, $F(5,115) = 3.84$, $p<.01$; see Table 2). Those that viewed from the perspective of the offending driver, attributed more responsibility to the driver of the truck.

Table 2. Significant Predictors of Skill, Riskiness, and Ratings of Victim Responsibility

Criterion	Predictor	b	t
Skill			
	Age	0.006	030
	Point of View	-1.388	-3.43*
	Age X Point of View	0.068	3.77*
	Intercept	1.955	
Riskiness			
	Gender	-0.247	2.52*
	Point of View	-0.361	-3.87*
	Intercept	3.348	
Responsibility			
	Point of View	-0.430	-3.85*
	Intercept	1.444	

n = 158.
* p<.05.

DISCUSSION

The present findings were generally in support of those of Storms (1973) rather than Martin and Huang (1984). As expected, the present study confirmed that visual perspective or point of view is an important determinant in the occurrence of the Actor Observer Bias. Those that viewed the simulated near collision from the perspective of a trailing motorist were more likely to focus on dispositional attributions of the offending driver (less skill and greater overall riskiness) and to subsequently discount situational attributions for the same event (less responsibility to the "victimized" truck driver). The opposite pattern was found for those that viewed the event from the perspective of the offending driver, in that they were more prone to discount dispositional attributions, but more likely to provide a situational attribution for the near collision.

As with attributions made in real driving settings, participants made quick evaluations under the authority of very minimal information, which generally downgraded the underlying ability the offending driver. For the trailing motorists, these evaluations were naively, and perhaps unnecessarily, critical of the overall ability of the offending driver given that the majority of their exposure involved error free driving. This tendency is consistent with the finding that drivers typically believe personal factors (inability, accident proneness, risk taking), rather than situational factors, are the predominant reasons for other drivers' collisions (McKenna, 1983; Rumar, 1988). It is also noteworthy that a dispositional enhancement was found for those that viewed the events from the perspective of the offending driver, despite the fact that they were not the actual driver. This further enhances the point of view argument to the extent that these raters would have been relatively free of personal history, social desirability, or self enhancement biases. Thus, the only difference from those in the trailing motorist group would have been the visual perspective of the events and not amount of information or knowledge.

Although not predicted, age was found to interact with point of view to impact attributions of skill. Specifically, older drivers that viewed the event as a trailing motorist were more likely to think the offending driver was unskilled. One possible explanation for this effect may have been the fact that attributions are partly based on personal experience and expectations in specific contexts (Green, Lightfoot, Bandy, & Buchanan, 1985). As drivers age, they likely have greater experience in observing undesirable or even dangerous actions from other drivers, which may reinforce a general schema of other drivers as "poorly skilled". According to Green et al. (1985), when confronted with the actions of a stranger, individuals generally rely on established schemas for that context, leading in this instance to a more automatic judgment of the offending driver as less skilled.

Limitations and Future Directions

While the present study found an age effect in attributions of skill, the age range was somewhat limited, with a maximum of 50 years. In some respects, skewing of predictor variables can be problematic in that they can overestimate relationships at the extremes. However, outliers were not present in the age distribution and the sample size was sufficiently large that this was unlikely an issue in the present regression analysis. From a more applied aspect, the limited range of ages excluded older drivers. According to Holland (1993), the experience and attribution processes of older drivers may be distinct from those of younger drivers. As a result, future research may benefit from a broader age representation. Another consideration is the validity of simulator use. While traffic simulators are an accepted research tool, there is always a concern regarding the perception of realism among users. Finally, the present study did not focus on personal factors that might alter the tendency to make such biased judgments, or the potential actions that might result from erroneous attributions of other drivers (e.g., vengeance, anger, aggression). Personality is an important factor in driving behavior and personality can influence cognitions, and hence evaluations, of the self and others (e.g.,Hennessy & Wiesenthal,1999; Matthews & Norris, 2002; Stradling & Meadows, 2000). Future research is needed in these areas to help gain a fuller understanding of the causes and consequences of attributions biases, and to ultimately minimize their negative influence on interdriver interactions.

References

Burger, J. M., & Cooper, H. M. (1979). The desirability of control. *Motivation & Emotion, 3*, 381-393.

Dalziel, J. R., & Job, R. F. S. (1997). Motor vehicle accidents, fatigue and optimism bias in taxi drivers. *Accident Analysis & Prevention, 29*, 489-494.

Green, S. K., Lightfoot, M. A., Bandy, C., & Buchanan, D. R. (1985). A general model of the attribution process. *Basic & Applied Social Psychology, 6*, 159-179.

Heider, F. (1958). *The psychology of interpersonal relations*. New York: Wiley.

Hennessy, D. A., & Wiesenthal, D. L. (1999). Traffic congestion, driver stress, and driver aggression. *Aggressive Behavior, 25*, 409-423.

_____ . (2001). Further validation of the Driving Vengeance Questionnaire. *Violence and Victims, 16*, 565-573.

Holland, C. A. (1993). Self bias in older drivers' judgments of accident likelihood. *Accident Analysis & Prevention, 25*, 431-441.

Jones, E. E., & Nisbett, R. E. (1971). *The actor and the observer: Divergent perceptions of the causes of behavior*. Morristown, NJ: General Learning Press.

_____ . (1972). The actor and the observer: Divergent perceptions of the causes of behavior. In E. E. Jones, D. E. Kanouse, H. H. Kelley, R. E. Nisbett, S. Valins, & B. Weiner (Eds), *Attributions: Perceiving the causes of behaviour* (pp. 79-94). Morristown NJ: General Learning Press.

Martin, D. S., & Huang, M. S. (1984). Effects of time and perceptual orientation on actors' and *Perceptual and Motor Skills, 58*, 23-30.

McKenna, F. P. (1983). Accident proneness: A conceptual analysis. *Accident Analysis & Prevention, 15*, 65-71.

McKenna, F. P., Stanier, R. A., & Lewis, C. (1991). Factors underlying illusory self assessment of driving skills in males and females. *Accident Analysis & Prevention, 23*, 45-52.

Matthews, B. A., & Norris, F. H. (2002). When is believing "seeing"? Hostile attribution bias as a function of self-reported aggression. *Journal of Applied Social Psychology, 32*, 1-32.

Rumar, K. (1988). Collective risk but individual safety. *Ergonomics, 31*, 507-518.

Storms, M. D. (1973). Videotape and the attribution process: Reversing actors' and observers' point of view. *Journal of Personality and Social Psychology, 27*, 165-175.

Stradling, S. G., & Meadows, M. L. (2000). Highway code and aggressive violations in UK drivers. Paper presented at *Aggressive Driving Issues Internet Conference*. [On-line]. Hosted by the Ontario Ministry of Transportation, October 16-November 30. Available at: http://www.aggressive.drivers.com.

Svenson, O., Fischhoff, B., & MacGregor, D. (1985). Perceived driving safety and seatbelt usage. *Accident Analysis & Prevention, 17*, 119-135.

Walton, D., & Bathurst, J. (1998). An exploration of the perceptions of the average driver's speed compared to perceived driver safety and driving skill. *Accident Analysis & Prevention, 30*, 821-830.

Wiesenthal, D. L., Hennessy, D. A., & Gibson, P. (2000). The Driving Vengeance Questionnaire (DVQ): The development of a scale to measure deviant drivers' attitudes. *Violence and Victims, 15*, 115-136.

PART 2
ANGER/AGGRESSION

In: Contemporary Issues in Road User Behavior and Traffic Safety ISBN:1-59454-268-6
Editors: D. Hennessy and D. Wiesenthal, pp. 23-36

Chapter 3

MOTORISTS' PERCEPTIONS OF AGGRESSIVE DRIVING: A COMPARATIVE ANALYSIS OF ONTARIO AND CALIFORNIA DRIVERS

Christine M. Wickens and David L. Wiesenthal
Department of Psychology, York University
Kathy Rippey
Professional Standards Bureau, Ontario Provincial Police

DRIVER STRESS AND MILD AGGRESSION

Driving is a common event that has been found to elicit elevated levels of stress and arousal, particularly in highly frustrating conditions, such as obstructed or congested traffic (Hennessy & Wiesenthal, 1997, 1999; Hennessy, Wiesenthal, & Kohn, 2000; Novaco, Stokols, & Milanesi, 1990; Rasmussen, Knapp, & Garner, 2000; Stokols & Novaco, 1981; Wickens, 2004; Wickens & Wiesenthal, 2004; Wickens & Wiesenthal, in press, Wiesenthal, Hennessy, & Totten, 2000). A recent survey found that rush hour congestion was the most prevalent source of daily stress among U.K. drivers (BBC News, 2000). Under stressful conditions, drivers are more likely to exhibit mild forms of driver aggression, including horn honking, swearing, and yelling at other drivers (Gulian, Matthews, Glendon, Davies, & Debney, 1989; Hartley & El Hassani, 1994; Hennessy & Wiesenthal, 1997, 1999, 2001a, 2002; Hennessy, Wiesenthal, Wickens, & Lustman, 2004; Wiesenthal, Hennessy, & Gibson, 2000). For many, the inability to deal effectively with stressful driving demands contributes to feelings of frustration, irritation, and anger, which can subsequently enhance the potential for driver aggression (Deffenbacher, Huff, Lynch, Oetting, & Salvatore, 2000; Ellison, Govern, Petri, & Figler, 1995; Joint, 1997; Matthews et al., 1998; Matthews, Tsuda, Xin, & Ozeki, 1999; Mizell, 1997).

Driver aggression has been defined as any behavior intended to physically, emotionally, or psychologically harm another within the driving environment (Hauber, 1980; Hennessy & Wiesenthal, 1999, 2001b). The study of aggressive driving has been performed in a variety of settings beyond North America with research conducted in Australia (Lupton, 2001; 2002; in press), Finland (Lajunen, Parker, & Summala, 1999), Germany (Krahé & Fenske, 2002), Great Britain (Lajunen & Parker, 2001; Lajunen, Parker, & Stradling, 1998; Lajunen et al., 1999), Israel (Shinar, 1998; Shinar & Compton, 2004; Yagil, 2001), Italy (Hennessy, Wiesenthal, & Piccione, 2001), and the Netherlands (Lajunen et al., 1999).

Due to its relationship with traffic violations (Novaco, 1991), driver aggression represents a potential danger, either directly or indirectly, to all roadway users. The American Automobile Association (Mizell, 1997) has estimated that incidents of roadway aggression in the United States increased by more than 50% between 1990 and 1996. Similarly, elevated levels of driver aggression have also been noted worldwide (Sleek, 1996; Taylor, 1997).

It has recently been argued that the increased incidence of roadway aggression can be linked to a steady increase in traffic volume, congestion, and resulting driver stress over recent decades (Donelly, 1998; Hennessy, 1999a; Taylor, 1997). The increase in the number of vehicles on the road has not been matched by a corresponding expansion of public roadways (Donelly, 1998). As a result of this greater competition for space, congestion level, frustration, irritation, anger, and aggression have escalated. As traffic congestion increases, other drivers represent obstacles that impede the attainment of specific goals, such as driving at a certain speed or arriving at a destination in a specific time frame (Novaco, Stokols, Campbell, & Stokols, 1979). In this respect, aggression is often directed toward other drivers who are perceived to be the source of frustration and irritation (Gulian, Debney, Glendon, Davies, & Matthews, 1989; Hennessy, 1999b). Vengeance represents one type of a reaction to stressful driving situations. The desire to retaliate against other drivers is often expressed as a wish to "teach the other driver a lesson". Wiesenthal, Hennessy, & Gibson (2000) constructed a scale (the Driving Vengeance Questionnaire, DVQ) to measure the desire to engage in vengeful driving and found that it assessed a single factor (vengeance), possessed high reliability, and produced no gender differences. Further research (Hennessy & Wiesenthal, 2001b) validated the DVQ as a predictor of aggressive and violent driving. Cross-national research has indicated that Italian drivers react with vengeful behaviors to the same situations that antagonize Canadian drivers (Hennessy et al., 2001).

Interaction Model of Driving Stress, Aggression, and Violence

As can be seen in Figure 1, the perception of driver stress is seen as arising from the interaction of both state and trait variables. State variables are those present in the situation and constitute the environment confronting the motorist. State variables are external to the driver and consist of the number of vehicles on the roadway, possible crowding of the roadway, heat, noise, vibration, time urgency, etc. Trait variables, internal to the driver, are individual differences and arousal due to various drive states. Driving history, age, level of risk preference, Type A behavior patterns, weak impulse control, and predisposition to stress (i.e., trait stress) would constitute such traits or individual difference variables. Both state and trait variables have been demonstrated to produce the perception of stress in drivers. These two categories of variables may also influence each other as indicated by the arrows. Individuals may change their trait stress levels through experience with commuting, or could increase in severity or extremity of response due to repeated exposure. Once stress is experienced, the driver will engage in a variety of both direct and indirect coping responses.

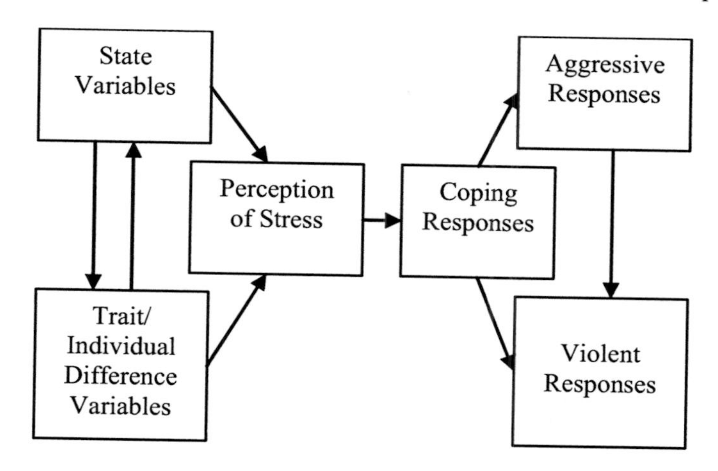

Figure 1. Interaction model of the causation of driving stress, aggression, and violence.

Direct versus Indirect Coping Behaviors

Direct coping behaviors are directed at removing or directly dealing with the source of stress. These responses typically are more persistent, more active, and more rational (Hennessy, 1995). Seeking information, changing behaviors, and planning (Carver, Sheier, & Weintraub, 1989) are examples of direct coping along with problem solving (Nakano, 1991). Indirect coping behaviors are emotion focused and deal with how a problem is experienced. Indirect coping responses tend to be more passive; denial, the need for emotional support, seeking comfort in religion, self-blame, and cognitive avoidance are all examples of this tactic (see Baum, Singer, & Baum, 1982; Carver et al., 1989; Nakano, 1991 Vingerhoets & Van-Heck, 1990; Nakano, 1991). Indirect coping behaviors have been linked to increased stress experiences, while direct coping behaviors have been associated with more effective stress reduction (Billings & Moos, 1981; Nakano, 1989, 1991).

Aggression directed against other motorists may be the emotional outcome of the unsatisfactory resolution of coping with driving stress. If milder aggression does not release the driver's tension, or if other drivers respond with hostility, then more extreme responses of a violent nature may ensue. The desire for vengeance may escalate the aggressive response very rapidly and, occasionally, to life-threatening levels.

Perceptions of Objectionable Driver Behavior: Public Concern with Aggressive Driving

Aggressive driving is currently perceived to be an increasingly pervasive problem in North America (Mizell, 1997; Wiesenthal, Wickens, & Rippey, 2002). The Ottawa-based

Traffic Injury Research Foundation (TIRF; Beirness, Simpson, Mayhew, & Pak, 2001) recently conducted over 1,200 telephone interviews with Canadians to sample public attitudes concerning road safety. Sixty-five percent of those interviewed perceived aggressive driving as a serious problem posing a greater danger than sleepy drivers, road conditions, or vehicle defects. Among acts of aggressive driving, 74% of Canadians view running a red light as a serious problem, with two thirds considering speeding to be a serious risk. Women were more threatened than men by aggressive driving, and Ontario drivers reported more encounters with aggressive drivers than did residents of any other province. In the TIRF's annual survey, it was seen that Canadian motorists experienced the following aggressive driving behaviors from most common to least common: speeding, tailgating, failing to signal, weaving in traffic, unsafe passing, driving slowly in the passing lane, failing to stop, and running red lights (Beirness et al., 2001). In contrast, courteous behaviors (e.g., waiting for pedestrians and allowing other drivers to merge) were reported to be less likely exhibited.

Toljagic (2000) indicated that 38% of Ontario drivers reported experiencing some form of abuse over the past year, while 90% of American Automobile Association members reported witnessing an aggressive driving incident in a year (National Conference of State Legislatures, 2000). Wald (1997) reported an estimated 28,000 highway fatalities in the United States attributable to aggressive driving. This increasing public concern over incidents of aggressive driving or "road rage" has led to a variety of initiatives by psychologists, automobile associations, police and regulatory agencies (see Rathbone & Huckabee [1999] for a survey commissioned by the American Automobile Association describing of these programs). One program adopted by a number of police forces across North America has involved encouraging drivers whose vehicles are equipped with cellular telephones to report incidents of dangerous driving. In southern Ontario, the Ontario Provincial Police focused on careless driving, especially when the motorist was driving with only one hand on the steering wheel, improperly using a cellular telephone, drinking coffee, reading, or driving too slowly (Mitchell, 1997). Citizen complaints averaged approximately 500 calls per week.

Rasmussen et al. (2000) surveyed drivers in Las Vegas and found that 21.6% had reported other drivers to the police. Drivers indicated that their major sources of annoyance were slow drivers, children not restrained in car seats, being followed too closely, tourists uncertain of their routes, and automobiles weaving from lane to lane. Somewhat less objectionable behaviors were cars passing in the parking lane, drivers conversing on cell phones, motorcyclists passing between lanes, bicyclists, drivers flashing headlights, and semi-tractor trailer trucks. The Las Vegas respondents overwhelmingly perceived drivers as more aggressive and dangerous than they were five years ago (75.8%) and indicated that drivers in their city were inferior to drivers elsewhere (57.8%).

Neighbours, Vietor, and Knee (2002) studied an ethnically diverse group of undergraduate psychology students at an urban university located in the American southwest. The vast majority of these students commuted to the university, with the median reported driving times from 8-12 hours per week. The sample produced a total of 775 driving anger records, with an average of 9.40 (SD = 4.18) per driver. When asked to indicate what provoked anger when they drove, the students indicated that discourtesy (59.1%) was the most frequent complaint. Since almost half the events fell into more than one category, 43.5% of events were described as due to dangerous driving, with slow driving causing jams (15.5%) constituting the third most anger provoking experience.

Underwood, Chapman, Wright, and Crundall (1999) equipped 100 drivers in England with tape recorders and instructed them to describe anger provoking situations that they encountered during their normal driving trips over a two-week period. Respondents recorded 293 near-accidents, 383 anger-arousing incidents, and 318 acts of courtesy. Anger was related to traffic congestion along with the number of near-accidents that were experienced, particularly when the driver reported that he/she was not at fault.

An association between aggressive roadway behaviors and the presence of firearms has been suggested by Miller, Azrael, Hemenway, and Solop (2002). Using random digit dialing, Arizona drivers were interviewed to determine how often they drove with firearms and whether they engaged in a variety of objectionable behaviors ranging from shouting obscenities to following or blocking other motorists. Engaging in aggressive roadway behaviors was more commonly reported by armed motorists, comprising 11% of the surveyed drivers, suggesting a possible priming effect of weapons. Whether aggressive drivers are disposed to carry weapons or whether the possession of weapons lowers inhibitions to aggressive behaviors remains a question for future research.

PERCEPTIONS OF CALIFORNIA MOTORISTS

Sarkar, Martineau, Emami, Khatib, and Wallace (2000) reported an analysis of 1,987 cellular telephone calls made to the California Highway Patrol in San Diego County during April, June, and September 1998. Improper lane usage constituted the largest category of complaints (27.1%) with speeding, in combination with other objectionable behaviors, accounting for a quarter of all complaints (24.6%). Speeding by itself, comprised another 19.8% of calls with "road rage" comprising another 16.1%. The results of this analysis may be problematic since there is no information on how the calls were categorized by the judges, nor is any reliability statistic presented. Sarkar et al. considered speeders to be inconsiderate drivers, but speeders may consider *slow* drivers to be inconsiderate. The categories were not defined, so readers may wonder just how close following vehicles had to be before they were labeled as tailgaters. Is tailgating during rush hour aggressive behavior? If following vehicles keep the recommended distance from a car in front of them, will another driver cut in front to fill the space? Is flashing a high beam serving as a signal of intent to pass, or a reminder rather than an aggressive response as described in their report? The use of classifications based upon a *combination* of categories (e.g., speeding plus some other behavior) produces an unclear categorization of the driving behavior. The Sarkar et al. data is unable to state the number of complaints involving weaving between lanes, tailgating, etc. Also problematic are assumptions made regarding the driver's intentions (e.g., "Forced vehicle off road" was classified as road rage, whereas the objectionable response might have been the result of driver inattention, defensive reactions, etc.). Sarkar et al. also included very minor incidents in their road rage category, producing a distorted view of driving violence (e.g., horn honking, preventing others from passing).

PERCEPTIONS OF ONTARIO MOTORISTS

Prior to the publication of the Sarkar et al. (2000) study, Wiesenthal et al. (2002) conducted a content analysis of the 14,406 telephone calls from concerned motorists made to the Ontario Provincial Police (OPP) in 2000. The OPP had recently concluded an initiative to encourage motorists equipped with cellular telephones to report dangerous or aggressive drivers. A sample of these reports was used to construct a content analysis coding system. These initial reports were taken from drivers using Highway 401 in the Greater Toronto Area. The data provided by the OPP was "cleaned", such that no identifying information was provided as to either the identity of the caller or the reported vehicle license plate number to ensure confidentiality and anonymity. A ten category system was developed as follows:

1. Improper speed (i.e., speeding/racing, unnecessarily slow driving, sporadic speed changes)
2. Tailgating or following too closely
3. Dangerous lane changes/improper lane usage
4. Improperly equipped or unsafe vehicles
5. Ignoring traffic signs/signals
6. Hostile driver displays (e.g., verbal or obscene gestures)
7. Erratic driving
8. Driver inattention
9. Hazardous road conditions not attributable to driver behavior (e.g., objects on road)
10. Other: behavior not matching the above categories.

Each telephone call could be placed in any one or more categories (e.g., "Possibly impaired, weaving, nearly forced me off the road, no headlights" would be coded as erratic driver [7], dangerous lane changes and lane usage [3], and improperly equipped and unsafe vehicle [4]). The category of erratic driving [7] was included to avoid making assumptions concerning driver intent. Some driver complaints gave descriptions of possibly impaired or sleepy drivers, or erratic, reckless, or aggressive driving behavior. Although in some cases there was some indication as to what form or category of behavior this described, it was decided to develop a category for these erratic driving complaints in order to avoid improperly distorting the data. Each month's telephone complaints were coded by two independent raters: a graduate student and one undergraduate student drawn from the research team. If there was a disagreement between the two primary coders, a third independent coder categorized the call. The final classification included the categories selected by at least two of the three coders. The coding system produced reasonably high interjudge reliability (Cohen's kappa = .84).

As can be seen in Figure 2, dangerous lane usage constituted 62% (N = 8,976) of the calls, 39% (N = 5,551) of the complaints concerned excessive speed, 24% (N = 3,433) of the complaints cited erratic driving, while 18% (N = 2,533) of the calls were complaints about tailgating. Figure 3 illustrates that lane usage, speeding, and tailgating constituted the three most common complaints for the category of erratic driving (comprising a total of 3,024 reports), representing 78%, 35%, and 23% of the calls (note that as the categories are not mutually exclusive, the total can exceed 100%). Although it is difficult to draw a direct comparison between the Ontario and California findings, due to the differences between the two coding schemes, there is some similarity in the findings. Ignoring their combination category, Sarkar et al. (2000) also identified dangerous lane usage, speeding, and tailgating, in this order, as the three most common complaints among California drivers. A noticeably larger proportion of Ontario than California drivers (62% versus 27%) identified lane issues as the source of their complaint; however, the use of a combination category and a "road rage" category in the California study is likely the cause of the reduced frequency in their weaving and cutting complaints.

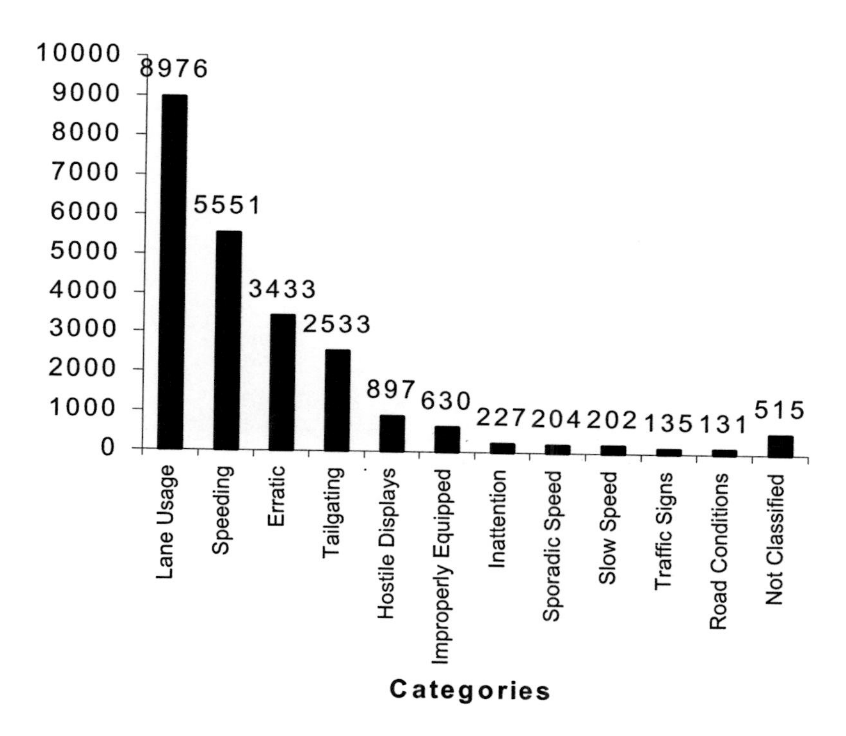

Figure 2. Frequency of driver complaints to the OPP.

Figure 4 indicates that Fridays (with 18% of the complaints) generated more complaints than other days, which replicates the findings of Sarkar et al. (2000). Mondays and Sundays represent periods of reduced complaints in Ontario (13% and 12% of the complaints respectively; $\chi^2(6) = 280.02$, $p<.001$), which is similar to California drivers who reported the fewest complaints on Sundays.

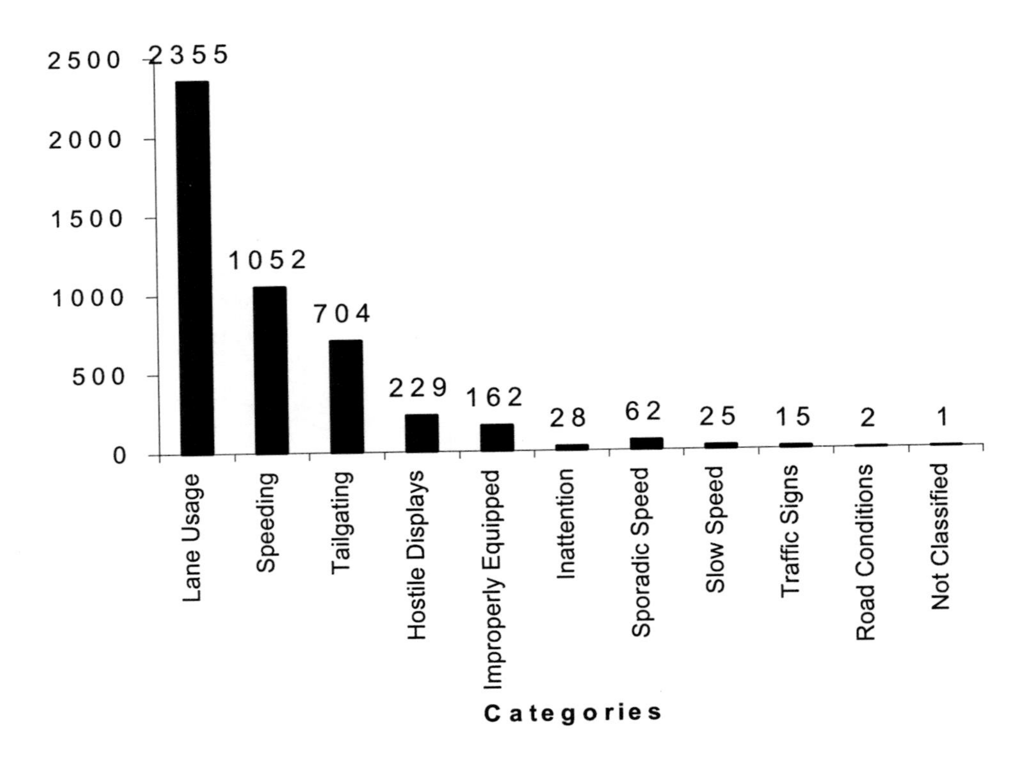

Figure 3. Frequency breakdown of driver complaints to the OPP classified as erratic driving and at least one other category.

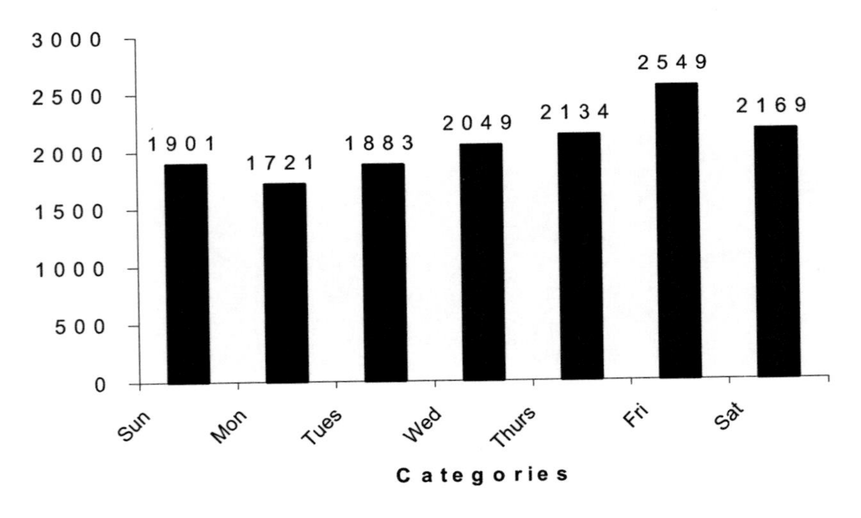

Figure 4. Frequency of driver complaints to the OPP by day of the week.

Figure 5 indicates that the afternoon rush hour (1500-1800 hours) produced the greatest number of complaints (21% of the complaints versus 9% of the complaints for the morning rush hour, 0600-0900; $\chi^2(7) = 3{,}723.08$, $p<.001$). Sarkar et al. (2000) also reported that the greatest number of complaints from California motorists came during this same afternoon rush hour. When each category of driving behavior in Ontario was analyzed separately, dangerous lane usage was most frequent between 1500 and 1800 hours, whereas speeding and tailgating were most frequent between 1200 and 1500 hours. For California drivers, weaving and cutting was also most commonly reported between 1500 and 1800 hours, but speeding plus some other behavior and tailgating alone were reported most commonly between 1200 and 1800 hours.

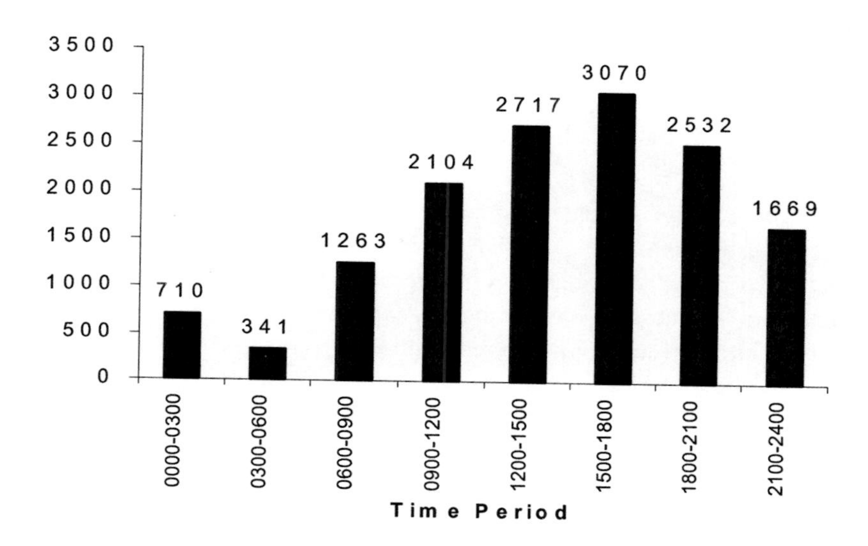

Figure 5. Frequency of driver complaints to the OPP by time of day.

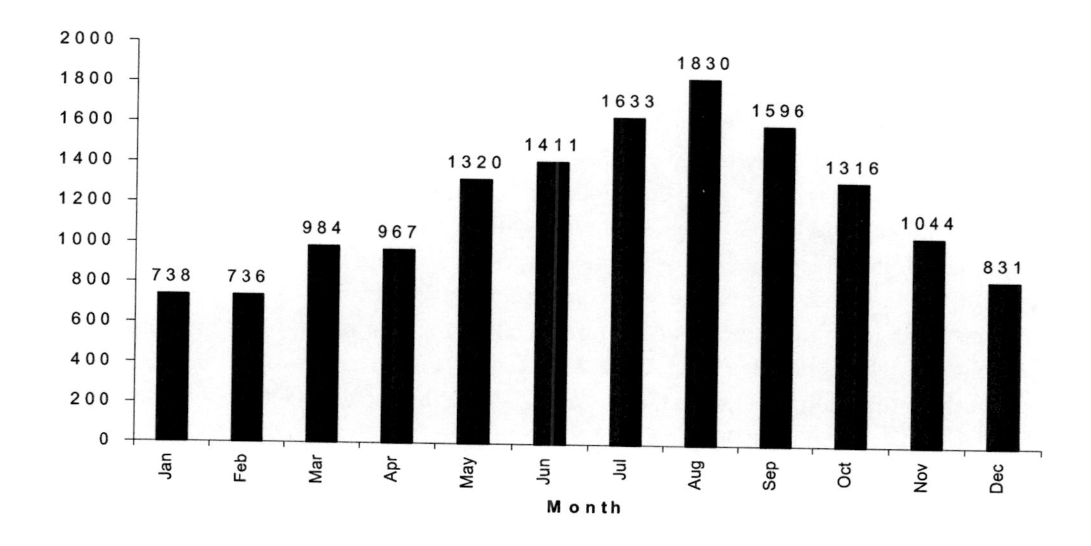

Figure 6. Frequency of driver complaints to the OPP by month of the year.

Figure 6 indicates that the total number of complaints was greatest in the summer months when more vehicles are on the road ($\chi^2(11) = 1,252.61$, $p<.001$). August represented the peak period of complaints (13%). For each category, the trends were reasonably stable over the course of the year. Unfortunately, because the California study only examined three months out of the year, a comparison of the seasonal variation in driver complaints between the two locales was not possible. Climate differences would likely have had a significant impact on any differences in seasonal variations in each community. It would be interesting to see if the same degree of seasonal variation that was identified in Ontario would arise in a warm climate area such as California.

Conclusion

These results clearly demonstrate that increased traffic congestion is related to the number of reported incidents of aggressive driving behavior. This is consistent with previous findings that have provided evidence for congestion-induced stress leading to aggressive roadway behaviors (Hennessy & Wiesenthal, 1997). The results also highlight the importance of drivers' discomfort when they perceive violations of crucial driving norms because these violations reduce the motorists' confidence in their ability to *predict* the behaviors of other drivers.

In total, the Ontario results are quite similar to the conclusions of the Sarkar et al. (2000) study of driver complaints in southern California. Supporting these studies of highway driving behavior, recent research has examined similar issues on local streets. Motorists in Toronto who routinely drove on local roadways were questioned regarding their perceptions of objectionable driving practices. Improper lane usage (58%) and speeding (30%) again produced the most complaints (Papalasarou, 2002). Overall, female drivers experienced more anger when observing improper driving, but male drivers were seen to endorse more vengeful driving tactics (Papalasarou, 2002).

A more thorough understanding of what drivers perceive to be objectionable driving behaviors does have some important potential applications. Knowing what these behaviors are, and when they are likely to occur, allows for greater efficiency for enforcement activities as well as for implementing educational campaigns. Likewise, the best way to educate is to deliver the message early in a driver's training, and this research identifies themes that can be incorporated into driver education and testing. The development and testing of a coding scheme offers simplicity for police to record and analyze driver complaints and could easily lend itself to assessing the success of enforcement and educational efforts aimed at reducing objectionable behaviors. The wide scale adoption of this system would facilitate studying these actions over a longer time frame and across differing locales. Geographical and cultural differences in perceptions of objectionable driving behavior, characteristics of driver complainants and offenders, features of their vehicles, and the effectiveness of grass-roots driving safety and enforcement campaigns, are all topics that could benefit from a more widespread use of this content analysis research, and all would contribute to improved safety on our roads and highways.

Acknowledgements

The authors wish to thank Faruk Gafic, Jonathan Belman, Margaret Chuong, Renée Desmond, Lisa Thompson, Wendi Perez, Debbie Ng, Kathy Mazurek, Christina Balioussis, Yvette Lobl, Fiorella Lubertacci, and Sara Ahadi for their assistance. Ms. Judy Manners assisted with her superlative word processing abilities. This research was partially supported by a scholarship awarded to Christine M. Wickens by the Canadian Transportation Research Forum and by Contract #540260 from the Ontario Ministry of Transportation awarded to David L. Wiesenthal. Opinions expressed in this report are those of the authors and do not necessarily reflect the views and policies of the Ministry.

References

Baum, A., Singer, J. E., & Baum, C. (1982). Stress and the environment. In G. W. Evans (Ed.), *Environmental stress* (pp. 15-44). New York: Cambridge University Press.

BBC News. (2000, November 1). *Commuting is "biggest stress"*. Retrieved November 1, 2000, from http://haddock.org/matt/ bbcnews.html

Beirness, D. J., Simpson, H. M., Mayhew, D. R., & Pak, A. (2001). *The Road Safety Monitor: Aggressive driving*: Ottawa, ON: Traffic Injury Research Foundation.

Billings, A. G., & Moos, R. H. (1981). The role of coping responses and social resources in attenuating the stress of life events. *Journal of Behavioral Medicine, 4,* 139-157.

Carver, C., Sheier, M., & Weintraub, J. K. (1989). Assessing coping strategies: A theoretically based approach. *Journal of Personality and Social Psychology, 56,* 267-283.

Deffenbacher, J. L., Huff, M. E., Lynch, R. S., Oetting, E. R., & Salvatore, N. F. (2000). Characteristics and treatment of high-anger drivers. *Journal of Counselling Psychology, 47,* 5-17.

Donelly, S. B. (1998, January 12). Road rage. *Time Magazine, 151,* 44-48.

Ellison, P. A., Govern, J. M., Petri, H. L., & Figler, M. H. (1995). Anonymity and aggressive driving behavior: A field study. *Journal of Social Behavior and Personality, 10,* 265-272.

Gulian, E., Debney, L. M., Glendon, A. I., Davies, D. R., & Matthews, G. (1989). Coping with driver stress, In F. J. McGuigan, W. E. Sime, & J. M. Wallace (Eds.), *In stress and tension control 3: Stress management* (pp. 173-186). New York: Plenum Press.

Gulian, E., Matthews, G., Glendon, A. I., Davies, D. R, & Debney, L. M. (1989). Dimensions of driver stress. *Ergonomics, 32,* 585-602.

Hartley, L. R., & El Hassani, J. (1994). Stress, violations, and accidents. *Applied Ergonomics, 25,* 221-234.

Hauber, A. R. (1980). The social psychology of driving behaviour and the traffic environment: Research on aggressive behaviour in traffic. *International Review of Applied Psychology, 29,* 461-474.

Hennessy, D. A. (1995). *The relationship between traffic congestion, driver stress, and direct versus indirect coping behaviors.* Unpublished master's thesis, York University, Toronto, Canada.

_____ . (1999a). *Evaluating driver aggression.* Unpublished major area paper, York University, Toronto, Canada.

_____ . (1999b). *The interaction of person and situation within the driving environment: Daily hassles, traffic congestion, driver stress, aggression, vengeance, and past performance.* Unpublished doctoral dissertation, York University, Toronto, Canada.

Hennessy, D. A., & Wiesenthal, D. L. (1997). The relationship between traffic congestion, driver stress and direct versus indirect coping behaviours. *Ergonomics, 40,* 348-361.

_____ . (1999). Traffic congestion, driver stress, and driver aggression. *Aggressive Behavior, 25,* 409-423.

_____ . (2001a). Further validation of the Driving Vengeance Questionnaire. *Violence and Victims, 16,* 565-573.

_____ . (2001b). Gender, driver aggression, and driver violence: An applied evaluation. *Sex Roles, 44,* 661-676.

_____. (2002). The relationship between driver aggression, violence, and vengeance. *Violence and Victims, 17,* 707-718.

Hennessy, D. A., Wiesenthal, D. L., & Kohn, P. M. (2000). The influence of traffic congestion, daily hassles, and trait stress susceptibility on state driver stress: An interactive perspective. *Journal of Applied Biobehavioral Research, 5,* 162-179.

Hennessy, D. A., Wiesenthal, D. L., & Piccione, G. (2001). Aggressive driving, stress and vengeance in Canada and Italy. Proceedings of the Canadian Multidisciplinary Road Safety Conference XII, June 10-13, London, ON.

Hennessy, D. A., Wiesenthal, D. L., Wickens, C. M., & Lustman, M. (2004). The impact of gender and stress on traffic aggression: Are we really that different? In J. P. Morgan (Ed.), *Focus on aggression research.* Hauppauge, NY: Nova Science Publishers.

Joint, M. (1997). *Road rage.* Washington, DC: AAA Foundation for Traffic Safety.

Krahé, B., & Fenske, I. (2002). Predicting aggressive driving behaviour: The role of macho personality, age, and power of car. *AggressiveBehavior, 28,* 21-29.

Lajunen, T., & Parker, D. (2001). Are aggressive people aggressive drivers? A study of the relationship between self-reported general aggressiveness, driver anger and aggressive driving. *Accident Analysis and Prevention, 33,* 243-255.

Lajunen, T., Parker, D., & Stradling, S. G. (1998). Dimensions of driver anger, aggressive and highway code violations and their mediation by safety orientation in UK drivers. *Transportation Research Part F, 1,* 107-121.

Lajunen, T., Parker, D., & Summala, H. (1999). Does traffic congestion increase driver aggression? *Transportation Research Part F, 2,* 225-236.

Lupton, D. (2001). Constructing "road rage"as news: An analysis of two Sydney newspapers. *Australian Journal of Communication, 28,* 23-36.

_____ . (2002). Road rage: Driver's understandings and experiences. *Journal of Sociology, 38,* 275-290.

_____ (in press). Pleasure, aggression, and fear: The driving experiences of young Sydneysiders. In W. Mitchell & R. Bunton (Eds.), *Young people and leisure.* London: Palgrave.

Matthews, G., Dorn, L., Hoyes, T. W., Davies, D. R., Glendon, A. I., & Taylor, R. G. (1998). Driver stress and performance on a driving simulator. *Human Factors, 40,* 136-149.

Matthews, G., Tsuda, A., Xin, G., & Ozeki, Y. (1999). Individual differences in driver stress vulnerability in a Japanese sample. *Ergonomics, 42,* 401-415.

Miller, M., Azrael, D., Hemenway, D., & Solop, F. I. (2002). 'Road rage' in Arizona: Armed and dangerous. *Accident Analysis and Prevention, 34,* 807-814.

Mitchell, B. (1997, November 22). "Road rage" on rise in GTA. *Toronto Star*, p. A2.

Mizell, L. (1997). *Aggressive driving*. Washington, DC: AAA Foundation for Traffic Safety.

Nakano, K. (1989). Intervening variables of stress, hassles, and health. *Japanese Psychological Research, 31,* 143-148.

_____ . (1991). Coping strategies and psychological symptoms in a Japanese sample. *Journal of Clinical Psychology, 47,* 346-350.

National Conference of State Legislatures, Environment, Energy and Transportation Program. (2000, January). *Aggressive driving: Background and overview report.* Retrieved April 7, 2001, from http://www.ncsl.org/programs/esnr/aggrdriv.htm

Neighbors, C., Vietor, N. A., & Knee, C. R. (2002). A motivational model of driving anger and aggression. *Personality and Social Psychology Bulletin, 28,* 324-335.

Novaco, R. W. (1991). Aggression on roadways, In R. Baenninger (Ed.), *Targets of violence and aggression* (pp. 253-325). Amsterdam, North-Holland: Elsevier.

Novaco, R. W., Stokols, D., Campbell, J., & Stokols, J. (1979). Transportation, stress, and community psychology. *American Journal of Community Psychology, 7,* 361-380.

Novaco, R. W., Stokols, D., & Milanesi, L. (1990). Objective and subjective dimensions of travel impedance as determinants of commuting stress. *American Journal of Community Psychololog*y, *18,* 231-257.

Papalasarou, E. (2002). *Driver perceptions of objectionable driving behaviors on Greater Toronto Area roadways.* Unpublished honors thesis, York University, Toronto, Canada.

Rasmussen, C., Knapp, T. J., & Garner, L. (2000). Driving-induced stress in urban college students. *Perceptual and Motor Skills, 90,* 437-443.

Rathbone, D. B., & Huckabee, J. C. (1999). *Controlling road rage: A literature review and pilot study.* Washington, DC: AAA Foundation for Traffic Safety.

Sarkar, S., Martineau, A., Emami, M., Khatib, M., & Wallace, K. (2000). Aggressive driving and road rage behaviors on freeways in San Diego, California. *Tranportation Research Record, 1724,* 7-13.

Shinar, D. (1998). Aggressive driving: The contribution of the drivers and the situation. *Transportation Research Part F, 1,* 137-160.

Shinar, D., & Compton, R. (2004). Aggressive driving: An observational study of driver, vehicle, and situational variables. *Accident Analysis and Prevention, 36,* 429-437.

Sleek, S. (1996). Car wars: Taming drivers' aggression. *APA Monitor, 27,* 13-14.

Stokols, D., & Novaco, R. W. (1981). Transportation and well being. In G. Altman, J. F. Wohlwill, & P. B. Everett (Eds.), *Human behavior and environment: Vol. 5. Transportation and behavior.* London: Plenum Press.

Taylor, B. (1997, August 25). Life in the slow lane. *Toronto Star*, p. D1.

Toljagic, M. (2000, October 28). Motorists tuning in to road rage. *Toronto Star,* pp. G1-G2.

Underwood, G., Chapman, P., Wright, S., & Crundall, D. (1999). Anger while driving. *Transportation Research Part F: 2,* 55-68.

Vingerhoets, A. J., & Van-Heck, G. L. (1990). Gender, coping, and psychosomatic symptoms. *Psychological Medicine, 20,* 120-135.

Wald, M. L. (1997, July 18). Temper cited as cause of 28,000 road deaths a year. *New York Times,* pA14.

Wickens, C. M. (2004). *Occupational stress, trait stress susceptibility, traffic congestion, and state driver stress.* Unpublished master's thesis, York University, Toronto, Canada.

Wickens, C. M., & Wiesenthal, D. L. (2004, June). Occupational stress, trait stress susceptibility, traffic congestion, and state driver stress. Paper presented at the Canadian Multidisciplinary Road Safety Conference XIV, Ottawa, ON, Canada.

______. (in press) State driver stress as a function of occupational stress, taffic congestion, and trait stress susceptibility. *Journal of Applied biobehavioral Research.*

Wiesenthal, D. L., Hennessy, D. A., & Gibson, P. M. (2000). The Driving Vengeance Questionnaire (DVQ): The development of a scale to measure deviant drivers' attitudes. *Violence and Victims, 15,* 115-136.

Wiesenthal, D. L., Hennessy, D. A., & Totten, B. (2000). The influence of music on driver stress. *Journal of Applied Social Psychology, 30,* 1709-1719.

Wiesenthal, D. L., Wickens, C. M., & Rippey, K. (2002, May). *Content analysis of driver complaints of aggressive behaviour.* Poster presented at the Canadian Psychological Association 63rd Annual Conference, Vancouver, British Columbia, Canada.

Yagil, D. (2001). Interpersonal antecedents of drivers' aggression. *Transportation Research Part F, 4,* 119-131.

In: Contemporary Issues in Road User Behavior and Traffic Safety ISBN:1-59454-268-6
Editors: D. Hennessy and D. Wiesenthal, pp. 37-46

Chapter 4

BEHIND THE WHEEL: CONSTRUCT VALIDITY OF AGGRESSIVE DRIVING SCALES

James Rotton, Paul J. Gregory and David L. Van Rooy
Florida International University

INTRODUCTION

It is probably safe to say that most people blame other drivers rather than environmental factors, such as congestion or road conditions, for aggression on the roadways. This assertion is based on results that Burns and Katovich (2003) obtained when they examined 390 accounts of "aggressive driving" and "road rage" in *The New York Times*, *The Dallas Morning News*, and *Los Angeles Times*. They found that the authors of 71.9% of the articles attributed the causes or aggressive driving and road rage to human factors; only 26.1% mentioned environmental factors. Although Burns and Katovich do not use the term, this finding is a good example of the fundamental attribution error: Individuals tend to explain other people's behavior in terms of dispositional factors even when situational cues are prominent.

The fundamental attribution error may explain why most researchers have addressed the problem of aggressive driving by adopting a trait approach. Table 1 summarizes recent measures of road rage and aggressive driving. It does not include early measures (e.g., Parry, 1968), which are described in Van Rooy, Rotton, and Burns (in press), nor does it list all of the quizzes and inventories that can be found in popular books (e.g., James & Nahl, 2000) and on the Internet (e.g., American Automobile Foundation for Traffic Safety, 2002). However, it might be noted that two of the instruments in Table 1 had their origins in books written for the general populace: the Road Rage Scale (Wells-Parker, Ceminsky et al., 2002) and the Driver Stress Profile (Larson, 1996). Finally, the table includes only one of the many versions of the Driver Behavior Questionnaire (DBQ), which was originally developed to measure three predictors of traffic accidents: lapses, errors, and traffic violations (Reason, Manstead, Stradling, Baxter, & Campbell, 1990). The DBQ has gone through a number of modifications and reincarnations (Meskin, Lajunen, & Summala, 2002). It appears that lapses and errors do not predict accident involvement (Parker, Reason, Manstead, & Stradling, 1995), only the DBQ's violations scale does so.

Table 1. Inventories and Scales Used to Assess Aggressive Driving

Inventory/Subscale	Author	k	Alpha
Aggressive Driving Behavior Scale (ADB)	Houston et al. (2003)	11	0.80
Speeding			
Conflict Behavior			
Aggressive Driving Scale (ADS)	Krahe & Fenske (2002)	24	0.85
Driver Anger Scale (DAS)	Deffenbacher et al. (1994)	14	0.80
Driving Anger Expression Inventory (DAX)	Deffenbacher et al. (2002)	62	0.90
Verbal Aggressive Expression		12	0.88
Personal Physical Aggressive Expression		11	0.81
Using Vehicle to Express Anger		11	0.86
Displaced Aggression		4	0.65
Adaptive/Constructive Expression		15	0.90
Driver's Angry Thoughts Questionnaire (DATQ)	Deffenbacher et al. (2003)		
Judgmental & Disbelieving Thinking		21	0.94
Pejorative Labeling & Verbally Aggressive		13	0.92
Revenge & Retaliatory Thinking		14	0.13
Physically Aggressive Thinking		8	0.93
Self-Instruction		9	0.83
Driver Vengeance Questionnaire (DVQ)	Wiesenthal et al. (2000)	15	0.79
Driver Behavior Inventory (DBI)	Glendon et al. (1991); Gulian et. al. 1989	37	0.80
Driving Aggression			
Overtaken irritation			
Alertness			
Dislike Driving			
Overtaking Frustration			
Driver Stress Profile (Larson, 1996)	Blanchard et al. (2000)	40	0.93
Impatience			
Anger			
Competing			
Punishing			
Dula Dangerous Driving Index (DDDI)	Dula & Ballard (2003)	31	0.92
Aggressive Driving			
Negative Emotional Driving			
Risky Driving			
Propensity for Angry Driving Scale (PADS)	DePasquale et al. (2001)	19	0.88-0.89
Road Rage Inventory (RRI)	Welles-Parker et al.	16	0.72
Violations Component of the DBQ	Lawton et al. (1997)	12	nr
Fast Driving			
Maintaining Progress			
Anger/Hostility			

k = number of items.
nr = not reported.

Developing scales to assess road rage and aggressive driving has become something of cottage industry, which makes it difficult to evaluate their construct validity. We use the term "cottage industry" to describe studies whose authors rarely mention earlier measures, let alone compare their scales with existing ones. For example, between the time when Van Rooy et al. (in press) undertook and completed their psychometric evaluation of three early scales, another six scales appeared in the literature. Referring to Table 1, it can be seen that one team of investigators (Deffenbacher, Lynch, Oetting, & Swaim, 2002; Deffenbacher, Petrilli, Lynch, Oetting, & Swaim, 2003; Deffenbacher, Oetting, & Lynch, 1994) has generated no fewer than three scales (and 11 subscales) to measure various aspects of angry driving.

The authors of the scales listed in Table 1 have been primarily concerned with the validity of their individual scales. This can be distinguished from the validity of aggressive driving--that is, the construct that the scales are meant to measure (Cronbach & Meehl, 1955). The latter requires a consideration of the convergent, discriminant, and criterion-related validity of multiple measures of the construct in question (Campbell & Fiske, 1959; Nunnally & Bernstein, 1994).

Convergent Validity

Van Rooy et al. (in press) expressed concern over the fact that correlations among the Driving Anger Scale (DAS), the Driver Vengeance Questionnaire (DVQ), and the General component of the Driver Behavior Inventory (DBI) were so high as to make the scales interchangeable. This concern is borne out by a recent analysis of data obtained from 291 (77 male, 214 female) undergraduates whose ages ranged from 16 through 49 years ($M = 22.42$; $SD = 4.65$). In addition to the DAS, DVQ, and the aggression component of the DBI, respondents in this more recent sample also completed the Aggressive Driving Scale (ADS), the Propensity for Angry Driving Scale (PADS), and the Road Rage Inventory (RRI). A principle components analysis indicated that the six scales loaded on a single factor that explained 58.6% of the variance.

Discriminant Validity

It is something on an irony that three of the measures in Table 1 were developed by a team led by an author whose previous work (Deffenbacher, 1992) is frequently cited as evidence that anger is a trait that shows consistency across situations. More than one investigator (Arnett, Offer, & Fine, 1997; Beirness, 1993; Lajunen & Parker, 2001) has suggested that aggressive driving is indicative of the more general trait that leads individuals to be aggressive in a variety of situations; that is, as Tillmann and Hobbes (1949, p. 329) asserted more than half of a century ago, "a man drives as he lives." This conclusion is consistent with results that Deffenbacher, Lynch, Oetting, and Yingling (2001) obtained in a validation study of the DAS. Their respondents completed four measures of driving-related aggression (e.g., "argument with passenger while you were driving") and three measures of general aggressive tendencies (e.g., "got into arguments when out with friends"). Applying Fisher's r-to-Z transformation, we determined that the DAS is more highly correlated with general aggressive tendencies (mean $r = .31$) than it is with driving-related aggression (mean $r = .24$).

Van Rooy et al. (in press) developed structural equation models to determine if the previously mentioned measures of aggressive driving (DBI, DAS, and DBQ) loaded on the same factor as scales from a measure of general aggression. They administered Buss and Perry's (1992) Aggression Questionnaire (AQ) to assess general aggression. The 29 items on this inventory load on four correlated factors: physical aggression ("If somebody hits me, I hit back"), verbal aggression ("I tell my friends openly when I disagree with them"), anger ("When frustrated, I let my irritation show"), and hostility ("I am sometimes eaten up by jealousy"). Contrary to our expectations, we found that the three measures of aggressive driving and four measures of general aggression did not load on a single factor. Instead, the best fit was obtained for a model that included separate factors for aggressive driving and general aggression; however, the correlation between the two factors was high (i.e., a phi coefficient of .71).

Although the results that Van Rooy et al. (in press) obtained are consistent with the idea that aggressive driving inventories measure something more than general aggression, it might be objected that this finding was based on an insufficient amount of evidence. Separate factors may have emerged because of the method variance that resulted from using a single instrument (AQ) to assess general aggression. To address this and other concerns, we had undergraduates in our follow-up study complete a second measure of general aggression as well as the six measures of aggressive driving described in the section on convergent validity. In addition to the AQ, our undergraduates completed Gladue's (1991) Aggression Inventory (AI). This 20-item inventory was designed to assess four aspects of aggressive behavior: physical aggression ("I get into fights with other people"), verbal aggression ("When a person is unfair to me, I get angry and protest"), impulsive aggression ("I seem to do things I later regret"), and avoidance of aggression ("I prefer to get out of the way and stay out of trouble whenever someone is harassing me"). We did not include scores on a fourth component (avoidance), because it is based on only two items of dubious reliability (Forrest, Banyard, & Shevlin, 2000).

Once again, we found that the six measures of aggressive driving and the seven measures of general aggression (from the AQ and AI) did not load on single factor. On the contrary, a highly significant chi-square statistic and disappointing fit indices were obtained when we attempted to develop a single factor model, $\chi^2 = 476.08$, $df = 65$, $p < .001$, normed fit index (NFI) = 0.78. Instead, as Figure 1 shows, the 15 variables loaded on two factors, which were highly correlated (phi = .79).

From the model in Figure 1, we can conclude that it is possible to discriminate between measures of general aggression and aggressive driving. However, the correlation between the two factors is as high as the reliability (i.e., alpha) coefficient for most inventories. We suspect that two separate factors emerge because respondents recognize that the items on the aggressive driving scales apply to a specific realm (namely, vehicles and roads). It should be cautioned that scores on these measures also share variance from having been obtained via a single method (i.e., self-report instruments).

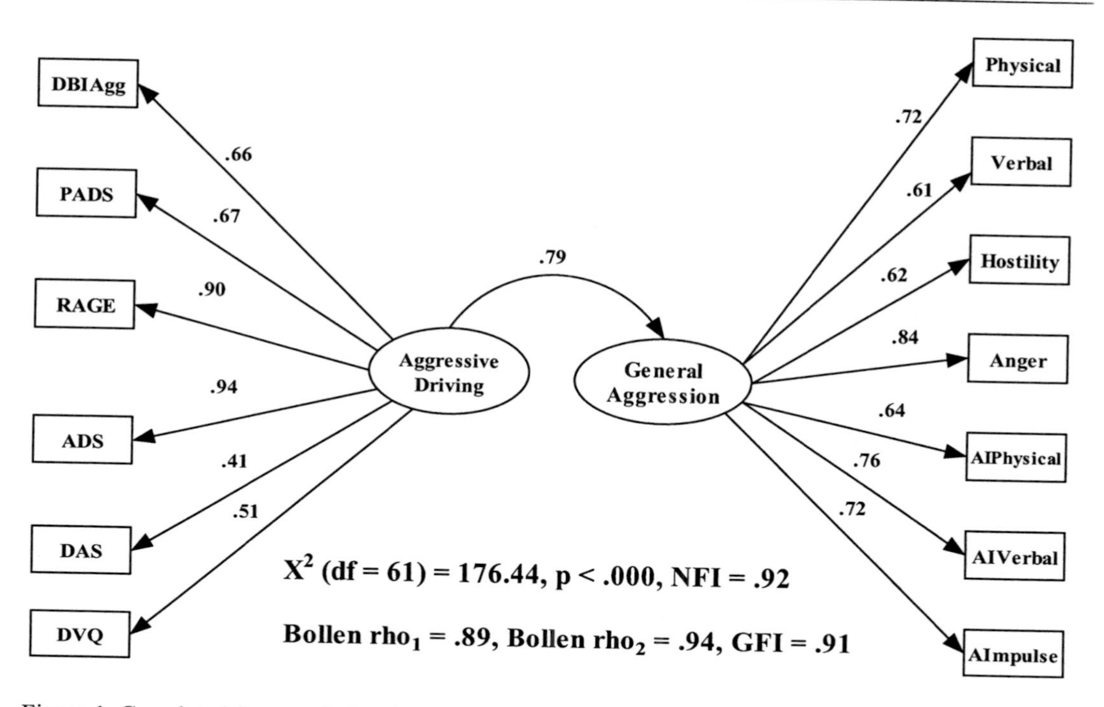

Figure 1. Correlated factor solution for aggressive trait and aggressive driving inventories. DBIAgg = Aggression scale of the Driver Behavior Inventory, DAS = Driving Anger Scale. PADS = Propensity for Angry Driving Questionnaire, RAGE = Road Rage Scale, ADS = Aggressive Driving Scale, DAS = Driving Anger Scale, DVQ = Driving Vengeance Questionnaire, Physical = Physical aggression on the Aggression Questionnaire (AQ), Verbal = Verbal aggression on the AQ, Anger = anger on the AQ, and Hostility = hostility on the AQ., AIPhysical = Physical aggression on the Aggression Inventory (AI), AIVerbal = Verbal aggression on the AI, and AIimpulse = impulsive aggression on the AI. (Error terms and correlations among errors are not displayed.).

In addition, it would be a mistake to conclude that respondents draw a distinction between aggressive driving and aggression in other realms (e.g., automobiles, home environments, stores, the work place). This limitation can be viewed as an opportunity because investigators (e.g., Hennessy, 2003) have begun to develop measures to assess aggression in other realms, such as the workplace. It will be interesting to find out if measures of workplace and driving aggression load on one or two factors. A single-factor solution would be consistent with the view that aggressiveness is a general trait that can be inferred from similar patterns of behavior in different places (i.e., vehicles and the work place).

Criterion Validity

Generally disappointing results have been obtained in studies that have examined correlations between the scales in Table 1 and crash-related variables, such as tickets for moving violations and number of major and minor accidents. For example, despite its frequent use, the DAS was not correlated with a person's frequency of minor and major accidents during the past year and during one's lifetime (Deffenbacher et al., 2001).

There are several reasons to believe that age and gender act as "common causes" that inflate correlations between aggressive driving and traffic violations. Taking age first, young people are much more likely to be involved in motor vehicle crashes than older individuals (Jonah, 1986a; 1986b). Age is also negatively correlated with aggressiveness (O'Connor, Archer, & Wu, 2001; Wilson & Herrnstein, 1985). Turning to gender, males are more likely to be involved in motor vehicle accidents (Lawton, Parker, Manstead, & Stradling, 1997; Simon & Corbett, 1996). Although there is some reason to believe that women score as high as men on measures of verbal aggression, there is ample evidence that men are more physically aggressive (Archer & Gartner, 1984). Thus, it is noteworthy that Welles-Parker et al. (2002) found that scores on the Road Rage Questionnaire predicted motor vehicle violations and crash involvement in analyses that controlled for age and gender. As previously noted, several investigators (e.g., Meskin et al., 2002) have uncovered reliable correlations between the Aggressive Violations component of the DBQ and crash-related variables, such as tickets for moving violations and number of minor and major accidents, after controlling for age and gender.

However, it is possible that correlations between measures of aggressive driving and crash-related variables reflect little more than the general trait of aggressiveness. To explore this possibility, we modified a scale that Cutler, Kravitz, Cohen, and Shinas (1993) developed to assess specific consequences of driving violations (e.g., tickets, accidents, attending Driving School to avoid points on license). After dropping some items and developing a scoring key for others, we obtained a scale that had an alpha coefficient of 0.80. We then performed a hierarchical regression analysis to determine if any of the measures of aggressive driving attained significance after we controlled for demographic variables (age, gender, miles driven) and the previously noted measures of general aggression (i.e., four scales from the AQ and three scales from the AI). As might be expected, age and gender were the best predictors of traffic violations. However, more importantly, two of the measures of aggressive driving emerged as predictors (i.e., as the next variable to be entered into the equation) in analysis that controlled for the demographic variables and aggressiveness. One of these was Krahé and Fenske's (2002) ADS (partial $r = .23$, $p < .001$), which is an amalgamation of items from earlier inventories. Interestingly, the other was the RRI (partial $r = .18$, $p < .003$), which was taken from a book written for the general public (James & Nahl, 2000).

It should be cautioned that these correlations are based on retrospective reports (i.e., concurrent validity). Only one investigation has been undertaken to determine if a scale designed to measure aggressive driving predicts behavior. Hennessy and Wiesenthal (2001) found that their measure of vengeance (the DVQ) predicted subsequent aggressive behavior when drivers got behind the wheel.

IMPLICATIONS

Results from our studies lead us to conclude that there is something unique about aggressive driving. Measures of aggressive driving can be distinguished from scales that assess general aggression, and two of the aggressive driving scales (ADS and RRI) emerge as significant predictors of traffic violations in analyses that control for demographic variables and general levels of aggressiveness. However, it should be cautioned that these results may reflect little more than the variance that self-report instruments share. Correlations between self-report measures and official accident rates tend to be low, with self-report measures accounting for less than 5% of the variance in violations and accident rates (Hatakka, Keskinem, Katila, & Laapotti, 1997). Of greater concern, Maycock (1997) found that a quarter of accidents are forgotten within a year's time.

However, assuming that scores on aggressive driving inventories do in fact predict crashes, we have to ask what there is about driving that elicits aggression. One possibility is that there is something unique about motor vehicles that lead individuals to behave aggressively; that is, as Parry (1968, p. 6) observed, "people who would be willing to support charities, scorn the growth of social violence, uphold the law in other respects, and generally try to live as good citizens, change into selfish, aggressive, and dangerous beings in the time it takes to get into and start a car." This observation is consistent with Berkowitz's (1990) cognitive neoassociationism theory, which grew out of his research on the "weapons effect." It has been found that individuals behave more aggressively after they have seen a weapon (Turner, Layton, & Simons, 1975). Just as the sight of a rifle or a shotgun acts as a prime, which leads to aggressive thoughts and behavior, some people may associate some kinds of vehicles with aggressive and reckless driving. This association may be yet another unintended consequence of the media (e.g., Johnson, 2002) which, in a recent commercial, shows a woman taming a monster that turns out to be the engine of a Mercedes Benz. Other commercials portray cars and trucks as "powerful" (Marsh & Collett, 1987) and, as one company puts it, "Dodge tough."

It would be an easy matter to design a laboratory experiment to determine if viewing movies and commercials that feature reckless driving leads individuals to engage in behavior that satisfies the most widely accepted definition of aggression: "behavior directed toward the goal of harming or injuring another" (Baron & Richardson, 1997, p. 7). We realize that questions can be raised about the construct validity of paradigms used to study aggression in laboratory settings (Tedeschi & Quigley, 1996; 2000). However, it should be possible to reduce suspicion and problems associated with demand characteristics by inserting commercials that feature aggressive and reckless driving into otherwise innocuous television programs. It would be interesting to assess the combined effects of type of commercial (aggressive vs. staid) and program content (violent vs. non-violent) on subsequent aggression.

REFERENCES

American Automobile Foundation for Traffic Safety. (2002, June 16). *Quiz: Are YOU an Aggressive Driver?* [Online]. Available on January 8, 2004: *http://www.aaafoundation.org/quizzes/index.cfm?button=aggressive.*

Archer, D., & Gartner, R. (1984). *Violence and crime in cross-national perspective.* New Haven, CT: Yale University Library.

Arnett, J. J., Offer, D., & Fine, M. A. (1997). Reckless driving in adolescence: "State" and "trait" factors. *Accident Analysis and Prevention*, 29, 57-63.

Baron, R. A., & Richardson, D. R. (1997). *Human aggression.* (2nd ed.). New York: Plenum.

Beirness, D. J. (1993). Do we really drive as we live? The role of personality factors in road crashes. *Alcohol, Drugs and Driving*, 9 (3), 129-143.

Berkowitz, L. (1990). On the formation and regulation of anger and aggression: A cognitive-neoassociationistic analysis. *American Psychologist*, 45, 494-503.

Burns, R. C., & Katovich, M. A. (2003). Examining road rage/aggressive driving: Media depiction and prevention suggestions. *Environment and Behavior*, 35, 621-636.

Blanchard, E. E., Barton, K. A., & Malta, L. (2000). The psychometric properties of aggressive driving: The Larson Drivers's Stress Profile. *Psychological Reports*, 87, 881-892.

Buss, A. H., & Perry, M. (1992). The aggression questionnaire. *Journal of Personality and Social Psychology*, 63, 452-559.

Campbell, D. T., & Fiske, D. W. (1959). Convergent and discriminant validation by the multitrait-multimethod matrix. *Psychological Bulletin*, 56, 81-105.

Cronbach, L., & Meehl, P. E. (1955). Construct validity in psychological tests. *Psychological Bulletin*, 52, 281-302.

Cutler, B. L., Kravitz, D. A., Cohen, M., & Schinas, W. (1993). The Driving Appraisal Inventory: Psychometric characteristics and construct validity. *Journal of Applied Social Psychology*, 23, 1196-1213.

Deffenbacher, J. L. (1992). Trait anger: Theory, findings, and implications. In C. D. Spielberger & J. N. Butcher (Eds.), *Advances in personality assessment* (Vol. 9, pp. 177-201). Hillsdale, NJ: Erlbaum.

Deffenbacher, J. L., Lynch, R. S., Oetting, E. R., & Swaim, R. C. (2002). The Driving Anger Expression Inventory: A measure of how people express their anger on the road. *Behaviour Research and Therapy*, 40, 717-737.

Deffenbacher, J. L., Lynch, R. S., Oetting, E. R., & Yingling, D. A. (2001). Driving anger: Correlates and a test of state-trait theory. *Personality and Individual Differences*, 31, 1321-1331.

Deffenbacher, J. L., Petrilli, R. T., Lynch, R. S., Oetting, E. R., & Swaim, R. C. (2003). The Driver's Angry Thoughts Questionnaire: A measure of angry cognitions when driving. *Cognitive Therapy and Research*, 27, 383-401.

Deffenbacher, T. L., Oetting, E. R., & Lynch, R. S. (1994). Development of a driving anger scale. *Psychological Reports*, 74, 83-91.

DePasquale, J. P., Geller, E. S., Clarke, S. W., & Littleton, L. C. (2001). Measuring road rage: Development of the Propensity for Angry Driving Scale. *Journal of Safety Research,* 32, 1-16.

Dula, C. S., & Ballard, M. E.. (2003). Development and evaluation of a dangerous, aggressive, negative emotional, and risky driving. *Journal of Applied Social Psychology*, 33, 263-282.

Forrest, S., Banyard, P., & Shevlin, M. (2000). An assessment of alternative factor models of the aggression inventory. *North American Journal of Psychology*, 2, 145-150.

Gladue, B. A. (1991). Qualitative and quantitative sex differences in self-reported aggressive behavioral characeristics. *Psychological Reports*, 68, 675-684.

Glendon, A. L., Dorn, G., Matthews, G., Gullian, E., Davies, D. R., Debney, L. M. (1991). Reliability of the Driving Behavior Inventory. *Ergonomics*, 36, 719-728.

Gulian, E., Matthews, G., Glendon, A. L., Davies, D., & Debney, L. M. (1989). Dimensions of driver stress. *Ergonomics,* 32, 585-602.

Hatakka, M., Keskinem, E., Katila, A., & Laapotti, S. (1997). Self-reported driving habits are valid predictors of violations and accidents. In T. Rothengaher & E. C. Vaya (Eds.), *Traffic and transport psychology: Theory and application* . New York: Pergamon.

Hennessy, D. A. (2003). *From driver stress to workplace aggression* . Toronto, Canada: Paper presented at the annual convention of the American Psychological Association.

Hennessy, D. A., & Wiesenthal, D. L. (2001). Further validation of the Driver Vengeance Questionnaire. *Violence and Victims*, 16, 565-573.

Houston, J. M., Harris, P. B., & Norman, M. (2003). The Aggressive Driving Behavior Scale: A self-report measure of unsafe driving practices. *North American Journal of Psychology*, 5, 269-279.

James, L., & Nahl, N. (2000). *Road rage and aggressive driving*. Amherst, NY: Prometheus.

Johnson, J. G. (2002). Television viewing and aggressive behavior during adolescence and adulthood. *Science*, 295, 2468-2471.

Jonah, B. A. (1986a). Accident risk and risk taking behaviour among young drivers. *Accident Analysis and Prevention*, 18, 255-271.

_____ . (1986b). Youth and traffic accident risk: Possible cause and potential solutions (Editorial). *Accident Analysis and Prevention*, 18, 253-254.

Krahe, B., & Fenske, I. (2002). Predicting aggressive driving behavior: The role of Macho personality, age, and power of car. *Aggressive Behavior*, 28, 21-29.

Lajunen, T., & Parker, D. (2001). Are aggressive people aggressive drivers? A study of the relationship between self-reported general aggressiveness, driver anger, and aggressive driving. *Accident Analysis and Prevention*, 33, 243-253.

Larson, J. A. (1996). *Steering clear of highway madness: A driver's guide to curbing stress and strain*. Wilsonville, OR: BookPartners.

Lawton, R., Parker, D., Manstead, A. S. R., & Stradling, S. G. (1997). The role of affect in predicting social behaviors: The case of road traffic violations. *Journal of Applied Social Psychology*, 27, 1258-1276.

Marsh, P., & Collett, P. (1987). *Driving passion: The psychology of the car*. Boston, MA: Faber and Faber.

Maycock, G. (1997). Accident liability--the human perspective. In T. Rothengatter & E. C. Vaya (Eds.), *Traffic and transport psychology* (pp. 65-76). New York: Pergamon.

Meskin, J., Lajunen, T., & Summala, H. (2002). Interpersonal violations, speed violations and their relation to accident involvment in Finland. *Ergonomics*, 45, 469-483.

Nunnally, J., & Bernstein, I. M. (1994). *Psychometric theory*. (3rd ed.). New York: McGraw-Hill.

O'Connor, D. B., Archer, J., & Wu, F. W. C. (2001). Measuring aggression: Self-reports, partner reports, and responses to provoking scenarios. *Aggressive Behavior*, 27, 79-101.

Parker, D., Reason, J. T., Manstead, A. S. R., & Stradling, R. G. (1995). Driving errors, driving violations and accident involvement. *Ergonomics*, 38, 1036-1048.

Parry, M. H. (1968). *Aggression on the road*. New York: Tavistock.

Reason, J., Manstead, A., Stradling, S., Baxter, J., & Campbell, K. (1990). Errors and violations on the roads: A real distinction. *Ergonomics*, 33, 1315-1332.

Simon, F., & Corbett, C. (1996). Road traffic offending, stress, age, and accident history among male and female drivers. *Ergonomics*, 39, 757-780.

Tedeschi, J. T., & Quigley, B. M. (1996). Limitations of laboratory paradigms for studying aggression. *Aggression and violent behavior*, 1, 163-177.

_____ . (2000). A further comment on the construct validity of laboratory aggression paradigms: A response to Giancola and Chermack. *Aggression and Violent behavior*, 5, 127-136.

Tillmann, W. A., & Hobbs, G. E. (1949). The accident-prone automobile driver. *American Journal of Psychiatry*, 106, 321-322.

Turner, C. W., Layton, J. F., & Simons, L. S. (1975). Naturalistic studies of aggressive behavior:

Aggressive stimuli, victim visibility, and horn honking. *Journal of Personality and Social Psychology*, 31, 1098-1107.

Van Rooy, D. L., Rotton, J., & Burns, T. M. (in press). Convergent, discriminant, and predictive validity of aggressive driving inventories: They drive as they live. *Aggressive Behavior.*

Wells-Parker, E., Ceminsky, J., Hallberg, V., Snow, R. W., Dunaway, G., Guiling, S., Williams,

M., & Anderson, B. (2002). An exploratory study of the relationship between road rage and crash experience in a representative sample of US drivers. *Accident Analysis and Prevention*, 34, 271-278.

Wiesenthal, D. L., Hennessy, D., & Gibson, P. M. (2000). The Driver Vengeance Questionnaire: The development of a scale to measure deviant drivers' attitudes. *Violence and Victims*, 15, 115-138.

Wilson, J. Q., & Herrnstein, R. J. (1985). *Crime and human nature*. New York: Simon & Schuster.

In: Contemporary Issues in Road User Behavior and Traffic Safety ISBN:1-59454-268-6
Editors: D. Hennessy and D. Wiesenthal, pp. 47-60

Chapter 5

UNDERSTANDING AND TREATING THE AGGRESSIVE DRIVER

Tara E. Galovski
University of Missouri – St. Louis
Edward B. Blanchard
University at Albany, State University of New York

DEFINING AGGRESSIVE DRIVING

There exists a myriad of definitions and severity levels of aggressive driving, ranging form the media coined phrase of "road rage" denoting aggressive driving in its most extreme form to merely "feeling angry" on the road. Most often, however, aggressive driving is defined as driving behaviors performed with the *intent* to punish, threaten, frighten, or harm another or his/her property. Harm can include physical and/or psychological damage. Commonly cited aggressive driving behaviors in the literature include slow driving with the intent to block another vehicle's passage, tailgating, improper passing (cutting other drivers off while passing, passing on the shoulder), failing to yield the right of way, horn honking, failing to keep right, flashing high beams, and failing to signal properly (Clayton & Mackay, 1972; Maiuro, 1998; Ross, 1940; Stradling & Parker, 1997). Personal attacks on fellow drivers are also included in the list of aggressive driving behaviors. These include obscene gestures, verbal insults, throwing objects, and vehicular assault. *Intent* is the key element in discriminating aggressive driving from driving error or lapses in judgment.

Estimates of the extent of aggressive driving on the roadways and the percentage of crashes attributed to aggressive behavior are considerable. The AAA Foundation for Traffic Safety estimates that an average of 1500 people died annually as the result of an escalation in aggressive driving behavior from 1990-1996 (Mizell, 1997). In a recent address to Congress, Martinez (1997) and Snyder (1997) have identified aggressive driving as a risk factor for MVA morbidity and mortality on a par with alcohol impaired driving. Martinez estimated that 33% of all MVAs resulting in personal injury and 66% of all MVAs resulting in fatalities were attributable to aggressive driving. More and more attention within the domain of traffic safety is being paid to aggressive driving. As a result, investigators have begun to conduct research aimed at understanding the aggressive drivers and changing their behaviors.

This chapter will review the available literature on the aggressive driver and the attempts at aggressive driving behavior change. Specific attention will be paid to the University at Albany, State University of New York research program.

PROFILING THE AGGRESSIVE DRIVER

In order to understand *how* to treat the problem, investigators must understand *who* is committing these types of behaviors on the roadways. Researchers have begun to fill in the gaps in a relatively sparse literature profiling the aggressive driver. Three related areas of research toward this end have been identified; the psychophysiological reactivity of aggressive drivers, psychological characteristics, and psychiatric comorbidity within the population.

PSYCHOPHYSIOLOGICAL REACTIVITY

Intuitively, aggressive driving occurs when drivers become angry. Empirically, Deffenbacher, Huff, Lynch, Oetting, and Salvatore (2000) support this supposition by demonstrating that aggressive drivers experience elevated levels of anger during provocative driving scenarios. Specifically, elevated levels of transitory state anger were correlated to aggressive driving behaviors in a sample of college students. Considering the potential role of anger and resultant physiological arousal in aggressive driving, Deffenbacher et al. (2000) included a relaxation component to their treatment program (as discussed below).

The Albany research program sought to specifically examine levels of physiological arousal in identified aggressive drivers. Towards this end, a group of college students identifying themselves as aggressive drivers were compared to their non-aggressive driving counterparts in an imaginal exposure to an anger-provoking driving scenario paradigm (Malta, Blanchard, Freidenberg, Galovski, Karl, & Holzapel, 2001). Physiological reactivity (heart rate, blood pressure, facial muscle activity, and skin resistance) was monitored in both groups while participants listened to idiosyncratic, anger-provoking driving vignettes, idiosyncratic fear-provoking non-driving vignettes, and a neutral stressor (a mental arithmetic task). The results indicated that the aggressive driving group experienced significantly greater physiological reactivity as measured by muscle tension and blood pressure than did their non-aggressive driving counterparts. The authors suggest that physiological arousal to provoking stimuli potentially contributes to aggressive driving behavior.

The second Albany physiological study of aggressive drivers improved on the first by including a sample of highly aggressive drivers identified by the courts ($N = 20$) and remanded to our program (Galovski, Blanchard, Malta, & Freidenberg, 2003). This group's driving offenses included felonies, misdemeanors, and multiple driving violations. Table 1 depicts the severity of the driving offenses of individuals diverted to our program. We also include a sample of self-referred drivers from the community ($N = 10$) who volunteered to participate in the program in exchange for free, psychological treatment aimed at remediating aggressive driving behavior (as explained below). Driving data gathered from our locally constructed interview is available for both court-referred (CR) and self-referred (SR) aggressive drivers in Table 1 as well. There were no significant differences between the CR and SR groups on any of these variables. A control, non-aggressive driving sample ($N = 14$) was recruited for comparison purposes. The goals of this study were twofold; to contribute to the available assessment literature and to assess the impact of a cognitive-behavioral intervention (described below) on physiological reactivity within this population.

Table 1. Driving Behavior by Referral Status

		Driving Behaviors				
Subject #	**Traffic Offense**	**Years Driving**	**# of MVAs**	**# Moving Violations**	**DWI Conviction**	**# annual driving violations**
		Court-Referred Sample (N = 20)				
100	Vehicular Assault	4	0	2	0	.75
101	Menacing	3	6	4	0	1.67
102	Menacing (with weapon)	2	2	4	0	2.5
103	Reckless driv Unsafe start (2x); Failure to comply	2	2	35	0	19.5
104	Speed (100/55); Failure to comply; Disorderly conduct	25	4	7	1	.4
105	Disorderly conduct; speed (10x)	3	2	10	2	4.33
106	Disorderly conduct; Run stop sign	10	4	10	4	1.3
107	Unsafe lane move; Speed (85/65)	16	2	4	0	.38
108	Criminal Harass. Failure to yield	55	2	6	0	.04
109	Menacing; Vehicular Assault	24	6	25	0	1.13
110	Disorderly conduct; Speeding	25	5	5	5	.28
111	2nd Degree Harassment	28	3	2	1	.11
112	Reckless Driving; Failure to yield	4	2	2	1	1.00
113	Criminal Harassment; Disorderly conduct	14	2	1	0	.21
114	Reckless endanger-ment; Reckless driv	12	0	1	0	.25
115	Unreasonable speed; Failure to keep right	14	3	1	0	.21
116	Disorderly conduct; Unsafe start; Speed	4	1	2	0	1.00
118	Menacing (with weapon)	4	1	1	0	.50
119	Reckless operation of vehicle; Passing on shoulder; Following too closely; Speeding; Crossing a hazard marking; Failure to keep right; Agg. Driv w/ suspended lic	17	1	50	1	3.24
120	Assault	9	0	4	0	.56
Mean		13.8	2.4	8.8	.65	1.96
SD		13	1.8	13	1.4	4.28

Table 1. Driving Behavior by Referral Status (Continued)

		Driving Behaviors				
Subject #	Traffic Offense	Years Driving	# of MVAs	# Moving Violations	DWI Conviction	# annual driving violations
		Self-Referred Sample (N = 10)				
200	X	34	5	1	0	.03
201	X	27	1	0	0	.00
202	X	33	2	1	1	.03
203	X	13	0	9	1	.69
204	X	24	5	6	0	.25
205	X	11	5	10	0	.91
206	X	40	7	6	1	.15
207	X	34	5	1	1	.15
208	X	6	1	3	0	.50
209	X	8	0	2	1	.25
Mean		23	3.1	3.9	.5	.30
SD		12.5	2.6	3.6	.5	.31

The same paradigm was used as described above and measures included heart rate, blood pressure and skin resistance. The entire psychophysiological assessment was conducted prior to treatment and two weeks following the conclusion of treatment. Prior to treatment, it was found that, when compared to non-aggressive drivers, the aggressive drivers were more reactive during the anger-provoking driving scenarios and less aroused during the neutral stressor as measured by systolic blood pressure. At post-treatment, the aggressive drivers showed significant decreases in reactivity to the anger-provoking driving scenario as measured by heart rate and blood pressure. No decreases from pre to post-treatment on any measures were observed during the neutral stressor, suggesting that the decreases observed in the driving scenario were not merely a habituation effect. In fact, arithmetically, heart rate increased from pre to post treatment in the neutral stressor phase. Furthermore, reactivity also did not decrease across time during the fearful scenario indicating a level of specificity of treatment effect to anger and related roadway provocation. These results suggest that dampening physiological arousal experienced on the roadways as a treatment component may be critical in any intervention dealing with this population.

Psychological Characteristics

There exists a rather sparse literature on the level and type of psychological distress within the aggressive driving population. Such information may help to identify potential causes of aggressive driving, thus informing treatment decisions. Larson (1996) developed a treatment program (described later) based on his observations that a substantial proportion of aggressive drivers endorsed Type A characteristics. He further developed a typology of aggressive drivers consisting of five categories including: The Speeder, The Competitor, The Passive-Aggressor, The Narcissist, and The Vigilante. The names of these driver categories give the reader some indication of the types of drivers included in each category. See Larson (1996) for a further description. These categories have particular value in helping the aggressive driver self-identify his/her own driving attitudes and beliefs as angry and aggressive.

Investigators have also taken a more empirical approach in identifying the psychological characteristics of aggressive drivers. In a series of studies, Matthews, Dorn, and Glendon (1991) sought to identify personality traits that may predispose a driver to become stressed and behave aggressively on the roadways. This group developed the Driving Behavior Inventory (Gulian, Matthews, Glendon, Davies, & Debney, 1989) to measure the amount of stress that drivers experience on the roadways. They found that traits such as hostility, neuroticism and, to a lesser extent, psychoticism were related to driver stress and the use of ineffectual coping strategies.

Deffenbacher, Oetting, and Lynch (1994) developed the Driving Anger Scale (DAS) which yields an overall driving anger score as well as six independent subscales including hostile gestures, illegal driving, police presence, slow driving, discourtesy, and traffic obstruction. These investigators found (using a sample of college students) that those drivers scoring in the top 25% of the DAS have experienced more motor vehicle accidents, including those resulting in personal injury, and endorsed significantly more anger and aggression on the roadways than their driving counterparts who score in the lowest 25% of the DAS. Furthermore, the high driving anger students also endorsed significantly higher levels of trait anger and anger directed outward on the State-Trait Anger Expression Inventory (STAXI; Spielberger, 1979) and indicated similarly elevated scores in trait anxiety on the State-Trait Anxiety Inventory (STAI; Spielberger, Gorusch, & Lushene, 1970). Thus it appears that both characterological anger and anxiety may be related to aggressive driving behaviors.

In the Albany aggressive driving program (Galovski & Blanchard, 2002a), thirty aggressive drivers (identified by the courts or self-identified as described above) were compared to 20 non-aggressive driving controls matched on age and gender. Measures of aggressive driving and driving anger included the Driver Stress Profile (DSP; Larson, 1996) yielding an overall score of driver stress as well as four subscales entitled anger, impatience, competing and punishing, and the DAS described above. In an effort to both quantify the extent of psychological distress within this population and identify a possible relationship between such distress and aggressive driving behavior, we also administered the Beck Depression Inventory (BDI; Beck, Ward, Mendelson, Mock, & Erbaugh, 1961) and the STAI to measure depressed mood and anxiety respectively. Measures of anger and hostility were also administered including the STAXI and the Buss-Durkee Hostility Inventory (BDHI; Buss & Durkee, 1957). Comparisons were conducted across three groups, court-referred aggressive drivers (CR), self-referred aggressive drivers (SR) and controls.

The results indicated that the SR group consistently scored higher in measures of psychological distress (see Table 2) than either of the other two groups. The CR group and the controls were essentially similar with the exception of an elevation in scores for the CR group in state anxiety. On measures of hostility and general anger, similar trends emerged on the STAXI and BDHI with the exception of the resentment and suspicion subscales (see Table 3).

Table 2. Psychological Distress Instruments by Referral Status

Psychological Distress Assessments: CR versus SR Between Groups' Comparisons					
Test	Condition: Mean (standard deviation)			Statistic	
	CR ($N = 20$)	SR ($N = 10$)	Control ($N = 20$)	F (2,47)	p
BDI	3.9^a (4.35)	8.8^b (6.97)	3.7^a (3.06)	4.83	0.01
State Anxiety (STAI)	43^a (17.51)	52.50^a (22.93)	26.45^b (5.23)	11.17	0.001
Trait Anxiety (STAI)	31.65^a (9.00)	44.60^b (13.74)	30.15^a (6.04)	9.11	0.001

* Post hoc multiple comparisons are conducted at the .05 alpha level. Superscripts denote significant differences (i.e. "a" differs significantly from "b".

Table 3. Anger and Hostility Instruments by Referral Status

Anger & Hostility Assessments: CR versus SR Between Groups' Comparisons					
Test	Condition: Mean (standard deviation)			Statistic	
	CR ($N = 20$)	SR ($N = 10$)	Control ($N = 20$)	F (2,47)	p
State anger (STAXI)	10.85^{ab} (3.34)	13.4^a (7.65)	10.1^b (.31)	2.35	0.10
Trait anger (STAXI)	15.55^a (6.29)	24.3^b (7.51)	14.05^a (2.8)	12.43	0.001
Angry temperament (STAXI)	6.05^a (2.86)	8.8^b (4.1)	5.15^a (1.27)	6.28	0.004
Angry reaction (STAXI)	6.05^a (1.85)	11.8^b (3.43)	6.45^a (2.09)	22.99	0.001
Anger In (STAXI)	11.8^a (2.35)	17.6^b (4.38)	15.2^b (4.98)	7.85	0.001
Anger Out (STAXI)	15.15^a (3.95)	19.6^b (5.15)	14.3^a (2.81)	6.77	0.003
Anger Control (STAXI)	24.35^a (4.96)	18.5^b (6.2)	24.95^a (3.38)	6.99	0.002
Buss-Durkee Hostility Inventory	25.7^a (12.74)	37.33^b (17.05)	21^a (7.71)	5.83	0.006
BDHI - assault	3.6^{ab} (2.46)	5.22^a (2.59)	2.3^b (1.63)	5.77	0.006
BDHI - indirect hostility	3.0^a (2.1)	5.33^b (2.74)	3.9^{ab} (2.2)	3.33	0.05
BDHI - irritability	2.65^a (2.28)	6.56^b (2.92)	3.25^a (1.71)	10.24	0.001
BDHI - negativism	1.6^a (1.57)	3.00^b (1.5)	1.5^a (1.32)	3.66	0.03
BDHI - resentment	2.0 (1.78)	2.22 (2.17)	1.35 (1.63)	0.99	0.38
BDHI - suspicion	3.25^a (2.15)	3.67^a (1.80)	1.25^b (.97)	9.58	0.001
BDHI - verbal hostility	6.9^{ab} (2.4)	8.56^a (3.43)	5.5^b (2.26)	4.61	0.02
BDHI - guilt	2.7 (2.05)	2.78 (2.91)	1.95 (1.54)	0.85	0.43

* Post hoc multiple comparisons are conducted at the .05 alpha level. Superscripts denote significant differences (i.e. "a" differs significantly from "b".

Finally, the aggressive driving scales showed very interesting results. Means and standard deviations of these measures as well as the comparisons across the groups are also portrayed in Table 4. As expected, the SR drivers reported significantly higher driving anger and driving stress as compared to non-aggressive driving controls on both the DAS and DSP as indicated by the overall scores and all subscales. Similarly, the SR group also showed significantly higher levels of driver stress and driving anger than the CR group on the DAS and DSP scales and subscales. Interestingly, the CR group indicated *less* driving anger and *less* driver stress than did the *controls* on every scale and subscale except DAS-police presence, DAS-traffic obstruction, DSP-competing, and DSP-punishing. On these four subscales, the CR group indicated almost identical scores to the controls. In essence, the CR group endorsed virtually no driving anger or stress. The responses on these psychometric tests clearly contradict the severity of the driving behaviors (objective samples of actual driving behavior) for which the CR group were arrested and remanded to the program (described in

Table?). The reader may hypothesize that the answers given on the DAS and DSP by the CR group may be less than accurate. Possible reasons for these inaccuracies are discussed later.

Table 4. Driving Anger instruments by Referral Status

Driving Anger Assessments: Between Groups' Comparisons					
Test	**Condition: Mean (standard deviation)**			**Statistic**	
	CR ($N = 20$)	**SR ($N = 10$)**	**Control ($N = 20$)**	**F (2,47)**	***p***
*Driving Anger Scale (DAS)**	59.9^{a} (16.44)	122.3^{b} (21.69)	68.75^{a} (10.66)	56.40	0.001
DAS - hostile gestures	4.7^{a} (1.89)	11^{b} (3.23)	6.6^{c} (1.9)	26.97	0.001
DAS - illegal driving	6.35^{a} (2.68)	14.5^{b} (3.66)	9.10^{c} (2.38)	28.52	0.001
DAS - police presence	5.7^{a} (2.05)	11.0^{b} (5.08)	5.75^{a} (2.1)	13.21	0.001
DAS - slow driving	10.95^{a} (3.15)	22.5^{b} (5.56)	11.05^{a} (2.86)	39.95	0.001
DAS - discourtesy	19.5^{a} (6.20)	39.20^{b} (4.96)	23.9^{c} (4.12)	48.75	0.001
DAS - traffic obstruction	12.75^{a} (4.74)	25.0^{b} (6.75)	12.35^{a} (2.7)	29.89	0.001
Driver Stress Profile (DSP)	15.15^{a} (12.68)	61.5^{b} (24.14)	18.95^{a} (8.65)	38.56	0.001
DSP - anger	5.45^{a} (3.95)	19.0^{b} (5.83)	7.75^{a} (2.92)	39.42	0.001
DSP - impatience	3.75^{a} (3.23)	17.0^{b} (7.48)	5.0^{a} (3.67)	31.67	0.001
DSP - competing	2.85^{a} (4.59)	12.40^{b} (8.25)	2.5^{a} (2.67)	15.51	0.001
DSP - punishing	3.10^{a} (3.16)	13.10^{b} (6.89)	3.6^{a} (2.5)	24.38	0.001

Superscripts denote significant differences (i.e. "a" differs significantly from "b".

In summary, investigators in the Albany program and beyond have reported elevated levels of neuroticism, driver stress, psychological distress (particularly anxiety), general anger, and overall hostility in the aggressive driving population.

Psychiatric Comorbidity

To our knowledge, the Albany program has been the only body of research that has systematically measured the incidence of diagnosable psychopathology in an aggressive driving population (Galovski, Blanchard, & Veazey, 2002). Using the Clinician-Administered Structured Clinical Interview for DSM-IV Axis I Disorders- Patient Edition (SCID-I/P; First, Spitzer, Gibbon, & Williams, 1995) to assess for Axis I disorders and the Structured Clinical Interview for DSM-IV Axis II Personality Disorders (SCID-II; First, Spitzer, Gibbon, & Williams, 1994) to assess for Axis II disorders, the full range of psychopathology was assessed and diagnosed in all 30 SR and CR drivers. Additionally, a SCID-like module was created by the investigator based on DSM-IV criteria for Intermittent Explosive Disorder (IED) and added to the assessment. IED is the only Axis I disorder to hold anger as its hallmark feature and is substantially understudied relative to disorders in the mood and anxiety spectrum (Felthous, Bryant, Wingerter, & Barratt, 1991; McElroy, Hudson, Pope, Keck, & Aizley, 1992; Monopolis & Lion, 1983). Thus we wanted to assess its presence in our aggressive population.

In summary, we found a substantial amount of psychopathology our sample. Table 5 indicates the prevalence of psychopathology in our sample. Interestingly, 33% of the overall population was diagnosed with IED. There was no difference in prevalence of this disorder between the SR and CR groups. It should be noted that aggressive driving behavior was not included diagnostically in assessing for this disorder meaning that all diagnoses were made based on examples of angry, destructive, impulsive behaviors made outside the context of driving.

Table 5. Psychiatric Diagnoses across Referral Source

Diagnosis	Groups (percentages) Court-referred (*N* = 20)	Self-referred (*N* = 10)	*Chi-Square*	*p*
Axis I Disorders				
Current Mood Disorder	10	10	0.000	1.00
Past Mood Disorder	15	50	2.33	0.13
Current Anxiety Disorder	10	50	5.96	.02
Past Anxiety Disorder	5	50	11.27	0.001
Intermittent Explosive Disorder	35	30	.08	.78
Current alcohol or substance abuse	15	30	0.94	0.33
Past alcohol or substance abuse or dependence	40	40	0.000	1.00
Somatoform Disorder	0	10	2.07	0.15
Eating Disorder	0	10	2.07	0.15
Any Axis I disorder	75	90	.34	0.56
Axis II Disorders				
Antisocial PD	30	10	1.49	0.22
Borderline PD	5	30	3.61	0.06
Narcissistic PD	15	0	1.67	0.2
Obsessive-Compulsive PD	5	30	3.61	0.06
Any Axis II	35	50	0.07	0.8

TREATMENT

Larson (1996) is considered a pioneer in aggressive driving treatment. His one-day intensive treatment program includes psychoeducation regarding anger, self-identification with the driver attitude categories mentioned above and their corresponding belief systems, cognitive techniques designed to aid a driver in challenging negative beliefs, and a variety of behavioral. Larson describes the success of this intervention as measured by pre- and post-treatment changes on the DSP as impressive (Larson, Rodriguez, & Galvan-Henkin, 1998).

Deffenbacher and colleagues have improved on Larson's work by conducting two controlled treatment outcome trials with multiple outcome measures in a college population. In the first treatment study, Deffenbacher et al. (2000) compared two active treatments versus an assessment only control condition. The two interventions were conducted in group format (7-10 people) and included eight, one-hour sessions conducted on a weekly basis. Self-managed relaxation coping skills (RCS) consisted of deep relaxation strategies coupled with four relaxation coping skills. The intervention aimed to teach subjects the skills necessary to relax in an anger-provoking or stressful driving situation in an effort to decrease driving anger. The second active treatment condition combined an abbreviated version of the relaxation component found in RCS with a cognitive component of challenging faulty thinking in driving situations. This intervention systematically used specific cognitive distortions (i.e. catastrophization, overgeneralization). Individuals in the control condition received no treatment.

Treatment outcome was measured by the DAS, subjective reports of anger on imagined provoking driving situations, a daily driving log, and the Driving Survey (querying accident related variables, risky and aggressive driving behaviors). Subjects were re-assessed at post-treatment and again at a four-week follow-up point. Both active treatment conditions resulted in improvements on the outcome measures compared to the control condition. There was very little differential improvement between the two active treatment conditions. RCS improved more on the DAS and CRCS resulted in greater treatment gains on the driving diaries.

Deffenbacher, Filetti, Lynch, Dahlen, and Oetting (2002) sought to further investigate the lack of differential effect between the two active treatment conditions in a follow-up study. They hypothesized that the weaker results seen in the CRCS condition on the DAS may be due to the implementation of the cognitive component of the intervention. Previously, the protocol dictated a specific cognitive distortion as the focus of each treatment session. The second study modified the cognitive component of CRCS to allow for more latitude in identifying and challenging cognitive distortions. The RCS treatment condition remained identical to the original study. The study also included additional outcome measures; the Personal Driving Anger Situations (a description of two angering driving scenarios) and the Driving Anger Expression Inventory (DAX; a dimensional measure of the expression of driving anger).

Both active treatments, RCS and CRCS, resulted in significant reductions in driving anger and aggressive driving and increases in adaptive driving behaviors as indicated by the DAX when compared to the untreated controls. One month following treatment, the treated drivers were re-assessed. Good maintenance of treatment gains was observed in the CRCS condition, specifically in decreases in hostile/aggressive expression of driving anger as measured by the DAX. There were no improvements seen in the no-treatment control group on any of the measures. Overall, the second treatment study replicated the positive outcome seen in the RCS condition and enhanced the treatment success of the CRCS condition.

Albany Treatment Trial

The Albany treatment trial offered a methodological improvement on previous research in its use of community participants and its inclusion of court-mandated subjects (Galovski & Blanchard, 2002b). Ten SR participants and 20 CR participants were treated with a cognitive-behavioral, multi-component intervention designed to specifically target aggressive driving behaviors. The treatment consisted of four group sessions, two hours in length, conducted once a week. The individual components consisted of education about the impact of aggressive driving and anger in general, identification of oneself as an angry driver, deep relaxation training, development of alternative coping skills, and cognitive restructuring.

A semi-crossover design was utilized such that within each referral condition, subjects were matched into pairs based on psychopathology, gender, age and socio-economic status. One member of each pair was randomly assigned to a control condition (6-week waitlist, no treatment) while the other member of the pair was assigned to immediate, active treatment.

Following the 6-week waitlist period, the control participant was crossed over into the active treatment condition. All participants thus eventually received the active treatment.

Twenty-eight aggressive drivers began the treatment program (two of the SR subjects were not able to join the program due to health and job difficulties). All twenty-eight participants completed treatment, but one SR subject did not return for a follow-up assessment. Therefore we have post-treatment data on twenty-seven participants (7 SR and 20 CR). There were no differences in demographic variables between the two groups, except that the CR was significantly younger. However, the two groups differed considerably on self-reports of driving histories (see Table 1). Clearly, the self-report of the CR group is less than consistent with the severity of the driving offenses for which the group was arrested (see Table 1).

The primary outcome measure was a locally constructed daily driving monitoring diary upon which the participant is able to record the extent to which he/she committed or experienced 13 overt AD behaviors and 4 covert AD behaviors on a daily basis for two weeks prior to treatment, for two weeks post-treatment and for one week at the two-month follow-up point. From these diaries a composite symptom reduction score was calculated following the method of Blanchard and Schwarz (1988). The CPSR score can be considered a percentage of improvement. In other words a CPSR score equaling .86 can be considered to indicate an 86% decrease in aggressive driving behavior. See Galovski and Blanchard (2002b) for information on calculating this score. The average CPSR score for all twenty-seven treated participants was 0.62 indicating that, on average participants decreased AD behaviors by approximately 62%. In comparing the waitlist control participants (average CPSR score = .007) to the treated participants (average CPSR score = .50), a significant difference was found. In other words, the treatment condition subjects had decreased aggressive driving behavior by an average of 50% while there had virtually been no change in the waitlist controls' driving behavior. The virtual lack of improvement in the waitlist control condition is an interesting finding as investigators often note some improvement in a waitlist condition, presumably a side effect of the assessment process and the symptom monitoring. Particularly in a study relying heavily on data from involuntary participants, one would normally expect to see change just from the punishment of having been sent to the program. These results give us more confidence that treatment gains seen in the active condition are not merely attributable to the effect of being remanded to the program, but are, in fact, treatment specific. When crossed over to treatment, this waitlist group averaged a 64% decrease in aggressive driving behavior. Thus it may be concluded that the intervention itself (and not necessarily the sentencing or the passage of time) directly impacted the aggressive driving behaviors.

These results were well-maintained at the two-month follow-up at which point the drivers averaged a 56% decrease in aggressive driving behaviors from their pre-treatment status. In other words, the drivers essentially cut their aggressive driving behaviors in half (felt less angry, engaged in half the dangerous behaviors, or engaged in less risky behaviors). Considering the relationship for between aggressive driving and motor vehicle accidents, property loss and societal cost, these results may be very meaningful from a public safety perspective.

Aggressive driving was also measured through the use of the DSP and DAS. The results across the entire aggressive driving sample indicated significant reductions in overall driving anger (DAS total score), and on several DAS subscales (hostile gestures, illegal driving, slow driving, discourtesy, and traffic obstruction). With respect to the DSP, significant decreases were realized in competing behavior on the roadways and trends in the expected direction were observed on the overall score as well as each of the remaining subscales. However, no differences were observed between the treatment group and controls on any of the driving measures. Possible reasons for this puzzling finding are discussed below.

We also sought to ascertain the extent to which treatment benefits may generalize to psychological distress and anger. Significant reductions in state anxiety, trait anger, angry temperament, angry reaction, and anger directed outward were realized across the entire treated sample. As expected, the aggressive drivers that received treatment showed more decreases in state anger, trait anger, and angry temperament than did the waitlist controls. Good maintenance of treatment gains in many of the self-report measures indicated stable trends and even continued improvement (specifically on measures of psychological distress and DSP subscales). The means, standard deviations, and statistics of all the paper and pencil measures are presented in Table 6.

Table 6. Entire Aggressive-driving Sample, Pre- to Post-treatment Comparisons of Self-Report Measures

Pre-Post Self-Report Measures: Entire Aggressive Driving Population ($N = 27$)				
Measure	**Pre-Treatment**	**Post-Treatment**	**Statistic**	
			***t* (25)**	**p**
General Psychological Distress				
BDI	4.69 (5.61)	3.08 (4.35)	1.49	.15
State Anxiety (STAI)	45.41 (17.98)	31.22 (10.54)	3.74	.001
Trait Anxiety (STAI)	35.3 (12.72)	32.7 (10.17)	1.52	0.14
General Anger				
State Anger (STAXI)	11.59 (5.3)	10.37 (1.39)	1.22	0.24
Trait Anger (STAXI)	17.74 (7.79)	15.07 (5.18)	2.58	0.02
Angry Temperament (STAXI)	6.74 (3.36)	5.3 (1.6)	2.86	0.008
Angry Reaction (STAXI)	7.33 (3.13)	6.52 (2.64)	2.16	0.04
Anger In (STAXI)	13.44 (4.27)	13.33 (3.17)	.19	0.85
Anger Out (STAXI)	16.44 (4.79)	15.04 (3.26)	2.33	0.03
Anger Control (STAXI)	23.04 (5.87)	22.67 (6.35)	.39	0.7
Driving Anger				
Driving Anger Scale (DAS)	75.74 (33.22)	63.85 (23.9)	3.43	0.002
Hostile Gestures (DAS)	6.26 (3.61)	4.89 (2.59)	2.61	0.02
Illegal Driving (DAS)	8.3 (4.45)	7 (3.42)	2.51	0.02
Police Presence (DAS)	6.81 (3.72)	6.41 (2.69)	.72	0.47
Slow Driving (DAS)	14.07 (6.74)	11.74 (4.9)	3.22	0.003
Discourtesy (DAS)	24.67 (10.7)	19.85 (7.48)	3.81	0.001
Traffic Obstruction (DAS)	16 (7.85)	14.07 (6.24)	2.62	0.02
Driver Stress Profile (DSP)	27.14 (26.56)	22.56 (17.42)	1.84	0.08
Anger (DSP)	8.96 (7.71)	7.89 (5.88)	1.12	0.28
Impatience (DSP)	7.11 (7.61)	6.22 (5.38)	1.01	.32
Competing (DSP)	5.3 (6.52)	3.59 (4.22)	2.31	0.03
Punishing (DSP)	5.78 (6.62)	4.52 (3.92)	1.73	.095

Limitations in Data

When one's population involves court-mandated individuals, the likelihood of untruthful or misleading responses is increased. The lack of anger and aggressive driving behavior endorsed by the CR participant was entirely incongruent with the severity of the behaviors objectively observed by law enforcement officers. In general, the CR group presented as much less forthcoming and more wary in their willingness to disclose information. These incongruencies clearly indicate inaccurate responding by the mandated, involuntary participants. One could certainly hypothesize a number of reasons for less than valid responding on these questionnaires including blatant dishonesty (this would be consistent particularly in the subset of antisocial personality disorder drivers seen in the study), fear of reprisal from the courts or disapproval/recrimination by the group leader or assessor, denial of the problem, or a general lack of insight. The rampant denial of any personal problems with aggressive driving within this population supports this overall lack of insight hypothesis. Almost 100% of the CR population came in to the study claiming that the police had gotten the "wrong guy/woman". Some drivers, in fact, felt entirely justified in committing serious and dangerous driving behaviors, especially in response to some perceived infraction by some other driver on the roadways. This lack of insight directly informs the administration of treatment within this population.

In anticipating similar responding styles by a court-mandated population, it may be helpful to administer a measure designed to detect "faking good" or inconsistent responding. It may also prove fruitful to employ the use of collateral sources (family members, friends) who drive often with the subject. Objective reports on the aggressive drivers behavior change could be gathered in this way.

Conclusion

It is becoming abundantly clear that aggressive driving is a significant public safety concern. Aggressive drivers appear to become more physiologically aroused in provoking driving situations than do their non-aggressive driving counterparts, perhaps contributing to their aggressive behavior. Aggressive drivers also appear to be more psychologically distressed and psychiatrically impaired. Thus their dangerous driving style might be, in part, a reaction to their inner turmoil or may be a sample of a larger repertoire of ineffectual coping strategies. The automobile affords the individual the perfect opportunity to relieve stress in a relatively anonymous fashion with the luxury of a quick getaway. Understanding the aggressive driver informs the clinical application of behavior change programs.

These research programs offer alternatives to merely ticketing and fining drivers for aggression on the roadways. Such interventions as described here directly target the problematic behavior, compel the drivers to spend some significant time acknowledging and learning about their behaviors and possible consequences and then offer alternative driving and coping strategies. These types of programs could potentially impact aggressive driving similarly to the impact of defensive driving courses or drunk-driving programs on their respective target behaviors. Thus the potential benefits on public health and safety is unknown at this point, but potentially substantial.

References

Beck, A.T., Ward, C.H., Mendelson, M., Mock, L., & Erbaugh, J. (1961). An inventory for measuring depression. *Archives of General Psychiatry, 4*, 561-571.

Blanchard, E.B. & Schwarz, S.P. (1988). Clinically significant changes in behavioral medicine. *Behavioral Assessment, 10*, 171-188.

Buss, A.H. & Durkee, A. (1957). An inventory for assessing different kinds of hostility. *Journal of Counseling Psychology, 21*, 343-349.

Clayton, A.B. & Mackay, G.M. (1972). Aetiology of traffic accidents. *Health Bulletin, 31*, 277- 280.

Deffenbacher, J. L., Filetti, L. B., Lynch, R. S., Dahlen, E. R., & Oetting, E. R. (2002). Cognitive-behavioral treatment of high anger drivers. *Behaviour Research and Therapy, 40*, 895-910.

Deffenbacher, J.L., Huff, M.E., Lynch, R.S., Oetting, E.R. & Salvatore, N.F. (2000). Characteristics and treatments of high-anger drivers. *Journal of Consulting Psychology, 47*, 5-17.

Deffenbacher, J.L., Oetting, E.R., & Lynch, R.S. (1994). Development of a driving anger scale. *Psychological Reports, 74*, 83-91.

Felthous, A.R., Bryant, S.G., Wingerter, C.B., & Barratt, E. (1991). The diagnosis of Intermittent Explosive Disorder in violent men. *Bulletin of the American Academy of Psychiatry and the Law, 19*, 71-79.

First, M.B., Spitzer, R.L., Gibbon, M., & Williams, J.B.W. (1994). *Structured Clinical Interview for Axis II Personality Disorders, Version 2.0* (July 1994 Draft). NY: Biometrics Research Department.

First, M.B., Spitzer, R.L., Gibbon, M., & Williams, J.B.W. (1995). *Structured Clinical Interview for DSM-IV Axis -I Disorders- Patient Edition (SCID-I/P) (Version 2.0)*, NY,NY: Biometrics Research Department.

Galovski, T.E. & Blanchard, E.B. (2002a). Psychological characterisitics of aggressive drivers with and without Intermittent Explosive Disorder. *Behaviour, Research, and Therapy, 40*, 1157-1168.

Galovski, T. E. & Blanchard, E. B. (2002b). The effectiveness of a brief, psychological intervention on aggressive driving. *Behaviour, Research and Therapy, 40,* 1385-1402.

Galovski, T. E., Blanchard, E. B., Malta, L. S., & Freidenberg, B. M. (2003). The psychophysiology of aggressive drivers: Comparison to non-aggressive drivers and pre- to post-treatment change following a cognitive-behavioural treatment. *Behaviour Research and Therapy, 41,* 1055-1067.

Galovski, T.E., Blanchard, E.B., & Veazey, C.S. (2002). Intermittent Explosive Disorder and other psychiatric co-morbidity among court-referred and self-referred aggressive drivers. *Behaviour, Research, and Therapy, 40*, 641-651.

Gulian, E., Matthews, G., Glendon, A.I., Davies, D.R., & Debney, L.M. (1989). Dimensions of driver stress. *Ergonomics, 32*, 585-602.

Larson, J.A. (1996). *Steering clear of highway madness: A driver's guide to curbing stress and strain.* Oregon: BookPartners, Inc.

Larson, J.A., Rodriquez, C., & Galvan-Henkin, A. (1998). *Pilot study: Reduction in "road rage" and aggressive driving through one day cognitive therapy seminar.* Represented at New York State Symposium on Aggressive Driving, May 13, 1998, Albany, NY.

Maiuro, R. (Summer, 1998). *Recovery: Rage on the road* [On-line]. Available online at http://www.icbc.com/oldrecover/volume9/number2/RageOnTheRoad/.

Malta, L.S., Blanchard, E.B., Freidenberg, B.M., Galovski, T.E., Karl, A., & Holzapfel. (2001).

Psychophysiological reactivity of aggressive drivers: An exploratory study. *Applied Psychophysiology and Biofeedback, 26*, 95-116.

Martinez, R. (1997). *The statement of the honorable Recardo Martinez, M.D.*, Administrator, National Highway Traffic Safety Administration before the Subcommittee on Surface Transportation and Infrastructure, U.S. House of Representatives.

Matthews, G., Dorn, L., & Glendon, A.I. (1991). Personality correlates of driver stress. *Personality and Individual Differences, 12*, 535-549.

McElroy, S.L., Hudson, J.I., Pope, Jr., H.G., Keck, P.E., & Aizley, H.G. (1992). The DSM-III-R impulse control disorders not elsewhere classified: Clinical characteristics and relationships to other psychiatric disorders. *American Journal of Psychiatry, 149*, 318-327.

Mizell, L. (1997). Aggressive driving. In AAA Foundation for Traffic Safety (Ed.), *Aggressive Driving: Three Studies* [On-line]. Available online at *http://www. aaafts.org/Text/research/agdrtext.btm.*

Monopolis, S. & Lion, J.R. (1983). Problems in the diagnosis of Intermittent Explosive Disorder. *American Journal of Psychiatry, 140*, 1200-1202.

Ross, H.L. (1940). Traffic accidents: A product of social-psychological conditions. *Social Forces, 18*, 569-576.

Snyder, D.S. (1997). *The statement of David S. Snyder, Assistant General Counsel*, American Insurance Association, representing advocates for highway auto safety before the Sub-committee on Surface Transportation, Committee on Transportation and Infrastructure, U.S. House of Representatives.

Spielberger, C.D. (1979). *State-Trait Anger Expression Inventory (STAXI) Professional Manua*l. Odessa, FL: Psychological Assessment Resources, Inc.

Spielberger, C.D., Gorusch, R.L., & Lushene, R.E. (1970). *STAI Manual for the State-Trait Inventory.* Palo Alto, CA: Consulting Psychologists Press.

Stradling, S.G. & Parker, D. (1997). Extending the theory of planned behaviour: The role of personal norm, instrumental beliefs, and affective beliefs in predicting driving violations. In J.A. Rothengatter & E. Carbonell (Eds.), *Traffic and transport psychology: Theory and application* (pp. 367-374). Amsterdam: Pergamon.

In: Contemporary Issues in Road User Behavior and Traffic Safety ISBN:1-59454-268-6
Editors: D. Hennessy and D. Wiesenthal, pp. 61-70

Chapter 6

ON THE ROAD: SITUATIONAL DETERMINANTS OF AGGRESSIVE DRIVING

David L. Van Rooy, James Rotton and Paul J. Gregory
Florida International University

Goin' places that I've never been.
Seein' things that I may never see again
And I can't wait to get on the road again.
On the road again
Willie Nelson

INTRODUCTION

Commercials often portray driving as an exhilarating experience for happy and carefree people who pilot their cars along mountain roads and through desert vistas (Marsh & Collett, 1987). People frequently name driving when they are asked to name activities that produce flow, which Csikszentmihalyi and LeFevre (1989) define as a positive emotional state that accompanies the effortless passage of time. These facts argue for drawing a distinction between driving and commuting (Parsons, Tassinary, Hebl, & Grossman-Alexander, 1998). Driving is something individuals do because they enjoy doing it (discretionary), whereas commuting is something individuals have to do in order to get to and from work (obligatory). As Kluger (1998) has pointed out, many drivers value the time spent behind the wheel; for some, it may be the only time they have to reflect and daydream.

The distinction between driving and commuting is useful for understanding an unanticipated result that Van Rooy (2003) obtained in an experiment that looked at the combined effects of commute length and congestion. Van Rooy varied commute length by having participants drive a set route for 6 or 18 miles (9.6 or 27 km). Congestion was varied by scheduling the drive at times when traffic volumes were light (8:00-11:58 am) or heavy (3:00-6:59 pm). The study's sample was composed of 136 undergraduates who received course credit for participating in an experiment whose protocol had been approved by the university's Institutional Review Board. Only females were recruited because women outnumbered men during the summer term when this experiment was conducted. Each was led to believe that the experiment's goal was to assess the physiological effects of driving

Following the commute, participants were told that there would be a delay while the equipment for testing physiological reactions was being set up. They were asked if, while waiting, they would use the time by helping a second experimenter doing research on personnel selection. They then rated three job applicants, which is a task that is frequently used to assess verbal aggression (Berkowitz, 1993). Based on results that Hennessy and Wiesenthal (1997; 1999) had obtained, Van Rooy (2003) hypothesized that ratings would be most negative after the participants had driven a long distance on congested streets. From the graph in Figure 1, it can be seen that the hypothesis was confirmed.

Figure 1. Candidate Ratings as a Function of Commute Length and Congestion. (Based on data in Van Rooy, 2003).

However, contrary to our expectations, candidates received favorable ratings when participants had driven a long distance on uncongested roads. In retrospect, we believe that this unexpected result stems from participants turning an uncongested commute into a leisurely drive; that is, it is likely that drivers leaned back in their seats and adopted a recreational frame of mind when they realized that they would not have to deal with a lot of traffic. This result may explain why drivers complain about commuting but choose to live far from where they work. Many commuters avoid congestion by relying on expressways that take them out of cities to their homes in the country. For example, the second author lives some distance from his place of employment, in Miami, which is known for its aggressive drivers (de Gale, 2002), but his commute ends with a drive through pastures, orange groves, and strawberry fields. There is a growing body of literature that indicates that natural scenery reduces the fatigue that results from the directed attention that driving entails (Cackowski & Nasar, 2003; Parsons et al., 1998).

ANGER

Van Rooy's (2003) experiment was designed to accomplish two goals. As noted, one was to assess the feasibility of measuring verbal aggression following a congested commute. The experiment's other goal was to clarify results from a prior study (White & Rotton, 1998) on commuter stress. Unlike previous investigators (e.g., Novaco, Stokols, Campbell, & Stokols, 1979; Novaco, Stokols, & Milanesi, 1990), White and Rotton obtained nonsignificant results when they used a reliable and common instrument (the Semantic Differential of Emotional States) to assess the effects of type of commute. Their disappointing results contrasted with ones that Hennessy and Wiesenthal had obtained when they used cell phone technology to obtain on-line measures of stress and aggression. This team of investigators found that congestion causes drivers to report stress (Hennessy & Wiesenthal, 1997) and to behave more aggressively (Hennessy & Wiesenthal, 1999) while driving on the busiest highway in North America. We reasoned that drivers might report anger and negative affect during a commute, but such effects would dissipate quickly following the commute. This hypothesis was based on laboratory studies (Doob & Climie, 1972; Konecni, 1975) that have found that the effects of frustration are short-lived.

Accordingly, the drivers in Van Rooy's (2003) experiment were asked several questions about their emotional states (e.g., anxiety, frustration, anger, stress) on four occasions. This was accomplished by having them complete a brief check list before their commute, use a cell phone to describe how they were feeling halfway into the commute, and filling out the same checklist immediately after their commute and then again after they rated job candidates. From the graph in Figure 2, it can be seen that the experimental groups showed equivalent levels of anger before the commute, and, as hypothesized, the highest levels of anger were reported by individuals given the task of driving a long distance on congested roads. However, differences between the experimental groups had largely disappeared by the time drivers completed their trip. As is the case with other stressors (Glass & Singer, 1972), it appears that individuals adapt to the stress of commuting. These results reconcile inconsistencies from prior research by showing that commute characteristics affect drivers while they are on the road (e.g., Hennessy & Wiesenthal, 1997), but individuals do not evidence changes in emotional reactions after they have completed a difficult commute (White & Rotton, 1998).

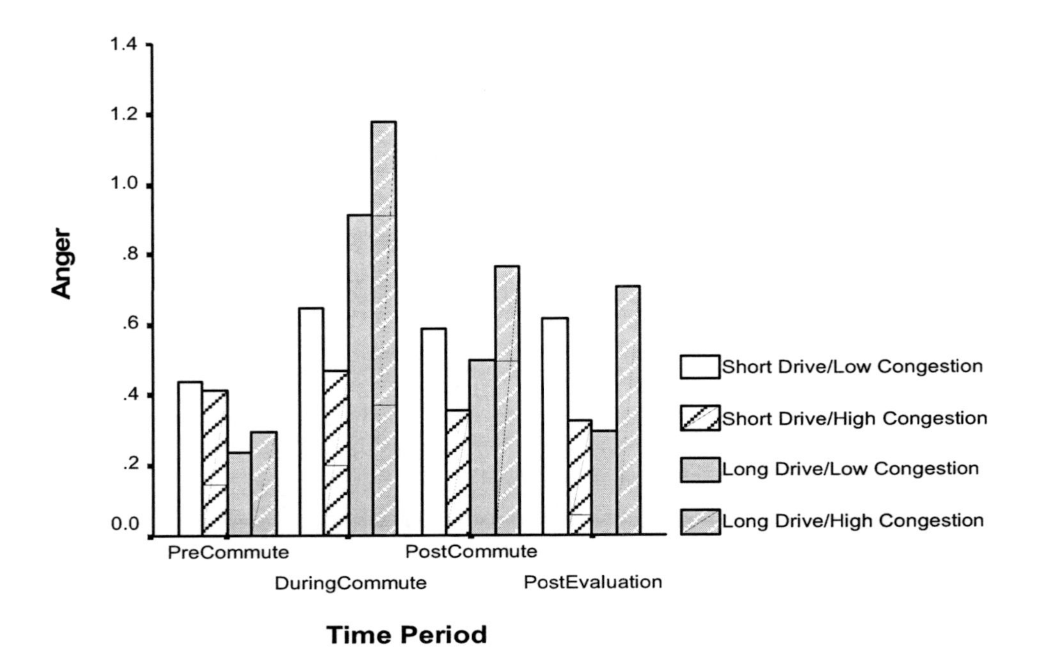

Figure 2. Mean Anger as a Function of Commute Length, Congestion, and Time Period. (Based on data in Van Rooy, 2003).

PROVOCATION

Research on aggressive driving has been dominated by the frustration-aggression hypothesis (Dollard, Miller, Doob, Mowrer, & Sears, 1939). Indeed, one author has gone so far as to define aggressive driving as "a syndrome of frustration-driven instrumental behaviors" (Shinar, 1998, p. 139). The frustration-aggression hypothesis has led investigators to focus on congestion: "blocked path of travel by other cars, red signal lights, insufficient number of lanes for the volume...and the tendency of many people to organize their work-recreation cycles around the same times and routes" (Shinar, 1998, p. 143).

However, frustration is only one of several factors that cause aggression. Others are provocation, anonymity, competition, and aversive stimuli, such as noise and high temperature (Berkowtiz, 1989; Ellison, Govern, Petri, & Figler, 1995; Kenrick & McFarland, 1986). Berkowitz (1989) reformulated the frustration-aggression hypothesis to propose that anything that makes us feel bad (i.e., causes negative affect) increases the probability of aggressive behavior. While it is possible to prepare a long list of stimulus factors that arouse negative affect, provocation has been found to be the most consistent and reliable cause of aggression (Berkowitz, 1993). This fact struck us when we examined the responses of 126 undergraduates (40 males, 86 females) who had been asked to "use single words and short phrases to describe all of the actions, behaviors, emotions, gestures, and expressions that come to mind when using the term 'aggressive driving'." We also asked this sample to prepare a similar list of words and phrases for "road rage." Our sample came up with 536 words and phases when asked for "aggressive driving"; they generated nearly as many (501) to describe "road rage." Table 1 lists the 20 words and phrases that were mentioned most often. These lists were prepared by separate pairs of coders, which explains why slightly different terms appear on each (e.g., "anger" under Aggressive Driving vs. "angry" under Road Rage). More than half of the words (11 out of 20) were used to characterize both terms. The overlapping terms are italicized. It appears "aggressive driving" and "road rage" are fuzzy sets that overlap in meaning. It is also interesting to note that none of the terms on the two lists refer to environmental factors, such as congestion. Instead, the terms that students associate most often with aggressive driving describe some sort of provocation: cutting off (48%), hand/body gestures (42%), cursing (29%), and honking (18%). The frustration-aggression hypothesis has led investigators to focus on congestion: "blocked path of travel by other cars, red signal lights, insufficient number of lanes for the volume...and the tendency of many people to organize their work-recreation cycles around the same times and routes" (Shinar, 1998, p. 143).

Table 1. Student Conceptions of Aggressive Driving and Road Rage

Aggressive Driving	Percentage	Road Rage	Percentage
Speeding	50.00	Angry	46.03
Cutting off	48.41	Rude gestures	42.86
Hand/body gestures	42.06	Aggressiveness	39.68
Cursing	29.37	Yelling	30.16
Anger	24.60	Cursing	17.46
Honking	18.25	Illegalities	16.67
Tailgating	15.87	Speeding	14.29
Dangerous	12.70	Dangerous	11.90
Impatience	12.70	Violent	10.32
Inconsiderate	11.90	Madness	9.52
Frustration	9.52	Fighting	8.73
Rudeness	8.73	Frustrated	8.73
Rushing	8.73	Honking	8.73
Carelessness	7.94	Rage	8.73
Traffic violations	7.14	Reckless	8.73
Irresponsible	5.56	Tailgating	8.73
Reckless	5.56	Impatience	7.94
Staring	5.56	Cutting off	7.14
Hurry	4.76	Screaming	5.56
Run red light	3.97	Death	4.76

N = 126.

It might be thought that it would be impossible to assess the role of provocation in a driving situation without placing research participants in danger. However, McGarva and Steiner (2000) recently developed a relatively safe and responsible procedure for studying how people react when they are the target of horn honking. Instead of using horn honking as a dependent variable, as has been done in the past (e.g., Ellison et al., 1995), this team had an observer ride along with participants on a test drive. The observer recorded how participants reacted when the driver in a car following their vehicle honked after they stopped at a sign. McGarva and Steiner found that drivers accelerated more rapidly when the person honking at them was driving a high than a low status vehicle; however, although nearly half of the drivers made gestures that were visible to the confederate in the other vehicle, status did not affect nonverbal gestures, verbalizations and horn honking.

McGarva and Steiner (2000) did not vary provocation, however. All of the drivers were subjected to a single honk of 95-100 dB intensity of two seconds duration. However, it would be an easy matter to manipulate provocation by including a no-honk control group and by varying either the duration of honks (e.g., 0.5-4.0 sec.) or their number (e.g., from 1 to 5 honks). It would also be interesting to observe how drivers responded when a confederate (either a pedestrian or another driver) gave them what is usually termed "the finger."

Assessing the effects of provocation would be of considerable theoretical interest, because very different predictions can be derived from two competing models of aggressive behavior. One of these is Anderson and Bushman's (2002) General Aggression Model (GAM), which grew out of research on temperature and aggressive behavior (for a review of the literature, see Rotton & Cohn, 2002). According to the GAM, aggression is a linear function of temperature: The hotter it is, the more individuals engage in aggression. This prediction is opposed by one that can be derived from the Negative Affect Escape (NAE) model. According to the NAE model (Baron & Bell, 1976), moderate levels of discomfort lead individuals to behave aggressively; however, when discomfort is very high, individuals choose to engage in alternative activities, such as escape and activities that minimize discomfort (e.g., staying inside an air-conditioned building). The latter behaviors conflict with people's desire to behave aggressively.

The NAE model's prediction is supported by results from archival analyses of criminal behavior, which indicate that assault is a curvilinear function of temperature. For instance, Cohn and Rotton (1997) found that more assaults were reported to police at moderately high than either low or very high temperatures in Minneapolis, MN. Rotton and Cohn (2000; 2004) obtained similar results when they examined assaults in Dallas, TX. Although these results are consistent with predictions that can be derived from the NAE model, they can be faulted for introducing an unmeasured variable (namely, escape) to explain why fewer assaults occur at high and low than moderately high temperatures. Unfortunately, it is difficult to derive a measure of escape from archival data. Interpretational difficulties are also encountered when investigators attempt to assess escape tendencies under laboratory conditions because of informed consent procedures, which let participants know that they have the right to withdraw from an experiment at any time (Asmus & Bell, 1999; Gardner, 1978).

However, McGarva and Steiner (2000) have identified a clear measure of escape tendencies (namely, acceleration). Their results also provide reason to believe that horn honking provokes one of the aggressive behaviors in Table 1 (nonverbal gestures). It would be possible to obtain more objective and reliable measures of nonverbal behaviors, such as expressions that communicate anger (cf. Cohen, Nisbett, Bowdle, & Schwarz, 1996), by mounting a web cam on dashboards, and an audio recording would provide an objective measure of verbal aggression.

SPILLOVER

Given the difficult task of assessing the effects of driving and commuting, investigators have substituted feelings (e.g., anger) for behavior (e.g., aggression). However, the responses described in the preceding section (facial grimaces, verbal remarks, nonverbal gestures) do not satisfy the most widely accepted definition of aggression: "behavior directed toward the goal of harming or injuring another living being who is motivated to avoid such treatment." (Baron & Richardson, 1997, p. 7). This and similar definitions (Berkowitz, 1993; Zillman, 1979) include words (e.g., goal) that require judgments of intention. For example, Baron and Richardson remind us that individuals may "strike one another with their automobiles quite by accident" (p. 9).

Rotton, Gregory, and Van Rooy (this volume) have found that scales designed to assess angry and aggressive driving load on a single factor. However, although the terms are frequently used interchangeably, anger and aggression are not the same thing. As Berkowitz (1993) has observed, anger is a feeling, and people do not always act on their feelings, especially if they fear retribution. Psychologists have developed standard procedures for assessing aggression under laboratory conditions. Like any methodology, these procedures have their disadvantages as well as advantages (Giancola & Chermack, 1998; Tedeschi & Quigley, 1996). Their principle advantage is that they remove doubts about an aggressor's intentions (e.g., Baron & Eggleston, 1972). Consequently, although questions have been raised about the external validity of laboratory experiments, they are regarded as "the gold standard" against which all other measures (e.g., verbal, indirect, passive) are judged.

Van Rooy's (2003) experiment was designed to explore the feasibility of assessing aggression after individuals had completed a commute and left their vehicles. It will be recalled that the effects of the manipulations had dissipated by the time the drivers had reached their destination. This finding might be thought to mitigate against finding effects after individuals have left their vehicle. However, while anger (and negative affect in general) increases the probability of aggressive behavior, Berkowitz (1978) has described anger and aggression as parallel processes: People do not always report anger when they aggress, and anger does not always lead to aggression (Averill, 1983). Indeed, not one of the many studies that have examined relationships between aversive stimulation and aggression has shown that anger (or negative affect) acts as a mediator (Rotton & Cohn, in press).

There are several reasons to hypothesize that commuting and congestion will increase the probability of aggressive behavior even when individuals do not report higher levels of negative affect. First, it has been found that workplace stress spills over to affect a person's mood and behavior at home (Burns & Rotton, 2004). Hennessy (2003) recently found that driving stress and gender combine to predict hostility and aggression in the workplace. Second, driving reduces an individual's willingness to persist when confronted with failure. This motivational deficit, which has been termed "a reduction on frustration tolerance" (White & Rotton, 1998), is notable because it occurs in the absence of any change in emotional states (Cohen, 1980). Third, individuals show elevated levels of arousal following a commute (for a review of the literature, see Koslowsky, Kluger, & Reich, 1995). Harrison (2003) has proposed that commuters might interpret their arousal as anger. According to Zillmann's (1994) excitation-transfer theory, individuals will behave more aggressively when they label their arousal as anger. This is most likely to occur if an individual has been provoked (e.g., insulted). Finally, provoked individuals have been found to behave aggressively *after* encountering environmental stressors, such as uncontrollable noises (Donnerstein & Wilson, 1976).

REFERENCES

Anderson, C. A., & Bushman, B. J. (2002). Human aggression. *Annual Review of Psychology*, 53, 27-51.

Asmus, C. L., & Bell, P. A. (1999). Effects of environmental odor and coping style on negative affect, anger, arousal, and escape. *Journal of Applied Social Psychology*, 29, 245-260.

Averill, J. R. (1983). Studies of anger and aggression: Implications for theories of emotions. *American Psychologist*, 38, 11451160.

Baron, R. A., & Bell, P. A. (1976). Aggression and heat: The influence of ambient temperature, negative affect, and a cooling drink on physical aggression. *Journal of Personality and Social Psychology*, 33, 245-255.

Baron, R. A., & Eggleston, R. J. (1972). Performance on the "aggression machine": Motivation to help or harm? *Psychonomic Science*, 26, 321-322.

Baron, R. A., & Richardson, D. R. (1997). *Human aggression*. (2nd ed.). New York: Plenum.

Berkowitz, L. (1978). Do we have to believe we are angry with someone in order to display "angry" aggression toward that person? In L. Berkowitz (Ed.), *Cognitive theories in social psychology* (pp. 455-463). New York: Academic.

Berkowitz, L. (1989). Frustration-aggression hypothesis: Examination and reformulation. *Psychological Bulletin*, 106, 59-73.

_____ . (1993). *Aggression: Its causes, consequences, and control*. Philadelphia: Temple University Press.

Burns, T., & Rotton, J. (2004, April). *Development of the Sphere Overlap Scale (SOS)*. Paper presented at the annual convention of the Society of Industrial and Organizational Psychology, Chicago, IL.

Cackowski, J. M., & Nasar, J. L. (2003). The restorative effects of roadside vegetation: Implications for automobile driver anger and frustration. *Environment and Behavior*, 34, 736-751.

Cohen, D., Nisbett, R. E., Bowdle, B. F., & Schwarz, N. (1996). Insult, aggression, and the southern culture of honor: An "experimental" ethnography. *Journal of Pesonality and Social Psychology*, 70, 945-960.

Cohen, S. (1980). Aftereffects of stress on human performance and social behavior: A review of research and theory. *Psychological Bulletin*, 88, 82-108.

Cohn, E. G., & Rotton, J. (1997). Assault as a function of time and temperature: A moderator-variable time-series analysis. *Journal of Personality and Social Psychology*, 72, 1322-1334.

Csikszentmihalyi, M., & LeFevre, J. (1989). Optimal experience in work and leisure. *Journal of Personality and Social Psychology*, 56, 815-822.

de Gale, A. (2002, June 18, 2002). Miami motorists are rudest -- again. *Miami Herald*, pp. 1E & 2E.

Dollard, J., Miller, N. E., Doob, L. W., Mowrer, G. H., & Sears, R. R. (1939). *Frustration and aggression*. New Haven, CT: Yale University Press.

Donnerstein, E., & Wilson, D. W. (1976). Effects of noise and perceived control on ongoing and subsequent aggressive behavior. *Journal of Personality and Social Psychology*, 34, 774-781.

Doob, A. N., & Climie, R. J. (1972). Delay of measurement anf the effects of film violence. *Journal of Experimental Social Psychology*, 8, 136-143.

Ellison, P. A., Govern, J. M., Petri, H. L., & Figler, M. H. (1995). Anonymity and aggressive driving behavior: A field experiment. *Journal of Social Behavior and Personality*, 10, 265-272.

Gardner, G. T. (1978). Effects of Federal Human Subjects Regulations on data obtained in environmental stressor research. *Journal of Personality and Social Psychology*, 36, 628-634.

Giancola, P. R., & Chermack, S. T. (1998). Construct validity of laboratory aggression paradigms: A response to Tedeschi and Quigley (1996). *Aggression and Violent Behavior*, 3, 237-253.

Glass, D. C., & Singer, J. E. (1972). *Urban stress: Experiments on noise and social stressors*. New York: Academic.

Harrrison, K. (2003). Fitness and excitation. In J. Bryant, D. Roskos-Ewoldsen, & J. Cantor (Eds.), *Communciation and emotion: Essays in honor of Dorf Zillmann* (pp. 473-489). Mahwah, NJ: Erlbaum.

Hennessy, D. A. (2003). *From driver stress to workplace aggression*. Paper presented at the annual convention of the American Psychological Association, Toronto, Canada.

Hennessy, D. A., & Wiesenthal, D. L. (1997). The relationship between traffic congestion, driver stress, and direct versus indirect coping behaviours. *Ergonomics*, 40, 348-361.

_____ . (1999). Traffic congestion, driver stress, and driver aggression. *Aggressive behavior*, 25, 109-123.

Kenrick, D. T., & MacFarlane, S. W. (1986). Ambient temperature and horn honking: A field study of heat/aggression relationship. *Environment and Behavior*, 18, 179-191.

Kluger, A. N. (1998). Commute variability and strain. *Journal of Organizational Behaviour*, 19, 147-165.

Konecni, V. J. (1975). The mediation of aggressive behavior: Arousal level versus anger and cognitive labeling. *Journal of Personality and Social Psychology*, 32, 706-712.

Koslowsky, M., Kluger, A. N., & Reich, M. (1995). *Commuting stress: Causes, effects, and methods of coping*. New York: Plenum.

Marsh, P., & Collett, P. (1987). *Driving passion: The psychology of the car*. Boston, MA: Faber and Faber.

McGarva, A. R., & Steiner, M. (2000). Provoked driver aggression and status: A field study. *Transportation Research: Part F*, 3, 167-179.

Novaco, R. W., Stokols, D., Campbell, J., & Stokols, J. (1979). Transportation, stress, and community psychology. *American Journal of Community Psychology*, 7, 361-380.

Novaco, R. W., Stokols, D., & Milanesi, L. (1990). Objective and subjective dimensions of travel impedance as determinants of commuting stress. *American Journal of Community Psychology*, 18, 231-257.

Parsons, R., Tassinary, L. G., Hebl, M., & Grossman-Alexander, M. (1998). The view from the road: Implications for stress recovery and immunization. *Journal of Environmental Psychology*, 18, 113-139.

Rotton, J., & Cohn, E. G. (2000). Violence is a curvilinear function of temperature: A replication. *Journal of Personality and Social Psychology*, 78, 1074-1081.

_____ . (2002). Climate, weather, and crime. In R. Bechtel & A. Churchman (Eds.), *The new handbook of environmental psychology* (pp. 461-498). New York: Wiley.

_____ (2004). Outdoor temperature, climate control, and criminal assault: The spatial and temporal ecology of violence. *Environment and Behavior*, 36, 276-306.

_____ (in press). The social escape avoidance (SEA) model of aggression and criminal violence. In F. Columbus (Ed.), *Psychology of aggression*. Hauppauge, NY: Nova Science.

Shinar, D. (1998). Aggressive driving: The contribution of the drivers and the situation. In J. A.

Rothengatten (Ed.), *Transportation research, Part B: Traffic psychology and behaviour* (pp. 137-160). New York: Pergamon.

Tedeschi, J. T., & Quigley, B. M. (1996). Limitations of laboratory paradigms for studying aggression. *Aggression and Violent Behavior*, 1, 163-177.

Van Rooy, D. L. (2003). *Effects of commute characteristics and cell phone use on affectand candidate evaluations: A field experiment*. Unpublished Masters Thesis, Florida International University, Miami, FL.

White, S. M., & Rotton, J. (1998). Type of commute, behavioral aftereffects, and cardiovascular activity: A field experiment. *Environment and Behavior*, 30, 763-780.

Zillmann, D. (1979). *Hostility and aggression*. Hillsdale, NJ: Erlbaum.

_____ . (1994). Cognition-excitation interdependencies in the escalation of anger and angry aggression. In M. Portegal & J. F. Knutson (Eds.), *The dynamics of aggression: Biological and social processes in dyads and groups* (pp. 45-71). Hillsdale, NJ: Erlbaum.

In: Contemporary Issues in Road User Behavior and Traffic Safety ISBN:1-59454-268-6
Editors: D. Hennessy and D. Wiesenthal, pp. 71-78

Chapter 7

FIELD METHODOLOGIES FOR THE STUDY OF DRIVER AGGRESSION

Andrew R. McGarva
Dickinson State University

INTRODUCTION

Simply put, there is something about driving an automobile that brings out the worst in people. Drivers yell and gesture, even lash out violently toward complete strangers with alarming regularity over things as insignificant as lane position or a parking space. Consider the relative frequency of pedestrian conflict, the likelihood of witnessing a dispute over an available seat on a park bench with one person shaking his fist at the other who responds with an enraged finger gesture. There is no comparison; in few other settings is human aggression more frequent and universal than between interacting drivers.

A study conducted by AAA Foundation for Traffic Safety (1999) reported that almost 90% of drivers in the United States have experienced at least one situation involving what they considered aggressive driving. It was also concluded that reported incidents involving driver aggression had recently increased by 51%, resulting in as many as 1,500 injuries or fatalities every year. In 1996 alone about 2,000 violent incidents were reported to police nationwide, a figure that likely represents only a fraction of its occurrence worldwide. It appears that to an alarming degree we become ogres at the wheel.

Perhaps the prevalence of driver aggression should be of little surprise. Driving incites anger, rage, and retaliatory intent in a context that appears to present minimal inhibitions against the expression of these feelings. The research literature in social psychology has identified numerous factors that influence human aggression, including crowding, anonymity, frustration, and provocation. Few commonplace situations present as many of these factors as does everyday commuter traffic.

Following Altman's (1975) territory theory, drivers view their vehicles as personal space, perhaps to be defended as one would their home or family (Wozniak, Miller, Bell, Troup, Szlemko & Benfield, 2004). The act of navigating one's personal space in a public space in conjunction with others navigating their own personal spaces invites perhaps a sort of territorial aggression. Furthermore, our vehicles both make us anonymous and act as equalizers, in that everyone's actions have equal effect; the behavior of and decisions made by individuals otherwise possessing little authority or influence (women, youth, low-level employees) are no less consequential than the actions of those who might otherwise have greater authority or influence (dominant males, parents, bosses). Driving may provide an opportunity for many whose aggressive actions have been stifled elsewhere to allow hostilities to run freely. Cars empower drivers, not only to travel quickly as an individual from point to point, but also to assert themselves anonymously over other drivers...aggressively if need be.

Aggression Research

Aggression is a common dimension of human behavior but can be a thorny subject of experimental research. While people can be observed acting in ways intended to harm one another on a daily basis, ethical and procedural issues limit the scientific study of aggression. It is unethical to inflict real pain and suffering on research volunteers as a means to instigate their retaliation. Nor can researchers give human subjects the opportunity to seriously aggress against even a bogus bystander without first gaining their informed consent and, arguably, it is fruitless to study high levels of aggression amongst participants who are aware of experimental manipulations and hypotheses.

Despite the ethical limitations to the empirical study of aggression and the particular difficulty of studying aggression between interacting drivers in real-life situations, some useful work has been done. The classic field study by Doob and Gross (1968) initiated an application of the social psychology of aggression to the behavior of automobile drivers. To satisfy a course requirement for a graduate seminar in social psychology—perform a study outside the laboratory based on Webb, Campbell, Schwartz, and Sechrest's 1966 text on the value of unobtrusive research methods in social science—Doob and Gross devised a test of the effects of frustrator status on driver responses in a naturally occurring context (Gross & Doob, 1976).

In an addendum to one of the study's several reprintings, Gross and Doob (1976) detailed the problems encountered in conducting their roadway manipulation. A black 1966 Chrysler Imperial was rented to serve as the high status vehicle. The car needed washing, but given that only the back of the car would be visible to participants, Gross polished only the rear half for the experiment. A rusted-out 1954 Ford station wagon and a drab, gray 1961 Rambler served as low status vehicles. The high- or low-status frustrating vehicle was to be positioned so that it was the first in a line of cars stopped at a red light. Intersections controlled by a stoplight were selected that provided the best opportunity to frustrate a driver. It was necessary to find lights that remained red long enough so that other drivers had time to pull up behind the experimental vehicle in a setting where frustrated drivers could not easily drive around the frustration. The status of the vehicle driven by the frustrated driver was of interest, but neither Doob nor Gross was adept at identifying the makes and models of cars, so two car savvy high school students were hired to ride around in the back seat of the frustrating vehicle as it was driven around downtown Palo Alto and surrounding residential areas, only to pop up at the termination of a trial to identify the car in question. Getting the paper published proved a larger problem than motion sickness, as the article was rejected by the *Journal of Personality and Social Psychology* and the *Journal of Personality* before being accepted by the *Journal of Social Psychology* in 1968.

In demonstrating the effect of the status of a vehicle that remained motionless at a green stoplight on the latency and duration of horn honks made by a frustrated driver, Doob and Gross (1968) initiated the naturalistic study of factors affecting aggressive behavior occurring on the roadways. For a time, horn honking became the staple dependent variable in research assessing the influence of gender (Bochner, 1971; Deaux, 1971), status of frustrating driver rather than vehicle (Unger, Raymond, & Levine, 1974), modeling (Harris, 1973), cell phone use of a frustrating driver (McGarva, Ramsey, & Shear, 2004), and aggression cues as well as deindividuation (Turner, Layton, & Simon, 1975) on driver aggression. Recently, Shinar, (1998) in demonstrating a relation between horn honking and congestion in city traffic, argued that the horn honk is largely ambiguous. It is not clear whether a driver has sounded his or her horn as a friendly "heads up" to another driver who is not attending to the task at hand or as a furious attack meant to rile offending drivers.

For the past several years a collection of student research assistants and I have been extending driver aggression research beyond the horn honk response. The effects of various traffic events on various aggressive driver reactions have been assessed with a series of experiments in which, under the guise of "a study on driver behavior" an experimenter is placed inside participants' vehicles. With this procedure, a range of responses to angering roadway scenarios can be measured. In one experiment hurriedness was manipulated as an independent variable with number of lane changes, vehicles passed, and top speed recorded as dependent variables (McGarva & Reis, 2001). For this study, undergraduates enrolled in general psychology at Dickinson State University, in southwestern North Dakota, were scheduled one at a time to arrive with their vehicle at the psychology building on campus where they met with an experimenter. Upon giving their consent, participants returned to their car accompanied by the experimenter who instructed them to drive from campus to a 4-lane, 25 mph (40 km/hr) major thoroughfare on which they traveled approximately 1.4 miles (2.3 km) to a gas station. During this time the experimenter asked a repeating series of questions unrelated to the experimental hypothesis, such as "without looking at the speedometer, how fast do you think you are going right now?" The purpose of these "distractor" questions was to lead participants to believe we were studying some sort of perceptual aspect of driving rather than driver aggression.

In the parking lot of the gas station, participants were informed that the experiment was finished and were then engaged a casual conversation regarding the participant's performance in general psychology. At this time, the experimenter commiserated with the other student about the difficulty of the course and participants in the hurried condition were told of an extra credit opportunity available to them back at the university only if they returned quickly. Those randomly assigned to the non-hurried condition were told of a similar opportunity occurring later in the day. While returning to campus the experimenter observed and recorded whether participants signaled while making lane changes, their top speed, number of lane changes, and number of cars passed. Once back at the university, all participants were debriefed.

Table 1. Means (and Standard Deviations) for Number of Lane Changes, Vehicles Passed, and Top Speed for Hurried and Non-hurried Drivers

	Lane Changes	Vehicles Passed	Top Speed
Hurried	2.8 (1.61)	6.2 (3.75)	34.46 (3.93)
Non-hurried	1.6 (0.95)	2.9 (1.63)	30.29 (3.39)

It was expected that people in a hurry would make more lane changes, pass more vehicles, and have a higher top speed than non-hurried drivers. Using ANOVA, hurriedness was found to significantly affect each of these dependent variables. These means and standard deviations are provided in Table 1. Drivers who felt the need to rush back to campus displayed their haste in the way they operated their motor vehicles.

Upon observing the effect of hurriedness on assertive driving maneuvers we decided to introduce a frustrating vehicle to the situation (McGarva, Reis, & Warner, 2002). By positioning a slow moving vehicle in traffic ahead of hurried participant drivers accompanied by an experimenter it became possible to measure a number of reactions to frustration that could not have been assessed in previous research. With a procedure similar to that described in the previous experiment, hurried and non-hurried participant drivers entered an intersection before turning left onto a 2-lane road on which they would travel for 2.1 miles (3.4 km) on their way back to the university. At this point a confederate—driving either a brand new Pontiac Grand Am or a shabby, dented, and impressively rusty 1983 Lincoln Continental Mark IV both with Montana license plates—timed her arrival to the intersection so as to position herself ahead of the participant vehicle and proceeded to travel well under the speed limit in a posted 35mph (57 km) passing zone.

The experimenter recorded passing latency and duration, counted the number of words spoken regarding the frustrating driver, and noted incidents of tailgating and nonverbal gesturing. Upon returning to the university, participants were debriefed and asked not to discuss the experiment with others. None of the participants reported suspecting the frustrating driver's authenticity.

As expected, hurried drivers exhibited more aggressive behavior than drivers who were not placed in a hurry. When hurried, participants were more likely to pass other drivers, make more lane changes, and drive faster than when unhurried. Additionally, hurriedness influenced drivers' reactions to frustration, encouraging more lengthy verbal lashings and more frequent gestures of the sort that can instigate retaliation from other drivers causing escalations toward more hostile actions. The two frustrating automobiles used as an attempted status manipulation did not have the expected effect. It was likely that the Pontiac was not of sufficiently high status. Admittedly, I was more interested in demonstrating the effects of frustration and hurriedness. Consequently, the data were collapsed across automobile status.

An unexpected gender difference was observed in that female drivers were *more* likely to express their anger toward a frustrating driver both verbally and non-verbally. Four of the 22 females, 2 in each hurried condition, made a nonverbal gesture (raising a hand or arm) that was visible to other drivers while there were no such gestures amongst the 28 males. This difference was found to be statistically significant using a 2 way chi-square. Females were also more inclined to verbalize about the confederate driver than males, either muttering to themselves or directly to the experimenter; however, using ANOVA this latter difference only approached statistical significance. Means and standard deviations are presented in Table 2.

Table 2. Mean (with Standard Deviations) Number of Words Spoken for Hurried and Non-hurried Female and Male Drivers

	Female Verbal	Male Verbal
Hurried	5.86 (7.03)	1.14 (1.95)
Non-hurried	1.71 (3.15)	.29 (0.76)

It is evident from both personal experience, as well as media reports, that frustration is not the sole determinant of roadway aggression. Perhaps induced by frustration or the perception that some other offense has been committed, drivers often provoke one another. In some cases it takes little more than lifting a finger to incite another driver to operate their vehicle in an aggressive way. Although provocation has been observed to invite aggressive responses in laboratory settings (see Bettencourt & Kernahan, 1997 and Bettencourt & Miller, 1996 for meta-analytic reviews), provocation has not received much representation in the research literature. To study the roll of provocation in driver aggression, a procedure was devised involving a confederate driver who honked his horn and made angry gestures at participant drivers who were stopped at a stop sign while waiting for directions issued by an experimenter (McGarva & Steiner, 2000).

Student volunteers drove with a female experimenter who issued instructions to drive out of the university lot, down a side street and onto a 4-lane road with a 25 mph (40 km/hr) speed limit. After traveling roughly ¼ mile (0.4 km), the experimenter directed the driver to pull into a parking lot where a driving questionnaire was administered orally. Participants were then directed to drive onto a side street to a stop sign where they were told to make a right turn onto another street with a 25-mph "40 km/hr") limit. A baseline recording of 0-20 mph (32 km/hr)-acceleration was made at this point.

At a second stop sign, the participant was made to wait for instruction while the experimenter appeared to be recording some information. This interval lasted several seconds. At this point a confederate, driving either a late model Nissan Pathfinder or an impressively filthy and roughed up pickup truck, pulled up behind the participant vehicle, and, not only honked his horn, but raised both his arms impatiently, waving the backs of his hands at the participant driver as though horribly frustrated by the brief delay. The experimenter then started a tape recorder, instructed the driver to make a right turn down a short block toward the 4-lane road and again recorded acceleration to 20mph (32 km/hr). Provocation continued onto the main 4-lane road where the confederate both shook his fist and made an irate facial expression as he pulled up beside the participant's vehicle before turning off the main road.

Upon independent review of the tape recordings, the length of time participants talked about the provoking driver was measured, as was a subjective judgment of the overall intensity of their response to provocation (ranging from "mild" to "intense"). The duration of their vocalizations ranged from 0 to 480 seconds with an average length of nearly one minute (56.6 sec.). Interestingly, while the majority of participant drivers made negative statements about the other driver (such as "what's with this jerk?" or "this guy behind me is really pissed!"), approximately one quarter of the participants appeared entirely unaffected by the provocation.

It was expected that drivers would accelerate more rapidly, make longer angry vocalizations, and be more likely to honk their horn and make visible, non-verbal responses when provoked by the low-status confederate than when provoked by the high status confederate. A significant effect of status on acceleration was observed in that participants were more inclined to speed away following low-status provocation than when provoked by a higher status driver. Length of vocalization was not related to status, but did vary across gender, with females talking an average of 63 seconds about the provoking driver and males exhibiting mean vocalization duration of only 43 seconds. Although this difference was not statistically significant—due largely to the overall variability in duration of talk—it supports the notion that women may, in some ways, be more aggressive than men when behind the wheel.

There are several advantages of these procedures over those used in the past. By placing an experimenter in participants' vehicles it is possible to observe and record a number of aggressive responses to roadway frustrations and provocations. An initial concern was that the presence of an experimenter might completely inhibit driver aggression, but the range of responses we observed suggests this not to be the case. For instance, in response to the provoking confederate driver some 0-20mph (32 km/hr) times were reduced by over 4 seconds; drivers occasionally made aggressive gestures; and, in some cases, voiced loud and angry comments for several minutes. Of the 51 participants, 26 made nonverbal responses that were visible to the confederate driver. For further details on the statistical outcomes please refer to McGarva & Steiner (2000).

The perseverance of angry responses in the presence of an experimenter underscores the serious nature of driver aggression. It appears that people will exhibit hostile driving behavior no matter who watches, as if this form of aggression is more socially acceptable than, say, throwing eggs at a person's house or waving a fist at someone who cuts in line at the delicatessen. Given the apparent persistence of driver aggression, researchers should be able to empirically measure any of a population of aggressive driver reactions to roadway events.

Ethical Concerns

The ethical treatment of research subjects is an ever-present concern in any type of experimental endeavor. When studying variables related to human aggression—even in the laboratory—ethical concerns intensify. Much of the research on driver aggression takes experimental methodologies to the streets and subjects participant/drivers to real-life roadway situations. There are costs to the methods employed: potentially very high costs for the volunteers, the experimenters, and the confederates in the case of an automobile accident caused or influenced by the experimental procedures. These costs, however, are surpassed by the potential benefits of research on what is popularly recognized as the growing road rage phenomenon. Driver rage is expressed far too often, affecting too many ordinary people, not to try to understand and minimize this form of behavior.

Although the experiments described herein involved relatively mild levels of driver aggression, it is not difficult to imagine scenarios where mild hostilities could escalate to more dangerous roadway behavior. Naïve research participants were placed in contexts resembling those that occasionally build toward acts of violence or even homicide (Michalowski, 1975). In this regard, the presence of an experimenter in participants' vehicles is essential to defuse excessive aggression should it occur. Luckily, circumstances requiring such intervention have not arisen during our research. Regardless, great care must be taken when conducting future research to maintain an environment where participants can be immediately informed of the nature of the experimental manipulation. We have found that, once debriefed, even the angriest participant was immediately pacified—some even entertained—by the experimental manipulations.

FUTURE RESEARCH

More research is necessary before any firm conclusion can be made about gender differences in mild forms of driver aggression. It may be that the responses we observed amongst women were a means of letting off steam, making women less likely to exhibit other forms of driver aggression. The observation that female drivers were more inclined to express their anger may merely reflect a female's expressive nature in general, rather than an inclination toward roadway aggression. It could also be that driving provides an avenue to express anger that may be inhibited elsewhere by perceived societal restrictions based on gender-role schemas. Only the results of future research will be able to shed light on these speculations.

It would be useful and of popular importance to investigate what inhibitory cues exist and are used in roadway interactions. To what degree does a friendly wave and a smile defuse a potentially hostile traffic incident? Are these gestures cross-cultural? Are there types of drivers that are more inclined to employ them? Do "Baby on board" tags have any effect? Perhaps drivers could be trained to identify instances when certain nonverbal cues should be displayed and what bumper stickers not to apply.

The effects of status on driver aggression are still not fully understood. Aggression appears influenced by perceptions drivers have of other drivers. It is not known what opinions people form of others based on cues present in driving situations. For instance, do commuters behave differently toward service drivers (taxi drivers, truck drivers) than toward other commuters or pleasure drivers? In order to curb anti-social driving behavior, we need to understand the impressions drivers make on one another and how these impressions influence aggressive roadway interactions. Driver aggression is a large and mounting problem; however, social scientists possess the tools to understand it and in time perhaps reduce it, making our roadways a safer place.

REFERENCES

AAA Foundation for Traffic Safety (1999). *Controlling road rage.* Washington, DC.

Altman, I. (1975). *The environment and social behavior.* Monterey, Ca: Brooks/Cole.

Bettencourt, B. & Kernahan, C. (1997). A meta-analysis of aggression in the presence of violent cues: Effects of gender differences and aversive provocation. *Aggressive Behavior, 23*, 447-456.

Bettencourt, B. & Miller, N. (1996). Gender differences in aggression as a function of provocation: A meta-analysis. *Psychological Bulletin, 119*, 422-447.

Bochner, S. (1971). Inhibition of horn-honking as a function of frustrator's status and sex: An Australian replication and extension of Doob and Gross (1968). *Australian Psychologist, 6*, 194-199.

Deaux, K. K. (1971). Honking at the intersection: A replication and extension. *Journal of Social Psychology, 84*, 159-160.

Doob, A. N. & Gross, A. E. (1968). Status of frustrator as an inhibitor of horn-honking responses. *Journal of Social Psychology, 76*, 213-218.

Gross, A. E. & Doob, A. N. (1976). Status of frustrator as an inhibitor of horn-honking responses: How we did it. In P. M. Golden (Ed.) *The Research Experience* (pp. 487-494), Itasca, Illinois: F. E. Peacock.

Harris, M. B. (1973). Field studies of modeled aggression. *Journal of Social Psychology, 89*, 131-139.

McGarva, A. R., Ramsey, M., & Shear, S. (2003, April). *Cell phone use and frustration-induced driver aggression.* Poster session presented at the annual meeting of the Western Psychological Association, Vancouver, B.C.

McGarva, A. R. & Reis, J. (2001). *The effects of hurriedness on aggressive driving behavior.* Unpublished manuscript. Dickinson State University, ND.

McGarva, A. R., Reis, J., Warner, T. (2002, March). *The effects of hurriedness on frustration-induced driver aggression.* Poster session presented at the annual Red River Undergraduate Psychology Conference, Moorhead, MN.

McGarva, A. R. & Steiner, M. (2000). Provoked driver aggression and status: A field study. *Transportation Research Part F: Traffic Psychology & Behaviour, 3,* 167-179.

Michalowski, R. J. (1975). Violence in the road: Crime of vehicular homicide. *Journal of Research in Crime and Delinquency, 12*, 30-43.

Shinar, D. (1998). Aggressive driving: the contribution of the drivers and the situation. *Transportation Research Part F: Traffic Psychology & Behavior, 1*, 137-160.

Turner, C. W., Layton, J. F., & Simons, L. S. (1975). Naturalistic studies of aggressive behavior: Aggressive stimuli, victim visibility, and horn-honking. *Journal of Personality and Social Psychology, 31*, 1098-1107.

Unger, R. K., Raymond, B. J., & Levine, S. (1974). Are women a minority group sometimes? *International Journal of Group Tensions, 4*, 71-81.

Webb E. J., Campbell, D. T. Schwartz, R. D., & Sechrest, L. (1966). *Unobtrusive Measures: Non-reactive research in the social sciences.* Chicago: Rand McNally.

Wozniak, R., Miller, R., Bell, P., Troup, L., Szlemko, W., & Benfield, J., (2004, April). Comparison of territorial influences on road rage across two campuses. In P.A. Bell (Chair), *Aggression and violence.* Symposium conduced at the meeting of the Rocky Mountain Psychological Association, Reno, NV.

PART 3
DRIVING VIOLATIONS AND COLLISIONS

In: Contemporary Issues in Road User Behavior and Traffic Safety ISBN:1-59454-268-6
Editors: D. Hennessy and D. Wiesenthal, pp. 81-100

Chapter 8

OBSERVING MOTORWAY DRIVING VIOLATIONS

A. Ian Glendon
WorkCover NSW Research Center of Excellence, School of Health Sciences,
University of Newcastle, New South Wales, Australia
Danielle C. Sutton
School of Psychology, Griffith University, Queensland, Australia

INTRODUCTION

Why is it important to study driving violations? Observational data from over 20 years ago established that close following and other violations are statistically related to drivers' crash involvement (Evans & Wasielewski, 1982; 1983). From questionnaire studies, Parker, Reason, Manstead, and Stradling (1995) report that driving violations are statistically associated with crash involvement. Golias and Karlaftis (2002) found that while speeding was associated with dangerous driving violations, neither was related to non-seat belt use or drink-driving violations. Kontogiannis, Kossiavelou, and Marmaras (2002) found that highway-code violations were related to speeding convictions. From telephone survey data, Jonah (1990) found that younger drivers reported higher accident and violation rates – including speeding and tailgating. Studies in several countries have found that males and younger drivers report significantly more violations than do females and older drivers (e.g., Kontogiannis et al., 2002; Lawton, Parker, & Stradling, 1997; Mesken, Lajunen, & Summala, 2002). Greater understanding of driving violations could serve to improve counter measures.

From survey data of drivers' reactions to speed camera installation, Corbett (2000) identified four driver types – "Conformers", "Deterred" (reduced speed to avoid detection), "Manipulators" (slowed down only when approaching a speed camera), and "Defiers" (undeterred by speed cameras). Although percentages of Manipulators and Defiers were relatively small, they were important categories of violating drivers. Two questions are pertinent. Would observational data on a range of driver violating behaviors produce similar findings? Would analysis of such data emerge with a similar driver typology?

Observing Driving Violations

In this chapter we report selected findings on driving violations from an observational study of nearly 2,700 driving cases on 2-, 3- and 4-lane stretches of motorway in Queensland, Australia. Data were collected from a moving vehicle during daylight hours in fine weather, with approximately equal numbers of cases in each direction, sampling mornings and afternoons on all days of the week. Total observing and recording time was approximately 20 hours. Jonah (1990) notes that direct observation studies of driver behavior are needed to validate self-reports, which tend to underestimate negative behaviors, such as violations. While making no claims for the broader representativeness of our findings, a prime contribution of this study is that data were collected on violations and other driving behaviors during "normal" driving while drivers were unaware that their behavior was being observed and recorded. This ensured that drivers' behavior was unaffected by the data collection method. All cases were anonymous – no individual drivers or vehicles could be identified.

In addition to basic driver demographics, data collection and coding for this study included classifying driving violations and other driving behaviors. Information was also recorded for vehicle type (14 categories), approximate vehicle age (newer, intermediate age, older vehicle) and other characteristics, as well as passenger details, driver gender and driver age group ("younger" – estimated as under 30 years of age, "intermediate age" – estimated at 30-50 years, and "older" as over 50 years of age).

Direct observation complements other methods of assessing driving performance (e.g., Parker, Manstead, Stradling, & Reason, 1992a). Its strengths include high ecological validity and ready availability of large samples for estimating frequencies of different violations and other driving behaviors. In free-flowing motorway traffic, as in this study, it is possible to obtain complete data on almost 100% of cases within the sampling frame. Real time data capture also provides data that are not available in any other way. Aspects of driving performance that cannot be measured by external observation include:

- Cognitions – e.g., percent attentional capacity devoted to driving task (Groeger, 2000).
- Motivations to violate – (Parker, Manstead, Stradling, Reason, & Baxter, 1992b; Parker et al., 1992a).
- Experience factors – e.g., vehicle and route driven, years since licence obtained, distance driven (Laapotti, Kesinen, Hatakka, & Katila, 2001; Lourens, Vissers, & Jessurun, 1999).
- Driver stress, both specific to that drive – (Gulian, Glendon, Davies, Matthews, & Debney, 1990), and general (Gulian, Glendon, Debney, Davies, & Matthews, 1989).
- Drugs or alcohol ingested – unless driving is thereby very severely and evidently impaired.
- Chronic and acute medical conditions.
- Listening to radio, CD, or using hands-free mobile phone.

Piloting indicated that three researchers optimized data quality and quantity. The driver was responsible for safely positioning the research vehicle to maximise data collection opportunities; the front-seat researcher video-recorded relevant aspects of the traffic environment; the rear-seat researcher tape-recorded relevant aspects of other vehicles, their drivers, passengers, and other information. When other vehicles were present on both sides of the research vehicle, the front-seat researcher could also record vehicle, driver and passenger details. The researchers performed the same roles throughout this study. As all tasks required high concentration, rest breaks were taken after approximately each hour of data gathering. Data were initially coded qualitatively using dedicated software (Noldus, Trienes, Hendriksen, Jansen, & Jansen, 2000) and subsequently entered into SPSS for quantitative analysis.

WHAT IS A DRIVING VIOLATION?

While some driving violations are almost universal (e.g., exceeding a signed speed limit), what constitutes the aggregate of all driving violations differs between jurisdictions, reflecting differences in history, legislation and culture. For example, in many jurisdictions "overtaking on the inside" (passing a vehicle while travelling in a "slower" lane) is a driving violation, but in Queensland it is allowed in some circumstances.[1] Some driving behaviors are "in transition" in respect of their violating status – for example, hand-held mobile phone use. Others are violations in some jurisdictions but not in others – for example, hands-free mobile phone use while driving. Stayer and Johnston (2001) indicate that both hand-held and hands-free mobile 'phone use significantly impair driving performance.

Lawton, Parker, Manstead, and Stradling (1997), and Parker, Lajunen, and Stradling (1998) categorised driving violations into three types: i) gaining advantage over other drivers, ii) maintaining progress, and iii) interpersonal aggression towards other drivers. Harrington and McBride (1970) used the extant Vehicle Code to classify violations into seven categories, only three of which (speed, equipment, passing) were identified in the current study. The basis for judging driving violations in this study was the extant version of the *Queensland driver's guide* (Queensland Transport, 2000[2]), complemented by the researchers' empirically-based assessments. Behaviors constituting driving violations in this study were:

- Speeding – exceeding the signed speed limit. In this study, vehicles were observed travelling in speed limit zones: 60, 70, 80, 90, 100, and 110 kph. To facilitate estimates of the extent to which other vehicles were speeding, the research vehicle was driven at, or as close to, the signed speed limit as was safely possible. Vehicle speeds were coded by consensus among the researchers according to whether a vehicle was being driven within the speed limit or was exceeding the speed limit by "up to 10%", "10-20%", "20%-30%", or "more than 30%". Videoed observations were used to verify initial judgements.

---

1 Blockey and Hartley (1995) include "Overtaking on the inside" as a violation in their Western Australia driver study. The descriptors "inside" and "outside" applied to lane descriptions are ambiguous. "Inside lane" generally refers to the lane in which the slowest vehicles normally travel – the usage employed in this chapter. However, it is also used to refer to the "fast" lane, which is "inside" in the sense that it is closest to the centre of a dual carriageway.

2 *The Queensland driver's guide* was replaced by: *Your keys to driving in Queensland* (Queensland Transport) in November 2002, a few months after the data for this study were collected.

- Tailgating – following a lead vehicle at a distance such that the time measured from a fixed point between the rear of the lead vehicle and the front of the following vehicle is less than two seconds. Consensus among the researchers provided an immediate judgement as to whether a following vehicle was tailgating. Videoed observations were used to verify initial judgements.
- Lane position – travelling in the outside lane of a 3- or 4-lane motorway when the lane immediately inside is safely available for use. Instances of vehicles travelling in the fourth lane when the third lane was vacant, or in the third lane of three when the second lane was vacant, were coded as violations, constituting 2.6% of all observed cases.
- Driver's view obstructed.
- Vehicle unsafely loaded.
- Vehicle in dangerous condition.
- Faulty lights.
- Inappropriate signalling.
- Increase speed when overtaken.
- Cutting in.
- Using hand-held mobile phone.
- Polluting vehicle.
- Driver/passenger aggression.
- Unruly passengers.

Frequencies and Predictors of Different Driving Violations

While speeding and tailgating were observed frequently in this study, other violations were much less common, and in some cases, almost no examples were observed. For example, only two cases of "Driver/passenger aggression" were observed, both younger males, and there were two cases of vehicles judged to be in a dangerous condition. Analyses relating only to the most frequently observed violations are considered here, although some analyses relate to all observed violations.

In reporting most findings, unless otherwise indicated, chi-squared (χ^2) summary statistics are provided where appropriate, while standardized residuals (SR) identify those cells that carry part of an effect. Where SR is either >2.0 or <-2.0 a significant effect is indicated (Hinkle, Wiersma, & Jurs, 1994, p. 540).

Driver's View Obstructed

Thirty cases of "Driver's view obstructed" were observed (1.1% of the total). While more cases of male (1.4%) than of female drivers (0.5%) having an obstructed view were observed, this difference was not significant. However, four out of 14 vehicle types accounted for virtually all cases of "Driver's view obstructed". These were (with percentages of these types): “Light van” (4.7%), “Station wagon” (3.4%), “Utility vehicle” (2.7%), and “4WD/camper” (1.9%). Apart from 0.2% of “Standard sedans”, all other vehicles were recorded with zero instances of this violation. Although the differences were significant, $\chi^2 = 55.36$, df 13, $p<.001$, many cells had zero counts, making further analysis problematic. When vehicle type was recoded as either “Commercial” or “Private-use” the significance became marginal, $\chi^2 = 4.98$, df 1, $p=.031$. However, while no cell could be clearly identified as carrying the observed effect, “Commercial” vehicles (1.8%, SR 1.9) had a higher percentage of obstructed views than did “Private-use” vehicles (0.8%). This evidence suggests that this particular violation is specific to certain categories of vehicle – mainly those involved in light commercial operations, with heavily loaded leisure vehicles being occasional transgressors.

Hand-Held Mobile (Cellular) Phone Use

Overall 1.6% (47 out of 2673) drivers were observed using a hand-held mobile (cellular) phone. Although fewer older drivers (0.6%) than intermediate age (2.0%) and younger drivers (1.8%) were observed using a mobile phone, the differences were not significant. No significant gender difference was found for mobile phone use, nor was it predicted by vehicle age.

However, there was a significant effect in respect of mobile phone use and “Vehicle occupancy”, $\chi^2 = 18.95$, df 3, $p<.001$. “Solo drivers” were more likely to be observed using a mobile phone (2.5%, SR 2.3), while “Drivers with one adult passenger” (0.1%, SR -3.2) were less likely than other vehicle occupancy combinations to be observed using a mobile phone. No cases of “Drivers with more than one adult passenger” were observed using a mobile phone, while “Drivers with child/ren in vehicle” were also relatively high users (2.3%).

While differences for the 14 vehicle types were not significant, when recoded into “Commercial” and “Private-use”, a significant difference was found, $\chi^2 = 13.45$, df 1, $p=.001$. “Commercial” vehicle drivers had higher mobile phone use (3.2%, SR 3.0) than did drivers of “Private-use” vehicles (1.2%, SR -2.0).

Vehicle Unsafely Loaded

Fifty vehicles (1.9%) were judged to be unsafely loaded. Intermediate age drivers were significantly more likely to be observed driving unsafely loaded vehicles, $\chi^2 = 6.76$, df 2, $p=.034$; 2.3% of intermediate age drivers, compared with 1.1% of older drivers and 0.5% of younger drivers were judged to be driving an unsafely loaded vehicle. Younger drivers’ low likelihood of driving an unsafely loaded vehicle accounted for the effect (SR -2.0). Unsafely loaded vehicles were more likely to be driven by males (2.4%) than by females (0.3%), $\chi^2 = 13.60$, df 1, $p<.001$. Vehicle loading was not predicted by “Vehicle occupancy”. Older vehicles were more likely (3.5%) to be observed with an unsafe load than were intermediate age (1.8%) or newer vehicles (1.0%), although this was of marginal significance, $\chi^2 = 4.76$, df 2, $p=.092$.

There was a significant difference between the propensity for different vehicle types to be observed with an unsafe load, $\chi^2 = 170.34$, df 13, $p<.001$. The effect was carried by 37 out of the 50 observations of unsafe load being either "Utility vehicle" (10.9%, SR 10.8) or "Small truck/van" (8.5%, SR 5.1), while "Standard sedans" (0.3%, SR -3.3) were very unlikely to be unsafely loaded.[3] Analysis of the recoded vehicle type variable confirmed a significant effect, $\chi^2 = 65.21$, df 1, $p<.001$, which resulted from 5.0% (SR 6.7) of "Commercial" vehicles being judged to be unsafely loaded, compared with 0.5% (SR -4.4) of "Private-use" vehicles.

Polluting Vehicle

While all road vehicles pollute, this category refers to those vehicles that were observed to have black smoke emanating continuously from their exhausts for at least ten seconds. Sixty vehicles (2.2%) were recorded as polluting. There was no significant likelihood that polluting vehicles would be driven by drivers in any particular age group – 2.6% of younger and 2.4% of intermediate age drivers were observed driving polluting vehicles, compared with 1.1% of older drivers. Males were significantly more likely than females to be observed driving polluting vehicles, $\chi^2 = 8.19$, df 1, $p=.004$; 2.8% of males compared with 1.0% of females. "Solo drivers" (3.2%) were significantly more likely, $\chi^2 = 14.52$, df 3, $p=.002$, to be observed driving a polluting vehicle than were "Drivers with two or more adults in vehicle" (1.3%), "Drivers with one adult passenger", or "Drivers with child/ren in vehicle" (0.5%). The higher likelihood of solo drivers polluting (SR 2.3) and the lower relative likelihood of vehicles with one adult passenger (SR -2.2) accounted for the effect.

As would be expected, older vehicles (4.3%) were significantly more likely than newer vehicles (0.2%) or intermediate age vehicles (2.1%) to be observed polluting, $\chi^2 = 27.82$, df 2, $p<.001$. The effect was accounted for by older vehicles' greater propensity to pollute (SR 3.8) and newer vehicles' lower polluting propensity (SR -3.5). There was a significant difference in the propensity for different vehicle types to be observed polluting, $\chi^2 = 313.20$, df 13, $p<.001$. Three-quarters (45/60) of all polluting vehicles were accounted for by three vehicle types: "Small truck/van" (18.9% of vehicles in this category, SR 11.2), "Medium truck/van" (17.3%, SR 10.3), and "Coach/bus" (16.7%, SR 5.7), while "Large truck/van" (6.3%, SR 2.1), also helped to account for the effect. At 3.7% of instances "Motorcycle" was the other vehicle type to be above the mean value, although numbers observed were too small for a reliable result. At the other end of the scale, "Standard sedan", "Station wagon", "Hatchback" and "Light van" all carried the effect with percentages of zero or near zero instances of polluting observed. Analyzing recoded vehicle type confirmed a significant effect, $\chi^2 = 87.18$, df 1, $p<.001$, which was accounted for by 6.4% (SR 7.7) of "Commercial" vehicles observed polluting compared with 0.5% (SR -5.1) of "Private-use" vehicles. It can be concluded that polluting is a highly vehicle-specific form of violating.

---

[3] On a subsequent driving observation study, tins of paint pigment fell off the back of a utility vehicle and struck the researchers' observation vehicle so that it was returned with a part colour change – the incident being video-recorded for posterity!

Aggregating Violations other than Speeding and Tailgating

When all driving violations apart from those separately identified as speeding and tailgating were aggregated, 10.3% of all vehicles observed were in the “Other violations” category. Driver age group produced a significant difference, $\chi^2 = 9.82$, df 2, p=.007, 11.0% of intermediate age drivers were observed violating, compared with 10.8% of younger drivers and 5.6% of older drivers. The effect was accounted for by older drivers violating less frequently (SR -2.8). Gender differences were significant, $\chi^2 = 9.59$, df 1, p=.002, with 11.6% of male and 7.5% of female drivers observed violating – female drivers' lower rate of violating accounted for the effect (SR -2.5)

“Vehicle occupancy” produced a significant effect, $\chi^2 = 15.50$, df 3, p=.001; “Solo drivers” (12.0%), “Drivers with child/ren in vehicle” (11.1%), and “Drivers with two or more adult passengers” (10.3%), were observed violating more frequently than were “Drivers with one adult passenger” (6.6%), the effect being attributable to the lower violation rate of drivers with one adult passenger (SR -3.2).

A significant difference was found in respect of vehicle age, $\chi^2 = 10.54$, df 2, p=.005; drivers of older vehicles (12.7%) and intermediate age vehicles (10.4%) were observed to be violating more frequently than were drivers of newer vehicles (7.5%). The effect was carried by the lower violation rate of newer vehicle drivers (SR -2.3) and the higher violation rate for drivers of older vehicles (SR 2.0). Vehicle type had a significant impact upon “Other violations”, $\chi^2 = 107.51$, df 13, p<.001. Carrying the effect were “Small truck/van” (28.3% of vehicles in this category observed violating, SR 5.7), “Medium truck/van” (23.6%, SR 4.3), and “Utility vehicle” (18.3%, SR 3.9) – “Coach/bus” also had a high percentage violating (19.4%), but the smaller number of cases meant that this category did not carry part of the effect. Least likely to be observed violating (speeding and tailgating excluded) were: “Standard sedan” (7.1%, SR -3.1), “Hatchback/5-door” (5.9%, SR -2.0), and “Small sedan” (4.0%, SR -3.2). Analysis using recoded vehicle type again produced a significant effect, $\chi^2 = 66.21$, df 1, p<.001, which was accounted for by “Commercial” vehicles being more likely to violate (17.8%, SR 6.4) compared with “Private-use” vehicles (7.4%, SR -4.2).

Tailgating (Close Following)

Twenty-two percent of vehicles (n = 588) were observed to be tailgating (<2 seconds to lead vehicle) while 19.2% of vehicles were observed travelling at =>2 seconds (“safe” distance) from a lead vehicle.

There was no significantly greater likelihood for drivers in any age group to tailgate, although younger drivers (24.5%) were slightly more likely to be observed tailgating than were intermediate age (21.7%) or older drivers (20.7%). Percentages for “safe distance” drivers in each age group were practically identical. Excluding the 58.7% of cases in which there was no lead vehicle (leaving 1102 valid cases) differences continued to be non-significant, with 56.2% of younger drivers, 53.0% of intermediate age drivers and 51.7% of older drivers observed tailgating, indicating that for this sample of drivers, tailgating is not influenced by age group.

From the complete data set there appeared to be a significant difference between males' and females' propensity to tailgate. However, the picture was confounded by the finding that while 61.4% of vehicles driven by males were observed to have no lead vehicle, this was true for only 52.5% of vehicles being driven by females. The reason for this difference is unclear. Excluding cases in which there was no lead vehicle, tailgating was observed in 53.6% of cases (n = 1116) for which driver gender was known and in which there was a lead vehicle. Percentages of tailgating male and female drivers were then almost identical (53.5% and 53.7%), indicating that tailgating cannot be attributed to gender differences.

There was a significant effect for "Vehicle occupancy", $\chi^2 = 9.46$, df 3, p=.024; "Solo drivers" were more frequently observed tailgating (in 56.7% of 1119 cases for which there was a lead vehicle) than were "Drivers with one adult passenger" (48.3%), or "Drivers with more than one adult passenger" (50.0%), or when there were "Child/ren in vehicle" (43.4%). However, the effect was not large enough to be attributable to any particular cell, suggesting that tailgating is only weakly influenced by vehicle occupancy. Parker et al. (1992a) found that the presence of a passenger of the same age and sex as the driver moderated drivers' stated intentions to violate, including tailgating, and that in daytime drivers indicated that the presence of a passenger would make them less liable to violate, while the opposite would be true for night driving.

Table 1. Percentages of Vehicle Types Observed Tailgating (Mean 53.4%)

Vehicle type	Percent observed tailgating
Medium truck/van	69.6
Large truck/van	63.6
Station wagon	61.8
Utility vehicle	57.7
Bus – all types	57.1
Small truck/van	57.1
Light van	53.8
Limo/luxury sedan	53.8
Hatchback/5-door	52.1
Standard sedan/coupe	52.0
Small sedan	51.5
Motorcycle	50.0
4WD/camper	41.4
(All other 4-wheeled vehicles* (n = 10)	70.0)

* Includes emergency vehicles.

There was no propensity for vehicles of any particular age to be observed tailgating. While the differences were not significant, some vehicle types were more likely to be observed tailgating in those cases (n = 1120) for which there was a lead vehicle. Percentages are shown in Table 1. Although differences were non-significant, a "gradient effect" may be seen from the ranked data. Ignoring the very small "Other 4-wheeled vehicles" category, all those tailgating above the 53.4% mean value could be considered to be "Commercial" vehicles, with the exception of "Station wagon" and "Limo/luxury sedan" categories (for which numbers were also small). It tends to be predominantly private-use vehicles that fall below the tailgating average, with 4WDs the lowest of all. When vehicle type was recoded, excluding cases in which there was no lead vehicle, there was still no significant difference in tailgating between "Commercial" and "Private-use" vehicles.

Findings on tailgating suggest that this is a general cultural phenomenon uninfluenced by driver demographic or vehicle-related variables. That there was no greater likelihood that drivers of tailgating vehicles would be observed violating in other ways ("Other violations" category) suggests that tailgating is independent of other types of violating (excluding speeding – see below).

Speeding

Data were initially coded on the signed speed limit for the road section on which observations were made, and whether a vehicle was judged to be travelling "At or below the signed limit", "Up to 10% above the limit", "10-20% above the limit", "20-30% above the limit" or "More than 30% above the speed limit". Less than 8% of observations were of vehicles in a speed limit zone of under 100 kph. Therefore the speed limit categories were reduced to three: "110 kph", "100 kph" and "Other speed limit", while retaining the five speeding variables. Findings are shown in Table 2.

Table 2. Percentages of Drivers Observed Driving within, and Exceeding Speed Limits

Driving behavior of all drivers	Estimated percent	Aggregate percentages
Drive within 110 kph limit	18.9	
Drive within 100 kph limit	39.9	62.4
Drive within other speed limit	3.6	
Exceed 110 kph limit by <=10%	5.0	
Exceed 100 kph limit by <=10%	13.4	20.1
Exceed other speed limit by <=10%	1.7	
Exceed 110 kph limit by 10-20%	5.3	
Exceed 100 kph limit by 10-20%	9.1	15.8
Exceed other speed limit by 10-20%	1.4	
Exceed 110 kph limit by 20-30%	0.4	
Exceed 100 kph limit by 20-30%	0.8	1.6
Exceed other speed limit by 20-30%	0.4	
Exceed 110 kph limit by >30%	0.1	
Exceed 100 kph limit by >30%	0.3	0.4
Exceed other speed limit by >30%	0.0	
Totals	100.3*	100.3*

* Rounding error.

Because the number of vehicles observed exceeding any signed speed limit by more than 10% was comparatively small, the speeding categories were reduced to three: "Drive within speed limit", "Exceed speed limit by up to 10%", and "Exceed speed limit by 10% or more". Although there was a number of empty and small cell counts, a highly significant result was obtained for the relationship between driver age group and speeding, χ^2 = 108.44, df 26,

$p<.001$. For all cases indicated by the SR values as contributing to this result, older drivers were more likely to be observed driving within the signed speed limit and younger drivers were more likely to be observed exceeding the speed limit.

Unreliability of analyses resulting from some small cell counts indicated repeating the analyses on recoded versions of the signed speed limit and speeding variables. The following contributed to the significant result, $\chi 2 = 88.54$, df 10, $p<.001$:

- *Drive within 110 kph speed limit:* Older drivers most likely to (25.5%, SR 2.8); (intermediate age drivers, 18.4%; younger drivers, 16.5%).
- *Exceed 110 kph speed limit by up to 10%:* Older drivers significantly less likely to (2.5%, SR -2 0); (younger drivers, 5.4%; intermediate age drivers 5.2%).
- *Exceed 110 kph speed limit by more than 10%:* Older drivers significantly least likely to (2.8%, SR -2.2); younger drivers significantly most likely to (8.0%, SR 2.0); (intermediate age drivers, 5.7%).
- *Drive within other speed limit:* Older drivers significantly most likely to (56.3%, SR 3.6); younger drivers significantly least likely to (34.0%, SR -2.9); (intermediate age drivers, 43.3%).
- *Exceed other speed limit by up to 10%:* Older drivers significantly least likely to (9.0%, SR -3.0); younger drivers significantly most likely to (22.4%, SR 3.7); (intermediate age drivers, 14.9%).
- *Exceed other speed limit by more than 10%:* Older drivers significantly least likely to (3.9%, SR -4.3); (intermediate age drivers, 12.6%; younger drivers, 13.7%).

With the original speeding variable, the only driver gender difference, producing a marginally significant result, $\chi^2 = 21.98$, df 13, $p=.056$, was that females were less likely than males (3.1% of females, SR -2.6: 6.1% of males) to exceed the 110 kph speed limit. When the speeding variable was recoded this finding was more robust, $\chi^2 = 15.29$, df 5, $p=.009$, with female drivers being less likely than male drivers to exceed the 110 kph speed limit (females 3.5% SR -2.5: males 6.6%).

Analysing the data using the recoded speed variable to determine the effect of vehicle occupancy, a significant result was found, $\chi^2 = 87.08$, df 15, $p<.001$. This was attributable to "Solo drivers" being significantly less likely to drive within the 110 kph speed limit (14.3%, SR -4.3), compared with "Drivers with two or more adult passengers" (28.4%, SR 2.7), "Drivers with child/ren in vehicle" (27.2%, SR 2.8) and "Drivers with one adult passenger" (25.2%, SR 3.8). However, somewhat paradoxically "Drivers with child/ren in vehicle" were *less* likely (33.2%, SR -2.3) to drive within another speed limit, compared with "Solo drivers" (45.8%), "Drivers with one adult passenger" (43.1%) and "Drivers with two or more adult passengers" (35.5%).

"Drivers with two or more adult passengers" were least likely to exceed some other speed limit by up to 10% (7.1%, SR -2.6), compared with "Solo drivers" (16.7%), "Drivers with child/ren in vehicle" (15.7%) and "Drivers with one adult passenger" (12.8%). "Drivers with one adult passenger" were least likely to exceed some other speed limit by more than 10% (8.0%, SR -2.8), compared with "Drivers with two or more adult passengers" (13.5%), "Drivers with child/ren in vehicle" (12.9%), and "Solo drivers" (12.8%). Baxter, Manstead, Stradling, Campbell, Reason, and Parker (1990), and Parker et al. (1992a) found similar moderating effects upon speeding intention for drivers with a single passenger compared with solo drivers' speeding intentions.

A significant difference was found for the predictor variable "Vehicle age" with the recoded speed variable, $\chi^2 = 105.35$, df 10, $p<.001$. The categories below accounted for this finding.

- *Exceed 110 kph speed limit by up to 10%*: Drivers of newer vehicles were most likely to (8.3%, SR 3.7); drivers of older vehicles were least likely to (2.3%, SR -3.5); (intermediate age vehicle drivers, 5.1%).
- *Drive within other speed limit*: Drivers of newer vehicles were significantly least likely to (32.7%, SR -4.1); older vehicle drivers were significantly most likely to (52.9%, SR 4.1); (intermediate aged vehicle drivers, 42.7%).
- *Exceed other speed limit by up to 10%:* Drivers of older vehicles were significantly least likely to (52.9%, SR -2.9); (drivers of newer vehicles, 17.9%; intermediate age vehicle drivers, 16.3%).
- *Exceed other speed limit by more than 10%:* Drivers of newer vehicles were significantly most likely to (17.5%, SR 4.3); older vehicle drivers were significantly least likely to (8.3%, SR -2.9); (intermediate vehicle age drivers, 11.0%).

With speeding recoded into: "Drive within speed limit", "Exceed speed limit by up to 10%", and "Exceed speed limit by more than 10%", a significant effect was found, $\chi^2 = 254.51$, df 26, $p<.001$. Vehicle types most likely to be observed travelling within the speed limit are shown in Table 3.

Table 3. Vehicle Types most Likely to be Observed Travelling within Signed Speed Limit

Vehicle type	% travelling within speed limit	SR
Medium truck/van	97.3	4.6
Coach/bus	94.4	2.4
Small truck/van	94.3	4.2
Large truck/van	89.1	2.7

Least likely to be observed travelling within a signed speed limit were "Standard sedan" (50.7% observed, SR -4.4) and "Motorcycle" (18.5% SR -2.9). Most likely to be observed exceeding the speed limit by up to 10% were: "Standard sedan" (25.3%, SR 3.5), "Small sedan" (26.7%, SR 2.3), and "Motorcycle" (37.0%, SR 2.0). Most likely to be observed exceeding a speed limit by more than 10% were: "Standard sedan" (24.0% observed, SR 4.6), "Limo/luxury sedan" (27.9%, SR 1.9), and "Motorcycle" (44.4%, SR 3.3). Least likely to be observed exceeding a speed limit by more than 10% were "Small sedan" (10.0%, SR -2.9) and "Light van" (9.9%, SR -2.4).

When vehicle type was recoded as either "Commercial" or "Private-use", the difference remained significant, $\chi^2 = 89.88$, df 2, $p<.001$. "Commercial" vehicles were most likely to be observed travelling within a speed limit (76.3% observed, SR 4.8), and were also least likely to be observed exceeding a speed limit by up to 10% (12.5%, SR -4.7) or by more than 10% (11.2%, SR -4.2). "Private-use" vehicles were least likely to be observed travelling within a signed speed limit (57.1% were observed doing so, SR -3.2) and were also most likely to be observed exceeding a signed speed limit by up to 10% (19.9%, SR 2.7). These findings are consistent with those of Wasielewski (1984), who observed that higher speeds were more likely for younger drivers, solo drivers, and newer vehicles, as well as being associated with crash-involved drivers and drivers who had prior traffic convictions.

Also recorded was whether a vehicle had its headlights on – a safety feature of daylight driving. A report (European Road Safety News, 1997) on 17 studies from nine counties on the effects of driving with dipped headlights during daylight hours, maintained that this practice had resulted in significant reductions in some categories of accidents. Overall, 4.3% of vehicles observed were using headlights. A significant association was found between speed and headlight use, $\chi^2 = 66.46$, df 5, $p<.001$. This was accounted for by drivers exceeding the 110 kph speed limit by up to 10% being significantly more likely to have their headlights on (16.9%, SR 6.0). Drivers driving within some other speed limit were significantly less likely to be using their headlights (SR -4.5), indicating that there is a group of drivers who exceed the speed limit by up to 10%, who are more likely to use their headlights during daylight driving.

Speeding, Tailgating and Other Violations

A significant but complex relationship between speeding and tailgating, $\chi^2 = 122.94$, df 10, $p<.001$, was accounted for by the findings described below.

- Drivers without a lead vehicle were significantly most likely to be driving within a 110 kph speed limit (23.0%, SR 3.8). Tailgating drivers were significantly least likely to be driving within a 110 kph speed limit (10.7%, SR -4.6); (non-tailgating drivers, 15.7%).
- Tailgating drivers were significantly least likely to be exceeding a 110 kph speed limit by up to 10% (3.0%, SR -2.2), compared with non-tailgating drivers (5.7%) and drivers with no lead vehicle (5.5%). The apparent inconsistency with the previous finding could be accounted for by these tailgating drivers "wanting to overtake" a vehicle that was adhering to the 110 kph speed limit.
- Non-tailgating drivers were significantly least likely to be observed exceeding a 110 kph speed limit by more than 10% (2.7%, SR -3.0); (tailgating drivers, 3.8%, SR -2.0). Drivers with no lead vehicle were significantly most likely to be exceeding this limit (7.6%, SR 2.9). Non-tailgating drivers who have recently overtaken a vehicle being able to achieve their preferred speed could account for the apparent inconsistency with the previous finding.
- Drivers with no lead vehicle were significantly least likely to be driving within some other speed limit (38.3%, SR -3.2), compared with non-tailgating drivers (51.5%, SR 2.7) and tailgating drivers (51.0%, SR 2.7). This finding might be accounted for by tailgating drivers being "held back" while drivers with no lead vehicle can "have their head". It could be that non-tailgating drivers effectively act so as to keep some other drivers' speeding behaviors "in check".
- Tailgating drivers were significantly most likely to be observed exceeding some other speed limit by up to 10% (21.2%, SR 3.9); (non-tailgating drivers, 15.9%).

Drivers with no lead vehicle were significantly least likely to be observed exceeding some other speed limit by up to 10% (12.5%, SR -2.7). A "traffic flow" phenomenon, whereby a line of traffic typically moves at a speed in excess of the signed limit – when this is less than 110 kph – but stays within 10% of that limit might account for this finding. This is an example of a driving culture phenomenon.

- Non-tailgating drivers were significantly least likely to be observed exceeding some other speed limit by more than 10% (8.4%, SR -2.2); (tailgating drivers, 10.2%; drivers with no lead vehicle, 13.3%).

There was a significant association between speeding and other violations, $\chi^2 = 104.59$, df 5, $p<.001$. This was accounted for by drivers driving within the 110 kph speed limit being significantly less likely to be observed committing at least one other violation (13.3%, SR -2.2), compared with drivers with no other violations (19.5%). Drivers with other violations were significantly more likely to be observed exceeding the 110 kph speed limit (18.2%, SR 8.7) compared with drivers with no other violations (4.4%, SR -3.0). However, drivers committing one or more "Other violations" were significantly *less* likely to be observed exceeding some "Other speed limit" by up to 10% (7.7%, SR -3.2), compared with drivers with no "Other violations" (16.0%).

Reducing the number of speed limit variables to two and the number of speeding variables to three, the highly significant effect, $\chi^2 = 55.20$, df 2, $p<.001$, was carried in two ways. First, drivers observed exceeding the speed limit by "More than 10%" were also more likely to be observed violating in other ways (19.6% of these drivers were observed violating in other ways, SR 5.9; 10.0% of drivers observed driving within the speed limit were observed violating in other ways). Second, a seemingly paradoxical finding was that drivers who were observed exceeding the speed limit by up to 10% were *least* likely to be observed violating in other ways (5.6%, SR -3.7). It seems that there is a sub-group of drivers whose only violation is to exceed the speed limit by up to 10%.

A dichotomous variable ("All violations") was created by including speeding and tailgating along with "Other violations" and identifying each driver as either violating or not violating. Overall, 55.9% of drivers (n = 1530) were observed violating one or more road rules, and 44.1% of drivers were seen not to be violating.

A significant association was found between "All violations" and driver age group, $\chi^2 = 63.40$, df 2, $p<.001$, which was explained by "Younger drivers" being significantly most likely to be observed violating (65.7%, SR 2.6) and "Older drivers" being significantly least likely to be observed violating (37.8%, SR -4.5). "Older drivers" were significantly most likely to be observed not violating (62.2%, SR 5.1), and "Younger drivers" were significantly least likely to be seen not violating (34.3%, SR -3.0). "Intermediate age drivers" were close to the mean values for this variable. Driver gender had no significant association with "All violations".

The significant effect found with respect to "Vehicle occupancy", $\chi^2 = 42.06$, df 3, $p<.001$, was explained by "Solo drivers" being most likely to be observed violating (60.6%, SR 2.5) and "Drivers with one adult passenger" being least likely to be observed violating (46.4%, SR -3.4). "Drivers with one adult passenger" were most likely to be observed not violating (53.6%, SR 3.8), while "Solo drivers" were least likely to be observed not violating (39.4%, SR -2.9).

The significant association between "All violations" and "Vehicle age", $\chi^2 = 30.98$, df 2, $p<.001$, was accounted for by drivers of newer vehicles being most likely to be observed violating (63.6%, SR 2.6), and drivers of older vehicles being least likely to be observed violating (49.2%, SR -2.6). Drivers of older vehicles were most likely to be observed not violating (50.8%, SR 3.0), and drivers of newer vehicles were least likely to be observed not violating (36.4%, SR -2.9). Drivers of intermediate age vehicles were close to the average for "All violations" (56.0%).

Including speeding and tailgating within the "All violations" category effectively switches the high violating category from older to newer vehicles. For the "Other violations" category (excluding speeding and tailgating) it is older vehicles that tend to carry the effect.

The significant effect for "All violations" and "Vehicle type", $\chi^2 = 94.25$, df 13, $p<.001$, was accounted for by the following categories being less likely to be observed violating: "Large truck/van" (37.5%, $SR_{All\ violations}$ -2.0; 62.5%, $SR_{No\ violations}$ 2.2), "Medium truck/van" (34.5%, $SR_{All\ violations}$ -3.0; 65.5%, $SR_{No\ violations}$ 3.4), "Small truck/van" (40.6%, $SR_{All\ violations}$ -2.1; 59.4%, $SR_{No\ violations}$ 2.4), "Coach/bus" (30.6%, $SR_{All\ violations}$ -2.0; 69.4%, $SR_{No\ violations}$ 2.3). The following vehicle categories were most likely to be observed violating: "Standard sedan" (61.2%, $SR_{All\ violations}$ 2.1; 38.8%, $SR_{No\ violations}$ -2.4), "Motorcycle" (88.9%, $SR_{All\ violations}$ 2.3; 11.1%, $SR_{No\ violations}$ -2.6), "Limo/luxury sedan" (73.8%, $SR_{All\ violations}$ 1.9; 26.2%, $SR_{No\ violations}$ -2.1).

When vehicle type was recoded into "Commercial" and "Private-use" categories, the significant result, $\chi^2 = 18.12$, df 1, $p<.001$, was accounted for by "Commercial" vehicles being least likely to be observed violating (49.4%, SR -2.4) and more likely to be observed not violating (50.6%, SR 2.7) compared with "Private-use" vehicles, of which 58.3% were observed violating. When speeding and tailgating are included in the "All violations" category, "Private-use" vehicles become the main category of offenders.

The correlation matrix shown in Table 4 summarises some of the relationships between variables in this study.

Table 4. Correlation Matrix of Selected Variables

Variable DV	1	2	3	4	5
Driver age	ns	$17^{‡}$	ns	$06^{†}$	$15^{‡}$
Driver gender	ns	ns	ns	$06^{†}$	ns
Vehicle age	ns	$-17^{‡}$	ns	$06^{†}$	$11^{‡}$
Vehicle category	ns	$18^{‡}$	ns	$16^{‡}$	$-08^{‡}$
Vehicle occupancy	09^{*}	$06^{†}$	$07^{‡}$	$07^{†}$	$12^{‡}$
1. Tailgating	-	ns	ns	ns	$07^{†}$
2. Speeding		-	$08^{‡}$	$14^{‡}$	ns
3. Lane violation			-	ns	$24^{‡}$
4. Other violations				-	ns
5. All violations					-

$^{*}p<.05$.

$^{†}p<.01$.

$^{‡}p<.001$.

Note: Correlations are Phi values, Cramer's V, or Spearman Rank Order Correlation Coefficients

Discussion

Driving is a complex phenomenon, involving multiple interactions between driver and traffic environment, moderated by vehicle characteristics as well as social features (e.g., passengers) and continuing cognitive/emotional appraisals. The picture to emerge from this observational study confirms some previous research on age differences but not on gender differences (e.g., Glendon, Dorn, Davies, Matthews, & Taylor, 1996). It has confirmed findings from studies using different methodologies that violations are more frequent among young drivers and male drivers (e.g., Jonah, 1990; Lawton et al., 1997; Parker et al., 1992a; Parker et al., 1995; Yagil, 1998). It has indicated that analysis of driving violations presents a rich picture of a driving environment in which drivers have substantial degrees of freedom in respect of how they choose to drive.

Just under 56% of drivers were observed to be violating at the time that their driving behavior was observed – in some cases these were multiple violations. The sample size makes this figure reliable. However, when the nature of these violations is considered in the light of other variables, it is clear that violating is not uni-dimensional. Aside from the general finding that younger drivers were more likely than older drivers to be violating, virtually no other variable consistently predicted drivers' violating behavior. One way of considering violating behaviors is to identify different driver groups on the basis of their violating characteristics, in similar fashion to the classification of driving styles and safety carried out some decades ago by Quenault (1967, 1968) or the stylised portraits of 'safe' and 'unsafe' young male drivers described by Rolls and Ingham (1992). On the basis of their observed violating behaviors, the driver groups described below emerged from this study.

Group 1: "Non-violating Drivers"

As 44% of drivers in this study were observed not violating, this figure represents an upper limit of the size of this group. The "true" size of the non-violating group of drivers is likely to be smaller than this, as at least some members migrate into one or more of the "violating" groups. However, a cross-sectional study such as this cannot elucidate further on these possibilities. Drivers in this group were likely to be older and to be travelling with one adult passenger.

Group 2: "Commercial Violators"

This appears to be a fairly distinct group, comprising around 10% of this study's driver sample. They comprise mainly younger and intermediate age drivers of older commercial vehicles – trucks, utility vehicles, buses, etc. Their violations are mainly "Obstructed view", "Unsafe load", "Using mobile phone", and "Polluting".

Group 3: "Tailgaters"

While 22% of vehicles were observed to be tailgating, this figure represents a lower limit of the percentage of drivers who tailgate – given that in nearly 55% of cases in which there was a lead vehicle, this was being tailgated. Thus, we might conclude that, given the opportunity, well over half of the users of this stretch of motorway will tailgate another vehicle. While this figure is unlikely to represent the upper limit of the proportion of drivers who tailgate, without further evidence an upper limit of drivers who tailgate cannot be determined. However, as in only just over 19% of cases was a vehicle following a lead vehicle at more than two seconds distance, it is possible that the potential tailgating driver group could be at least 80% of all drivers in this jurisdiction. A marked feature of this particular violation is that it is predicted by virtually no other variable, although solo drivers seem to be slightly more prone to tailgate. As a broad driving cultural phenomenon (Zaidel, 1992) tailgating is found more or less equally among all age groups, both genders and among drivers of all vehicle types and ages. Tailgating has a complex relationship with speeding – see findings outlined above.

Group 4a: "Speedsters"

In this study, 38% of drivers were observed to be exceeding a signed speed limit, again indicating a widespread phenomenon, although again one that is not readily encompassed within a single violating category. Drivers in this group were more likely to be male and younger. They were also much more likely to be driving newer private-use vehicles, particularly sedans – standard sized sedans were likely to be observed speeding at all rates over signed speed limits, while small sedans were more likely to be observed speeding up to 10% over the signed limit – reflecting the relative capacity of these vehicle types. While solo drivers were more likely to be observed exceeding the top speed limit – 110 kph in this study, less intuitively, drivers with one or more children in their vehicle were more likely to be observed exceeding other speed limits. The respective risks selected by these sub-groups might reflect on the one hand pressure upon "professional" drivers to reduce the amount of time spent travelling between appointments or between work and domestic locations. For drivers involved with childcare, their speeding behavior could reflect the urgency of transporting children to and from school, fitting such activity within an already full schedule, or other motivations. Multiple antecedents can generate the "same" objective behavior.

Group 4b: "Judicious Speeders"

While a substantial proportion of speeding drivers were also observed violating in some other way, a small group – around 6% of all drivers, could be separately identified on the basis that they were driving within 10% in excess of the signed speed limit but were not violating in any other way. In other respects they were similar to Group 4a. For drivers in this group it is possible that an implicit (sub-)cultural driving behavior is that it is "acceptable" to drive up to 10% over a signed speed limit. While cognitive states can only be inferred from a purely behavioral study, it might be that drivers in this group consider that they can safely drive within 10% above a signed speed limit, and are alert to the presence of police radar or speed cameras such that they can quickly reduce their speed to within the signed limit should this be necessary. Drivers in this group, while essentially "law abiding", in the case of driving may make a reasoned case for accepting risks associated with exceeding the speed limit by a certain amount in order to obtain benefits of faster travel. As noted by Reason, Manstead, Stradling, Baxter, and Campbell (1990), risk/benefit trade-offs are among motivational factors likely to play a significant role in committing violations. Drivers in this group were most likely to be using their headlights, perhaps indicative on the one hand of a sense of urgency, and on the other as a safety feature.

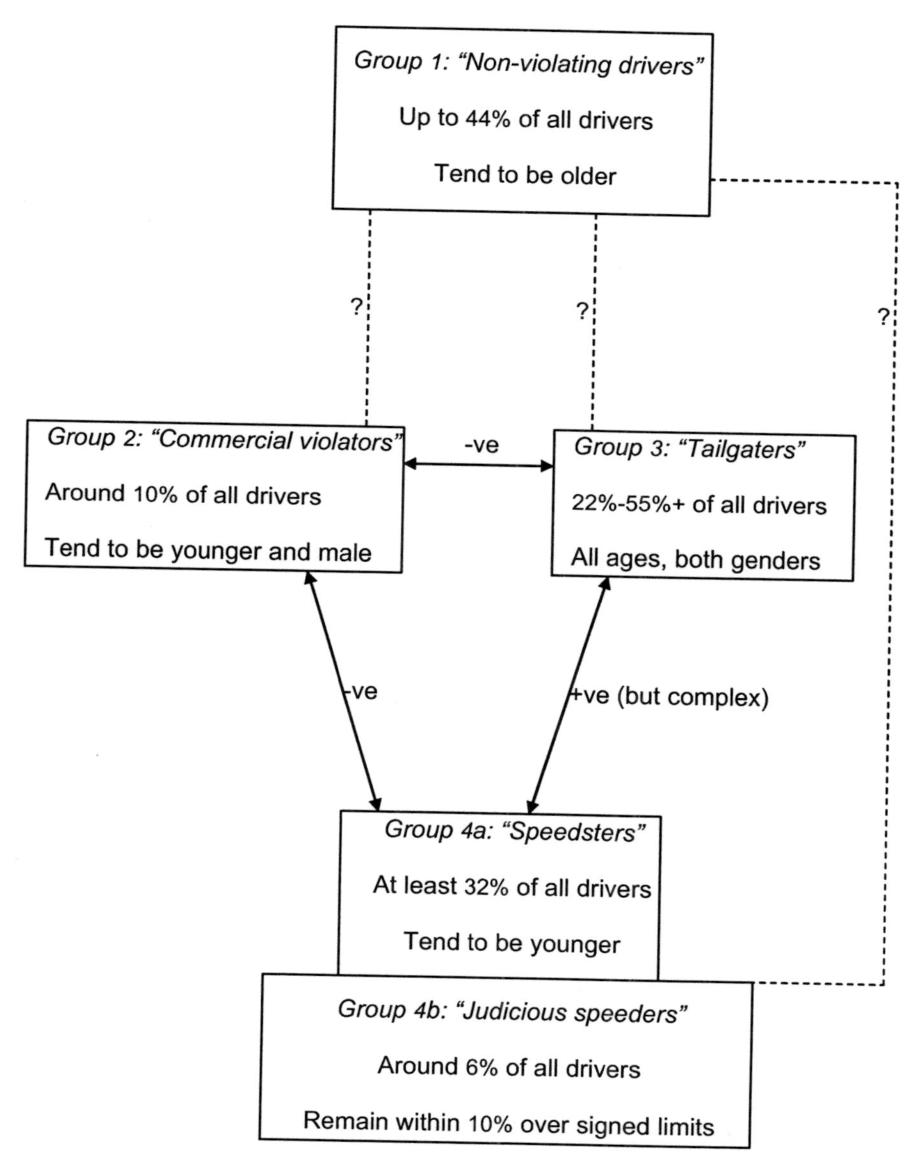

Figure 1. Possible relationships between driving population sub-groups

Relationships between these postulated driver groups are shown in Figure 1. These groups bear some limited comparisons with those identified by Corbett (2000). While "Non-violating" drivers (44%) in this study match Corbett's "Conformers" (29-54%), because this study observed a wider range of violating behaviors and did not access drivers' motivations to speed or to violate in other ways, further comparisons are problematic. It is likely that Corbett's "Defiers" (1-7%) will be included within the "Speedster" category (38% of the current study sample), while it is also possible that some of Corbett's "Deterred" (36-61%) will also be found within this group. There is also likely to be overlap between Corbett's "Manipulators" (3-14%) and the "Speedster" and "Judicious speeder" driver categories (6%) observed in the current study.

The widespread observation of drivers' violating behaviors in this study indicates that violating is a cultural norm for a substantial proportion of the motorway driving population in Queensland. This is consistent with the notion of driving culture, espoused by Zaidel (1992), and also the proposition that violations such as speeding and tailgating are imitative in respect of other drivers' behavior (Connolly & Åberg, 1993; Groeger & Chapman, 1997).

ACKNOWLEDGEMENTS

Thanks to Lucinda Gavin for her excellent data collection and to the Australian Research Council for funding the research *Exploring Dimensions of Driving Culture* under the Small Grants Scheme.

REFERENCES

Baxter, J. S., Manstead, A. S. R., Stradling, S. G., Campbell, K. A., Reason, J. T., & Parker, D. (1990). Social facilitation and driver behavior. *British Journal of Psychology, 81*, 351-360.

Blockey, P. N., & Hartley, L. R. (1995). Aberrant driving behaviour: Errors and violations. *Ergonomics, 38*, 1759-1771.

Connolly, T., & Åberg, L. (1993). Some contagion models of speeding. *Accident Analysis and Prevention, 25*, 57-66.

Corbett, C. (2000). A typology of drivers' responses to speed cameras: Implications for speed limit enforcement and road safety. *Psychology, Crime & Law, 6*, 305-330.

European Road Safety News. (1997). Driving with day-time lights: The advantages for road safety are obvious. *European Road Safety News,* December, *10*, 16.

Evans, L., & Wasielewski, P. (1982). Do accident-involved drivers exhibit riskier everyday driving behavior? *Accident Analysis and Prevention, 14*, 57-64.

Evans, L., & Wasielewski, P. (1983). Risky driving related to driver and vehicle characteristics. *Accident Analysis and Prevention, 15*, 121-136.

Glendon, A. I,, Dorn, L., Davies, D. R., Matthews, G., & Taylor, R. G. (1996). Age and gender differences in perceived accident likelihood and driver competences. *Risk Analysis, 16*, 755-762.

Golias, I., & Karlaftis, M. G. (2002). An international comparative study of self-reported driver behavior. *Transportation Research Part F: Traffic and Transport Psychology, 4*, 243-256.

Groeger, J. A. (2000). *Understanding Driving: Applying cognitive psychology to a complex everyday task.* Hove, UK: Psychology Press.

Groeger, J. A., & Chapman, P. R. (1997). Normative influences on decisions to offend. *Applied Psychology: An International Review, 40*, 265-285.

Gulian, E., Glendon, A. I., Davies, D. R., Matthews, G., & Debney, L. M. (1990). The stress of driving: A diary study. *Work & Stress, 4*, 7-16.

Gulian, E., Glendon, A. I., Debney, L. M., Davies, D. R., & Matthews, G. (1989). Dimensions of driver stress. *Ergonomics, 32*, 585-602.

Harrington, D. M., & McBride, R. S. (1970). Traffic violations by type, age, sex, and marital status. *Accident Analysis and Prevention, 2*, 67-79.

Hinkle, D. E., Wiersma, W., & Jurs, S. G. (1994). *Applied statistics for the behavioral sciences* (3rd ed.). Boston: Houghton Mifflin.

Jonah, B. A. (1990). Age differences in risky driving. *Health Education Research: Theory & Practice, 5*, 139-149.

Kontogiannis, T., Kossiavelou, Z., & Marmaras, N. (2002). Self-reports of aberrant behavior on the road: errors and violations in a sample of Greek drivers. *Accident Analysis and Prevention, 34*, 381-399.

Laapotti, S., Keskinen, E., Hatakka, M., & Katila, A. (2001). Novice drivers' accidents and violations – a failure on higher or lower hierarchical levels of driving behavior. *Accident Analysis and Prevention, 33*, 759-769.

Lawton, R., Parker, D., Manstead, A. S. R., & Stradling, S. G. (1997). The role of affect in predicting social behaviors: The case of road traffic violations. *Journal of Applied Social Psychology, 27*, 1258-1276.

Lawton, R., Parker, D., & Stradling, S. G. (1997). Predicting road traffic accidents: The role of social deviance and violations. *British Journal of Psychology, 88*, 249-262.

Lourens, P. F., Vissers, J. A. M. M., & Jessurun, M. (1999). Annual mileage, driving violations, and accident involvement in relation to drivers' sex, age, and level of education. *Accident Analysis and Prevention, 31*, 593-597.

Mesken, J., Lajunen, T., & Summala, H. (2002). Interpersonal violations, speeding violations and their relation to accident involvement in Finland. *Ergonomics, 45*, 469-483.

Noldus, L. P., Trienes, R. J., Hendriksen, A. H., Jansen, H., & Jansen, R. G. (2000). The Observer Video-Pro: New software for the collection, management, and presentation of time-structured data from videotapes and digital media files. *Behavior Research Methods, Instruments, & Computers, 32,* 197-206.

Parker, D., Lajunen, T., & Stradling, S. G. (1998). Attitudinal predictors of interpersonally aggressive violations on the road. *Transportation Research Part F: Traffic and Transport Psychology, 1*, 1-14.

Parker, D., Manstead, A. S. R., Stradling, S. G., & Reason, J. T. (1992a). Determinants of intention to commit driving violations. *Accident Analysis and Prevention, 24*, 117-131.

Parker, D., Manstead, A. S. R., Stradling, S. G., Reason, J. T., & Baxter, J. S. (1992b). Intention to commit driving violations: An application of the theory of planned behavior. *Journal of Applied Psychology, 77*, 94-101.

Parker, D., Reason, J. T., Manstead, A. S. R., & Stradling, S. G. (1995). Driving errors, driving violations and accident involvement. *Ergonomics, 38*, 1036-1048.

Queensland Transport. (2000). *The Queensland driver's guide*. Brisbane: Queensland Transport.

Quenault, S. W. (1967). *Driver behaviour: Safe and unsafe drivers*. Report LR 70. Crowthorne, UK: Road Research Laboratory.

_____ . (1968). *Driver behaviour: Safe and unsafe drivers*. Report LR 146. Crowthorne, UK: Road Research Laboratory.

Reason, J. T., Manstead, A. S. R., Stradling, S. G., Baxter, J., & Campbell, K. (1990). Errors and violations on the road: A real distinction? *Ergonomics, 33*, 1315-1332.

Rolls, G., & Ingham, R. (1992). *'Safe' and 'unsafe' – a comparative study of younger male drivers*. Basingstoke, UK: AA Foundation for Road Safety Research.

Stayer, D. L., & Johnston, W. A. (2001). Driven to distraction: Dual-task studies of simulated driving and conversing on a cellular telephone. *Psychological Science, 12*, 462-466.

Wasielewski, P. (1984). Speed as a measure of driver risk: Observed speeds versus driver and vehicle characteristics. *Accident Analysis and Prevention, 16*, 89-103.

Yagil, D. (1998). Gender and age-related differences in attitudes toward traffic laws and traffic violations. *Transportation Research Part F: Traffic and Transport Psychology, 1*, 123-135.

Zaidel, D. M. (1992). A modelling perspective on the culture of driving. *Accident Analysis and Prevention, 24*, 585-597.

In: Contemporary Issues in Road User Behavior and Traffic Safety ISBN:1-59454-268-6
Editors: D. Hennessy and D. Wiesenthal, pp. 101-112

Chapter 9

TRAFFIC SAFETY IN HONG KONG: CURRENT STATUS AND RESEARCH DIRECTIONS

J. P. Maxwell
Institute of Human Performance, University of Hong Kong

HONG KONG SAR: THE AREA AND ITS TRANSPORT SYSTEM

Hong Kong Special Administrative Region, located on the south-eastern coast of China, consists of four distinct areas: Hong Kong Island, Kowloon Peninsula, New Territories, and the outlying Islands such as Lantau and Lama. Hong Kong was described as a 'barren rock' when gifted to Queen Victoria some 150 years ago, but has since developed into one of the most important financial centers in Asia. The area was under British rule for a century and a half until it retroceded to China on 1st July, 1997, at which time it was designated as a Special Administrative Region (SAR) allowing it to maintain existing economic, legal, and social structures until 2047. The British rule of Hong Kong had a profound effect on these structures, including its transport policy and infrastructure, which continue to closely replicate the UK system. Driving laws in both countries are similar, traffic moves on the left and signage closely resembles the UK standard except that all written signs and street names are dual language.

The 1102 square kilometers of land that constitutes Hong Kong is populated by approximately 6.82 million people. The average population density is 6,300 per square kilometer concentrated mostly on Hong Kong Island and the Kowloon Peninsula (total 128 square kilometers and densities of 16,290 and 43,220 persons per square kilometer, respectively) making Hong Kong Island and Kowloon two of the most densely populated areas in the world. This high populace density is replicated in transport; there are approximately 525,551 licensed vehicles on 1,924 kilometers of road in Hong Kong, which includes Tsing Ma Bridge, the world's longest road-rail suspension bridge. The number of vehicles per kilometer of road is almost four times that of the UK and eight times that of the

USA (Hong Kong Transport Department Annual Report, 2002[1]); however, the number of vehicles relative to the total population is much lower in Hong Kong than both the UK and USA. There are approximately 77 licensed vehicles per 1,000 population in Hong Kong, 488 in the UK and 800 in the USA. This relationship between population and number of vehicles has an important impact on the calculation of accident rates and implications for road safety.

Public transport is used by the majority of the population; roughly 90% of the population makes a total of 11.7 million journeys on public transport daily. The public transport system is amongst the best in the world, consisting of railways (including an underground Mass Transport Railway, MTR), trams (Hong Kong's oldest form of public transport, the Peak Tram, has been in operation since 1889 and connects Central with Victoria Peak), buses, taxis and ferries. Public buses and the MTR are the most used transport systems; buses carry 4.3 million passengers and over 2 million passengers cram into MTR trains daily. The high population density is also responsible for concentrating pedestrian traffic, exacerbating traffic congestion and bringing vehicles and pedestrians into close proximity.

Traffic Safety in Hong Kong: Accident Rates and Comparison with the UK

A number of recent high profile accidents involving public transport buses have highlighted the issue of improving transport safety in Hong Kong. On Tuen Mun Road (22nd July, 2003) 22 people died and 20 were injured when a bus collided with a container truck and broke through a guardrail before plummeting down a steep hill. While on the West Kowloon Highway (18th October, 2003), 37 people were injured when a bus overturned traveling on the southbound highway. Public buses and bicycles are the only classes of vehicle that have evidenced increased accident involvement over the past decade; all other classes have shown decreased or stable accident involvement rates despite increasing numbers of vehicles on the road.

There has been an increase in the number of traffic injury accidents in Hong Kong during recent years. In the five years from 1998 to 2002 the total number of accidents increased by about 300 per year from 14,014 to 15,576 following a drop by the same amount per year from 1993 to 1997 (see Figure 1). Thus, over the past decade, the number of accidents has remained relatively unchanged in Hong Kong compared with a slight drop in the UK. Casualties are not limited to passengers and drivers; during the first six months of 2003, 55 pedestrians were killed and 2,110 injured, a 31% increase over the same period in 2002 (reported at www3.news.gov.hk, 12th August, 2003).

[1] All statistical data were collected from the Hong Kong Transport Department Annual Report, (2002) and UK Transport Department Road Traffic Accident Statistics (2002) unless otherwise stated.

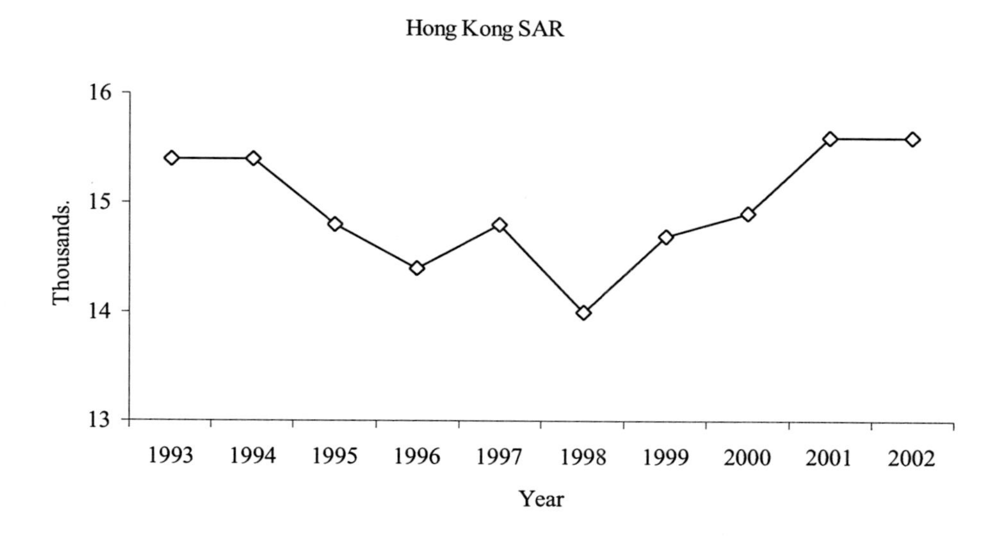

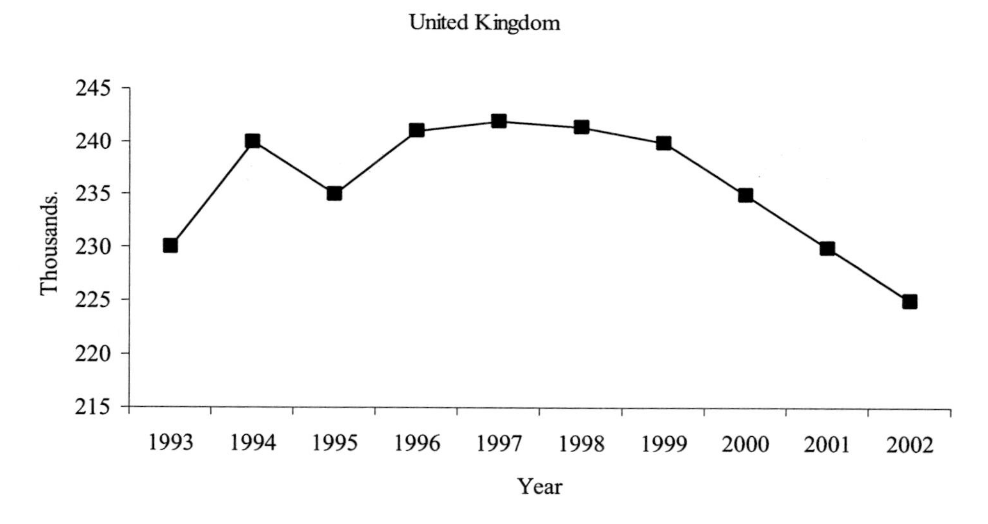

Figure 1. Yearly number of injury accidents reported in Hong Kong SAR and the UK.

In Hong Kong during 2002, 1% of all traffic injury accidents resulted in one or more fatalities, 22.1% led to serious injury and 76.9% to slight injury during 2002. These statistics do not compare favorably with the corresponding statistics from the UK where 1.4%, 13.8% and 84.8% resulted in fatal, serious and slight injuries, respectively (Department for Transport, UK, 2002). It is possible that the higher proportion of accidents in Hong Kong involving serious injury is purely a reflection of definitional disparity. Unfortunately, this is not the case. Table 1 details the definitions used by the UK and Hong Kong traffic departments when reporting accident statistics; they are almost identical suggesting that the higher prevalence of serious injury in Hong Kong is a reflection of some other local factor.

Table 1. Definitions of Slight, Serious and Fatal Injury Accidents

	United Kingdom*	Hong Kong SAR†
Slight injury	An injury of a minor character such as a sprain, bruise or cut not judged to be severe or slight shock requiring roadside attention. This definition includes injuries not requiring medical treatment.	An injury of a minor character such as a sprain, bruise or cut not judged to be severe, or slight shock requiring roadside attention and detention in hospital is less than 12 hours, or not required.
Serious injury	An injury for which a person is detained in hospital as an in-patient, or any of the following injuries whether or not the casualty is detained in hospital: fractures, concussion, internal injuries, crushing, severe cuts and lacerations, severe general shock requiring medical treatment and injuries causing death 30 or more days after the accident.	An injury for which a person is detained in hospital as an 'in-patient' for more than twelve hours. Injuries causing death 30 or more days after the accident.
Fatal injury	Human casualties who sustained injuries which caused death less than 30 days after the accident.	Sustained injury causing death within 30 days of the accident.

Source: *www.transtat.dft.gov.uk (10/24/2002); †www.info.gov.hk/td/eng/transport/ (10/23/2002).

Accident prevalence is affected by traffic density, that is, the available road space and the number of vehicles trying to use it. When calculating fatality rates, for example, it is common to represent the number of deaths per million capita. Number of deaths per million in Hong Kong is 24; in the UK it is 53, suggesting that UK roads are less safe. However, this can be a very misleading statistic, particularly in high density populations who have a high dependence on public transport, as is the case in Hong Kong. Number of fatalities can also be affected by the available road space; low space availability suggests that vehicles will be in closer proximity and more likely to collide. A more appropriate figure would be reached by calculating number of fatalities relative to traffic density, in other words, the number of vehicles per kilometer of road. When calculated in this way, Hong Kong drivers are three times more likely to be involved in a traffic accident than are UK drivers. Another surprising observation is that the area of lowest traffic density, the New Territories, is the site of over a third of the total number of traffic accidents in Hong Kong and is the only area to experience an increase in accidents over the past decade.

Naturally, other factors influence accident rates, but have yet to be considered. For example, driver age, traffic speed, distance traveled, driver education, concentration, road layout, signage, weather, driver personality, traffic law obedience, and vehicle quality may all influence accident involvement. Young drivers are involved in a greater number of accidents than are older drivers (Jonah, 1986; Summala, 1987; Ulleberg, 2002). It is possible that Hong Kong has a higher proportion of younger drivers, explaining the higher frequency of accidents to some extent. Unfortunately, age related accident data is unavailable, preventing a comparison with the percentage of young drivers in the UK, but considering the expense of driving in Hong Kong, it seems unlikely that there is a larger proportion of young drivers. A number of other factors would be expected to alleviate the accident rate in Hong Kong compared to the UK. High density traffic usually moves slower than low density traffic, which one would assume reduces risk of serious injury (typical "rush hour" driving speeds are 22.2 km/h and 27.6 km/h on Hong Kong Island and in Kowloon, respectively) and may be a partial explanation of the high accident rate in the New Territories where traffic density is low and driving speeds higher.

A brief inspection of cars in Hong Kong reveals a high proportion of new, top brand cars which are likely to have state of the art safety devices fitted, again reducing fatality risk. Driving a private car is relatively expensive in Hong Kong; car ownership often involves status symbolism rather than pure convenience. The top five contributing vehicle factors in 2002 were defective brakes, unidentified vehicle (involved in a hit-and-run type accident), defective body work, tire blow out and illegal tires. They accounted for 216 road injury accidents in 2002 with 81 hit-and-run cases. Weather in Hong Kong is generally bright and when rain storms and rare typhoons do strike, traffic almost disappears from the roads. In 2002, weather conditions were cited for fewer than 100 of the total number of traffic injury accidents in Hong Kong. Other environmental factors, such as potholes, objects in the road, road layout and slippery surfaces (not caused by weather) were listed as the cause of less than 250 accidents during 2002. It seems that we are left with human factors, such as driver education, individual differences in law obedience, personality and skill, as the major contributory factors for approximately 12,000 of the remaining traffic injury accidents (the cause of about 3,000 accidents was listed as unknown).

Human factors are reported to cause or contribute to 90-95% of traffic accidents (Lajunen, Parker & Summala, 2004; Rumar, 1985). Human factors include failing to pay attention, losing control of the vehicle, driving under the influence of drugs or alcohol, speeding, driving too closely to other vehicles and erratic lane changing. The top five factors involving the fault of a vehicle driver or pedestrian are shown in Table 2.

Reason, Manstead, Stradling, Baxter and Campbell (1990), developed a taxonomy of aberrant driving behaviors based on the principals of volition and neglect. Three categories were described -- violations, errors and lapses. Violations measure volitional actions that contravene traffic laws or safe driving practices. For example, driving above the speed limit, or when intoxicated, would be classified as violations. Errors are actions that are inappropriate for the situation (e.g. indicating left when turning right), whereas lapses are omissions such as forgetting to signal. Research indicates that young, male drivers are more likely to report high violation levels than older drivers and female drivers; females are more likely to report higher incidences of lapses. High violation score, as measured by the Driver Behavior Questionnaire (DBQ; Parker, Reason, Manstead & Stradling, 1995), has also been linked to accident involvement (Parker et al., 1995; Xie & Parker, 2002).

Table 2. Top Five Human Factors Contributing to Accident Involvement

Contributory factor	Number of accidents
Driving too close to the vehicle in front	2,132
Pedestrian not paying attention to traffic	1,982
Careless lane changing	1,256
Losing control of vehicle	1,251
Turning negligently	950

Table 3. Number of Accidents Caused by Violations, Errors or Lapses

Contributory factor	Number of accidents		
	Fatal	Serious	Slight
Violation	40	555	2,996
Error	71	1,306	5,461
Lapse	3	65	212

Source: www.info.gov.hk/td/eng/ (10/24/2003).

The Hong Kong Transport Department listed in their 2002 annual report 43 causes of accidents that reflect human contributory factors together with severity of injury to those involved. Table 3 lists the number of accidents by severity for each class of error. The majority of accidents were caused by unintentional errors, such as losing control of the vehicle, or turning negligently. This seems to contradict previous findings that suggested a link between violations and accident involvement (Parker et al., 1995). This trend possibly reflects a more law abiding nature of Hong Kong Chinese compared to other cultures (but see Xie & Parker, 2002, below); alternatively, negligence may include intentional aspects that argue for a violation classification (e.g. attempting to pull out when not enough space is available). The imprecision of police reports make it almost impossible to readily classify accident cause and hamper the ability of scientific researchers to develop effective countermeasures (Evans & Courtney, 1985).

TRAFFIC SAFETY RESEARCH IN HONG KONG

The need for socio-culturally specific research is evident; but to what extent has this already been achieved? Very little traffic safety research about Chinese drivers appears in English language publications and even less has been conducted in Hong Kong. A literature search of peer reviewed scientific psychological and sociological journals using the keywords driver, driving, safety, accident(s), transport, traffic, road, and their multiple combinations reveals 22 published articles (mostly written in Chinese with English abstracts) in the past 130 years when China is stipulated and only five when Hong Kong is stipulated. Unfortunately, two of the five involved either Singapore motorcyclists (Wong, Lee, Phoon, Yiu, Fung & McLean, 1990) or British cyclists and pedestrians (Bagley, 1992). The former found an inverse relationship between experience and accident involvement; the latter concluded that individual vulnerability is a factor in child accident involvement in some urban areas. Both were written by researchers based in Hong Kong.

Of the remaining three, the first paper examined the accident involvement of public buses in Hong Kong (Evans & Courtney, 1985). The study reported higher road accident rates compared with Japan, the UK, and the USA. Evans and Courtney reported a three to four times higher accident rate involving injury in Hong Kong compared with the UK during the years 1975 to 1980. Focusing on public buses, the accident rate in Hong Kong was double that of the UK and was twice as likely to involve pedestrians; although, it must be recalled that pedestrian traffic is much higher in Hong Kong than the UK. Typically, public bus accident rates peak at around 0900 hrs and 1800 hrs in the UK, corresponding to peak traffic flow; however, in Hong Kong accidents peak at 0800 hrs and remain high until 1900 hrs. Evans and Courtney suggest fatigue as a possible causative factor together with speeding and negligence. Hong Kong bus drivers typically fulfilled 9 hour shifts, with accident rates increasing severely after 7 hours of driving. This research remains pertinent given the rising number of accidents involving public buses over the past decade.

The second article described the use of eye movement desensitization reprocessing for the treatment of post-traumatic stress disorder in a 40 year old male following a traffic accident (Wu, 2002). The case study found that the treatment was effective and suggested application to the Chinese population. The most recently published article describes risk factors affecting the severity of single vehicle accidents (Yau, 2004). Yau examined the effect of geographical, human, vehicle, safety, environmental and site factors using data from the Traffic Accident Data System (TRADS), developed by the Hong Kong government departments and Police Force. Male gender, young (<25 years) and old age (>55 years), old car age (>5 years), good street lighting or daylight, and seat belt use were all associated with an increased probability of being involved in a serious injury or fatal accident; although, not all factors were influential for all three vehicle types. For example, seat belt usage predicted severity of accidents for goods vehicles, but not private cars, possibly reflecting goods vehicle drivers' reluctance to wear seatbelts when making multiple deliveries in local areas.

Regrettably, the paucity of research to date does not allow us to clearly describe the psychological characteristics of Hong Kong drivers, but research carried out in mainland China may provide some speculative insights that are applicable to Hong Kong Chinese even though they are typically more westernized than their mainland counterparts. Xie and Parker (2002) used a modified version of the Manchester Driver Behavior Questionnaire (DBQ, Parker, et al., 1995) to analyze the aberrant driving behaviors of drivers from two Chinese cities, Beijing and Cheng-de. Factor analysis revealed a four factor structure to the modified DBQ relating to lapses and errors, aggressive violations, inattention errors, and maintaining progress violations. A significant effect of age was found for lapses and errors, and aggressive violations. The oldest drivers reported fewer lapses and errors, and aggressive violations than the youngest. Males reported fewer lapses and errors, and inattention errors than females. The aggressive violations subscale was positively related to accident involvement, which concurs with previous findings in western populations (Parker et al., 1995).

Xie and Parker (2002) proposed that cultural factors may play an important role in determining the behavior of Chinese drivers. They developed the Chinese Driving Questionnaire (CDQ) to test this possibility. The CDQ contained four subscales that measure sense of social hierarchy, potential road safety countermeasures, belief in interpersonal networks and challenging legitimate authority. The sense of social hierarchy sub-scale measures the driver's belief that their personal status affects police decisions and the likelihood of punishment after committing a traffic offence. The social hierarchy subscale is closely related to the third and fourth subscales, belief in interpersonal networks and challenging legitimate authority, in that personal influence and status are believed to reduce the risk of prosecution by police. These three subscales were moderately inter-correlated (Pearson's product moment correlation coefficients between .34 and .43 were reported), but were weakly or uncorrelated with the second subscale, road-safety countermeasures. Xie and Parker interpreted their results as confirmation of Hofstede's (1980, 1984) observation that mainland Chinese society tends to self-concern and social ordering that is associated with 'poor rule obeying in general and a high rate of traffic accidents in particular' (p. 306). The social structure of Hong Kong Chinese bears some similarity to their mainland counterparts despite the Western influence afforded by the extended British rule. It remains to be determined to what extent Xie and Parker's (2002) results can be generalized to Hong Kong Chinese. It is also difficult to speculate how hierarchical social structure impacts on Chinese drivers when a similar study has not been completed in countries with less defined social hierarchies. It is possible that the attitudes measured by the CDQ are prevalent in all cultures and that other factors motivate attempts to influence police decisions.

SAFETY INITIATIVES IN HONG KONG[2]

The preceding text may give the impression that Hong Kong is an unsafe city in which to drive; however, this is far from the case. Law enforcement and government legislation has helped to maintain relatively low accident levels despite ever increasing population and vehicle densities. Various countermeasures have been adopted in Hong Kong based on their success in other countries, particularly the UK. Speed enforcement cameras and red light cameras are now common with over 20 cameras rotated over 136 sites. The scheme has led to a 23% reduction in speeding related accidents at one site (the other sites have not been evaluated) and red light jumping has been reduced by 43-55% at the junctions where the system is installed with a 34% reduction in accidents. The use of seatbelts is now mandatory for both front and rear seat passengers in all vehicles except public buses. New legislation will introduce the use of seatbelts in public buses (with 14 seats) in 2004, reflecting similar trends in the UK. Alcohol consumption is relatively low compared to Western countries such as the UK and USA. This is reflected by the low incidence of alcohol related accidents. The legal limit is 50 mg of alcohol per 100 ml of blood; only 444 convictions for driving over this limit were reported in 2002, 81 cases were registered as the causes of injury accidents, only one of which was fatal. In the UK, where the legal limit is 80 mg (the bill to reduce the limit to 50 mg was rejected by the British Parliament in March 2002), the figure was approximately 80,000 convictions, 7,560 injury accidents and approximately 500 deaths. To what extent these figures represent drink driving rates or efficiency of law enforcement is unknown.

The use of hand held communication devices by drivers was banned in July 2000 and extended to the use of radio phones in taxis on 1st July 2001. To date, no data have been collected to evaluate the effectiveness of this intervention, but data from several experimental studies suggest that it is an appropriate course of action and is in line with current practice in many other Western and Eastern countries (Alm & Nilsson, 1994, 1995; Haigney, Taylor & Westerman, 2000; Reed & Green, 1999).

The education of drivers has also received considerable attention from the Hong Kong Government. Publicity material, focusing on new legislation, behavioral and attitudinal modification and risk awareness, are regularly produced and are intended to influence accident occurrence, as well as educate. A driver improvement scheme was introduced in 2002 to "promote road safety and make drivers more law-abiding through better understanding of their driving attitude and behavior" (Transport Department Annual Report, HK, 2002; p. 52). Drivers may join the scheme voluntarily or be ordered to attend by the

[2] Material concerning government safety initiatives in this section are cited from Leung (2002).

court. Satisfactory course completion allows the removal of three penalty points from the driver's license. Again, the benefits of this scheme have not been appraised.

Probationary licenses were introduced for motorcyclists in October, 2001. The introduction of this scheme led to a drop in number of accidents involving inexperienced motorcyclist during 2002 relative to 2001 (although accident levels have remained stable over the past decade). The scheme is similar to the graduated driver licensing (GDL) schemes described by McKnight & Peck, (2003) that have also been associated with better driving and lower accident rates. GDL schemes withhold the award of a full driving license through a variable period of provisional licensing, lengthening the learning period or requiring an extended period of supervised driving. In essence, the learner is guided through more driving experiences under supervision from a qualified driver and is more likely to successfully negotiate hazards encountered for the first time.

FUTURE RESEARCH DIRECTIONS

Cross-cultural psychological research, particularly between Western and Eastern cultures, is rather sparse in the driving literature, but is clearly a valuable line of enquiry. Culture refers to the "shared attitudes, beliefs, categorizations, expectations, norms, roles, self-determinations, values and other such elements of subjective culture found among individuals whose interactions are facilitated by shared language, historical period and geographic region" (Triandis, 1972, p. 3). Clark (1987) proposed culturally grounded research as an essential element in the understanding of human behavior. Chinese culture has distinct and unique elements that may have profound influences on the way its people behave, instantly precluding the bootstrapping of Western findings onto Chinese society (particularly if the trend towards global Westernization is to be avoided).

One of the major obstacles facing cross cultural researchers is that of cultural equivalence. Most comparative research involves the distribution of a translated scale developed in one culture to members of another culture (Sue & Sue, 1987); for example, Xie and Parker's (2002) application of the DBQ to Chinese. Typically, established scales are translated and back translated, modified by the addition of intuitively apparent cultural idiosyncrasies, distributed to the target population then subjected to exploratory or confirmatory factor analysis in an attempt to identify common factors and discrepancies (e.g. Lajunen et al., 2004; Xie & Parker, 2002).

Translation of scales provides the first point of difficulty for cross-cultural research. Often, direct translation is impossible, particularly between phonetic and pictorial written languages. For example, there is no direct translation in Chinese of the words 'Institute of Human Performance', the title is translated as 'Physical Activity and Potential Development Institute'. This change in wording does not fully capture the concept of 'human performance' because it is often interpreted as relating only to sport. Add to this the further complication that there is often little agreement about the definition of terms such as 'aggressive driving' between speakers of the same language, and the requirement for linguistic equivalence becomes a thorny problem. Back translation is often employed to overcome the issue of linguistic equivalence (Brislin, 1980), but may not be sufficient (Duda & Hayashi, 1998). Additional measures, such as continuing the process of translation until equivalence is reached, or performing a test-retest with bilingual individuals responding to both variations of the scale at different time points, may enhance confidence in the translation but fail to address other difficulties such as psychometric, testing, and sampling equivalence. These difficulties can only be overcome by an appreciation of the cultures in question and careful application of rigorous research and statistical methodologies.

To make cross-cultural comparisons, the researchers must be confident that they are asking the same question of both cultures. It is tempting, when conducting cross-cultural research, to make speculative statements about cultural differences that are based on personal observation rather than rigorous measurement. Whilst this process is necessary for the development of relevant questions, it makes comparisons with previous findings almost redundant. Factor structures may be broadly replicated, but the items that constitute them become so different as to provide little insight into cultural differences. Culturally specific differences can only be identified when identical scales are used on multiple populations rather than the generation of multiple versions of the same scale that each differs from the others.

Conclusions

The need for a comprehensive driving research program in Hong Kong is unquestionable; Hong Kong remains very much "a barren rock" in this respect, but the directions this research should take are debatable. The psychological, behavioral, and attitudinal characteristics of Hong Kong drivers are currently unknown; the high incidence per vehicle on the road and pattern of accident causes, reported above, suggests that they may differ considerably from Western populations. The incidences of aggression, anger, stress, frustration, negligence, violation, and traffic law conformity require detailed description and quality of driver education should be assessed. Whilst overseas research can inform policy, local and cultural factors need to be accounted for so that the implementation of inappropriate legislation, without benefit for road safety, is avoided.

Acknowledgements

The author would like to thank staff at the Hong Kong Transport Department, particularly T.F. Leung, for providing copies of Annual Reports dating from 1993-2002 and Marie Claire Rösiö for support during the production of this chapter.

References

Alm, H., & Nilsson, L. (1994). Changes in driver behavior as a function of hands-free mobile phones – a simulator study. *Accident Analysis and Prevention, 26,* 441-451.

Alm, H., & Nilsson, L. (1995). The effects of a mobile telephone task on driver behavior in a car following situation. *Accident Analysis and Prevention, 27,* 707-715.

Bagley, C. (1992). The urban environment and child pedestrian and bicycle injuries: Interaction of ecological and personality characteristics. *Journal of Community and Applied Social Psychology, 2,* 281-289.

Brislin, R.W. (1980). Translation and content analysis of oral written material. In H.C. Triandis & J.W. Berry (Eds.), *Handbook of cross-cultural psychology: Vol. 2. Methodology* (pp. 389-444). Boston: Allyn & Bacon.

Clark, L. A. (1987). Mutual relevance of mainstream and cross-cultural psychology. *Journal of Consulting and Clinical Psychology, 55,* 461-470.

Department for Transport, UK (2002). *Transport Statistics*. Available: www.dft.gov.uk.

Duda, J.L., & Hayashi, C.T. (1998). Measurement issues in cross-cultural research within sport and exercise psychology. In J.L. Duda (Ed.), *Advances in Sport and Exercise Psychology Measurement* (pp. 471-483). Morgantown, WV: Fitness Information Technology, Ltd.

Evans, W.A., & Courtney, A.J. (1985). An analysis of accident data for franchised public buses in Hong Kong. *Accident Analysis and Prevention, 17,* 355-366.

Haigney, D.E., Taylor, R.G., & Westerman, S.J. (2000). Concurrent mobile (cellular) phone use and driving performance: Task demand characteristics and compensatory processes. *Transport Research: Part F, 3,* 113-122.

Hofstede, G. (1980). *Culture's consequences: International differences in work-related values.* Beverly Hills, CA: Sage.

Hofstede, G. (1984). *Culture's consequences: International differences in work-related values (Abridged edition).* Beverly Hills, CA: Sage.

Jonah, B.A. (1986). Accident risk and risk-taking behaviour among young drivers. *Accident Analysis and Prevention, 18,* 255-271.

Lajunen, T., Parker, D., & Summala, H. (2004). The Manchester Driver Behaviour Questionnaire: A cross-cultural study. *Accident Analysis and Prevention, 36,* 231-238.

Leung, T.F. (2002). *Road safety work in Hong Kong.* Paper presented at the International Symposium on Road Safety, University of Hong Kong, Hong Kong.

McNight, A.J., & Peck, R.C. (2003). Graduated driver licensing and safer driving. *Journal of Safety Research, 34,* 85-89.

Parker, D., Reason, J.T., Manstead, A.S.R., & Stradling, S.G. (1995). Driving errors, driving violations and accident involvement. *Ergonomics, 38,* 1036-1048.

Reason, J.T., Manstead, A.S.R., Stradling, S.G., Baxter, J.S., & Campbell, K. (1990). Errors and violations on the road: A real distinction? *Ergonomics, 33,* 1315-1332.

Reed, M.P., & Green, P.A. (1999). Comparison of driving performance on-road and in a low-cost simulator using a concurrent telephone dialling task. *Ergonomics, 42,* 1015-1037.

Rumar, K. (1985). The role of perceptual and cognitive filters in observed behaviour. In L. Evans & R.C. Schwing (Eds.), *Human Behaviour and Traffic Safety* (pp.151-165). New York: Plenum Press.

Sue, D., & Sue, S. (1987). Cultural factors in the clinical assessment of Asian-Americans. *Journal of Consulting and Clinical Psychology, 55,* 479-489.

Summala, H. (1987). Young driver's accidents: risk taking or failure of skills? *Alcohol, Drugs, and Driving, 3,* 79-91.

Transport Department, HK (2002). *Transport Department Annual Report 2002: Putting our hearts into it.* Hong Kong: Transport Department Press.

Triandis, H.C. (1972). *The analysis of subjective culture.* New York: Wiley.

Ulleberg, P. (2002). Personality subtypes of young drivers: Relationship to risk taking preferences, accident involvement, and response to a traffic safety campaign. *Transportation Research Part F, 4,* 279-297.

Wong, T.W., Lee, J., Phoon,W., Yiu, P.C., Fung, K.P., & McLean, J.A. (1990). Driving experience and the risk of traffic accident among motorcyclists. *Social Science and Medicine, 30,* 639-640.

Wu, K.K. (2002). Use of eye movement desensitisation and reprocessing for treating post-traumatic stress disorder after a motor vehicle accident. *Hong Kong Journal of Psychiatry, 12,* 20-24.

Xie, C., & Parker, D. (2002). A social psychological approach to driving violations in two Chinese cities. *Transport Research (Part F), 5,* 293-308.

Yau, K.K.W. (2004). Risk factors affecting the severity of single vehicle traffic accidents in Hong Kong. *Accident Analysis and Prevention, 36,* 333-340.

In: Contemporary Issues in Road User Behavior and Traffic Safety ISBN:1-59454-268-6
Editors: D. Hennessy and D. Wiesenthal, pp. 113-123

Chapter 10

SPEEDING BEHAVIOR AND COLLISION INVOLVEMENT IN SCOTTISH CAR DRIVERS

Stephen G. Stradling
Transport Research Institute, Napier University, Edinburgh, UK

INTRODUCTION

Recent reviews (Lancaster & Ward, 2002; Stradling et al., 2003) have summarized research showing that for car drivers in highly motorized societies levels of speeding behavior vary systematically with driver gender (e.g., Brook, 1987; Buchanan, 1996; French, West, Elander, & Wilding, 1993; Meadows & Stradling, 2000; Shinar, Schechtman, & Compton, 2001; Waterton, 1992), with driver age (e.g., Boyce & Geller, 2002; Ingram, Lancaster, & Hope, 2001; Parker, Manstead, Stradling, & Reason, 1992; Quimby, Maycock, Palmer, & Grayson, 1999; Shinar et al., 2001; Stradling, Meadows, & Beatty, 2000), with driver exposure measures such as reported annual mileage (e.g., Stradling et al., 2003), with trip purpose and time pressure (e.g., Rietveld & Shefer, 1998; Silcock, Smith, Knox, & Beuret, 2000), with vehicle performance and size (e.g., Evans & Herman, 1976; Horswill & Coster, 2002; Wasielewski & Evans, 1985) and with indices of driver personality (e.g., Elander, West, & French, 1993; Lajunen, 1997; Rimmo & Aberg, 1999; Stradling & Meadows, 2000; Sumer, 2003; Ulleberg, 2002).

Research has also shown that road traffic accident involvement is associated with having been detected speeding for both car drivers (Cooper, 1997; Gebers & Peck, 2003; Rajalin, 1994; Stradling et al., 2000, 2003) and powered two wheeler riders (Ormston, Dudleston, Pearson, & Stradling, 2003; Stradling & Ormston, 2003).

This chapter combines data sets from two recent studies of Scottish drivers, one conducted for the Scottish Executive and one for the Strathclyde Safety Camera Partnership. Both surveys involved in-home interviews with quota samples of drivers. Full details of sampling strategy and sample demographics are given in Stradling et al. (2003) and Campbell and Stradling (2003). Data from 1088 drivers from the first survey and 1101 from the second, who held a current driving license, had driven within the previous year and who cited 'car' as their main vehicle when driving are combined here to give a picture of the current speeding behaviors and collision involvement of Scottish car drivers. The chapter concludes with a suggestion for the remediation of speeding behaviors.

DETECTED SPEEDERS

How Many of These Drivers had Been Detected Exceeding the Speed Limit?

Respondents were asked how many times they had been stopped by the police for speeding during their driving career and how many times they had been flashed by a speed camera in the past three years. Twenty seven percent had been stopped by the police for speeding during their driving careers and 20% had been flashed by a speed camera. Overall 37% had been stopped or flashed or both and were labeled 'speeders'.

Does Having Been Detected Speeding Vary with Gender and with Age?

The final column of Table 1 shows that half of the male drivers and a quarter of the female drivers had been detected speeding. Table 1 also shows how the proportion of 'speeders' varies across the age range, separately for male and female drivers. Chi-square tests showed that the proportions were significantly different between males and females overall and within each age range and that, within both genders, the proportions differed with age (all $p < .001$).

Table 1. Proportions of Speeders by Age Group, Separately for Male and Female Drivers

	Age in years	17-24	25-34	35-44	45-54	55-64	65+	Total
Male	% speeders	32	54	56	66	52	44	49.9%
Female	% speeders	13	35	33	26	23	18	25.4%

Does Having Been Detected Speeding Vary with Reported Annual Mileage?

Table 2 gives the proportions of speeders amongst those currently driving different annual mileages. Chi-square analyses showed that for both male and female drivers the proportion that are speeders increased as annual mileage increased and that the proportion of male speeders exceeded that of female speeders at each mileage band (all $p < .001$). The proportions that had been detected speeding varied from 13% of those female drivers currently driving less than 5,000 miles per annum to 62% of those male drivers currently driving more than 12,000 miles per annum.

Table 2. Proportions of Speeders by Annual Mileage Group, Separately for Male and Female Drivers

	Annual mileage (miles)	< 5,000	5,000-9,999	10,000-12,000	> 12,000	Total
Male	% speeders	30	44	53	62	49.9%
Female	% speeders	13	24	28	46	25.4%

Does Having Been Detected Speeding Vary with Current Engine Size?

Table 3 shows the variation in proportion of speeders with engine size of car currently driven. Chi-square analyses showed that for both males and females the proportion who were speeders tended to increase as current engine capacity increased, while the proportion of males who have been detected speeding exceeded that for females in each engine size band (all $p < .001$). Approaching two thirds of males (64%) currently driving large cars of 2 liters or above had been detected speeding, compared to below 1 in 5 (18%) of females currently driving small cars.

Table 3. Proportions of Speeders by Engine Capacity Group, Separately for Male and Female Drivers

	Engine capacity (liters)	<1.3	1.3-1.6	1.6-1.9	>2.0	Total
Male	% speeders	39	45	57	64	49.9
Female	% speeders	18	28	29	39	25.4

Speeding and Recent Collision Involvement

Does Having Been Detected Speeding Covary with Having a Recent Collision History?

Fifteen per cent of the sample (18% of the male drivers and 13% of the female drivers) reported having been involved in one or more road traffic accidents as a driver within the last three years. While 13% of male and 11% of female non-speeders reported recent road traffic accident (RTA) involvement, the proportions reporting recent RTA involvement were significantly elevated ($p < .001$) to 22% for both male and female speeders.

While overall half (50%) of males and a quarter (25%) of female drivers in the sample were speeders, these proportions were significantly elevated ($p < .001$) to 64% for males and 41% for females amongst those who reported recent RTA involvement.

This section has shown, using data from two recent large-scale surveys of Scottish car drivers and defining 'speeders' as drivers who had been ever stopped by the police for speeding or had been flashed by a speed camera in the previous three years, that twice as many male as female drivers are speeders, that the incidence varies with age, with annual mileage and with engine size, and that approaching twice as many speeders as non-speeders had been recently involved in a road traffic accident as a driver.

SITUATIONAL INFLUENCES ON SPEED CHOICES

Under What Circumstances Do Drivers Vary Their Speed on the Road?

Respondents were asked to indicate whether they would drive faster, slower or the same as usual in a number of situations. They rated 18 scenarios asking how they would drive 'Compared to how I normally drive on my own …', on a 7-point scale from 1 'Much slower', through 4 'Much the same as usual' to 7 'Much faster'. Table 4 shows the distribution of responses collapsed to Slower (scale points 1-3), Same as Usual (4) and Faster (5-7), with the situations arranged in descending order of percentage of drivers saying they would drive faster.

Half of drivers (55%) say they would speed up when late and a third (30%) would speed up if the traffic around them were moving faster than they normally drive. Almost all drivers would slow down when driving in fog (98%) and heavy rain (96%). Many drivers indicated they would make 'no change' in their speed choice in some of the circumstances.

Table 4. Influence of Journey Conditions on Driver Speed Choices

[Row percentage]	Under this condition, I drive...		
	Slower	Same as usual	Faster
When you are late for a meeting or appointment	1	44	55
When the traffic ahead is moving faster than you normally drive	3	67	30
When feeling stressed	23	56	21
When someone is driving close behind you	34	54	12
When listening to music	4	88	8
When the weather is hot	10	85	6
With people your own age in the car	6	90	4
When the traffic is moving more slowly than you normally drive	69	27	4
When driving under streetlights	34	65	1
When driving in the dark	66	33	1
On unfamiliar roads	88	11	1
With older people in the car	37	62	0
When driving in light rain	42	57	0
With children in the car	57	42	0
When you see speed camera warnings	58	41	0
When you spot a speed camera	65	35	0
When driving in heavy rain	96	4	0
When driving in fog	98	2	0

Most of the variables were skewed to one pole or the other (faster or slower) with only two variables showing substantial bi-polar differentiation across drivers: 23% would drive more slowly when feeling stressed while 21% would drive faster; 12% of drivers would speed up when someone was driving close behind them, while one third (34%) say they would slow down when being tailgated.

The data was submitted to principal components analysis (KMO .756; Bartlett's test $p < .001$). Three factors had eigenvalues greater than unity and passed the scree test. The three factors were labeled Adverse driving conditions, Responsibilities to others, and Arousal.

Variables loading on the first factor had *Adverse driving conditions* as a common link, with driving in heavy rain (.71), in the dark (.59), on an unfamiliar road (.57), in fog (.56), in light rain (.55), under streetlights (.54) and in traffic moving more slowly than the respondent normally drives (.34) loading positively on this factor. From Table 4 it may be seen that all the variables that loaded on this factor tended to make the respondent drive more slowly than their usual speed and they may thus all be seen as constraining a driver's opportunity to speed.

Variables loading on the second factor covered *Responsibilities to others*. These obligations included compliance with enforcement authorities and the law when spotting a speed camera (.79) or speed camera warning signs (.73); duty of care to vulnerable present others, both older people (.58) and children (.55) in the car; and undertakings made to distant others at one's destination when running late for a meeting or appointment (.39). While the latter tended to make drivers increase their speed, the other items tended to make respondents drive slower than their usual speed (see Table 4).

The variables which loaded positively on the third factor were having people of the driver's age in the car (.62), driving when the weather is hot (.60), driving while listening to music (.59), when feeling stressed (.42), being late (.39), when someone is driving close behind you (.36) and when the traffic ahead is faster than the respondent's usual speed (.35), all situations in which feelings of *Arousal* or stimulation are likely to be present. This factor consisted of variables tending to make respondents drive faster than their usual speed (see Table 4).

These three factors may be seen as sorting influences on driver in-journey speed choices into three groups: those influencing, by constraining, the opportunity to speed; those influencing obligation to refrain from speeding; and those driving the inclination to speed. This pattern of results is consistent with the claim that transport choices are driven by the interaction of opportunity ('Can I do it?'), obligation ('Should I do it?') and inclination ('Do I want to do it?') (Stradling, 2002; 2003).

SPEED CHOICES AND SPEEDERS

Are Any of These Speed Choices – Driving Faster or Slower than Normal – More Prevalent Amongst Those Drivers Who have been Detected Speeding?

Table 5 shows there were statistically significant differences ($p < .10$) between speeders and non-speeders on 8 of the 18 speed choice scenarios.

More of those drivers who had been detected speeding reported driving faster when late, when the traffic ahead is moving faster or slower than they normally drive, and when feeling stressed. Fewer of them drive slower on unfamiliar roads, and more of them drive slower when they spot a speed camera or speed camera warning signs and when they have older passengers in the car.

Table 5. Speed Choices under Different Scenarios for Drivers Detected Speeding and not

	% faster		% slower		p for chi-square
	Not speeder	Speeder	Not speeder	Speeder	
Late for meeting or appointment	49	64			.000
Traffic ahead faster than you normally drive	26	35			.005
Feeling stressed	18	24			.091
Traffic ahead slower than you normally drive	3	6			.009
On unfamiliar roads			89	86	.000
You spot a speed camera			59	74	.000
See speed camera warning signs			56	63	.001
With older passengers			34	44	.000

Speed Choices and Recent Collision Involvement

Are Drivers Who have been Recently Collision Involved More Likely to Speed up or Slow down under Any of these Scenarios?

Table 6 shows there were significant differences in behavior ($p < .10$) on 6 of the 18 speed choice scenarios between those who had and those who had not been RTA involved as a driver in the previous 3 years. More of the recently RTA-involved drivers say they would drive faster compared to how they normally drove when late for an appointment, while listening to music, when feeling stressed, and when the traffic ahead was moving faster than they normally drove. More say they would drive more slowly when they spot a speed camera, suggesting they know they will likely be exceeding the speed limit, and with children in the car, suggesting they know their normal rate of progress would be inappropriate when transporting child passengers.

Table 6. Proportions of non RTA-involved and RTA-involved Drivers who Report Varying their Speed in Particular Circumstances

	% faster		% slower		p for chi-square
	No RTAs	Some RTAs	No RTAs	Some RTAs	
Late for meeting or appointment	53	65			.000
Listening to music	7	15			.000
Feeling stressed	19	28			.005
Traffic ahead faster than you normally drive	29	34			.038
You spot a speed camera			64	71	.001
With children in the car			56	65	.083

CONCLUSIONS AND SUGGESTIONS

From a large survey of car drivers in Scotland a group designated 'speeders' were identified. Thirty seven per cent of Scottish car drivers had ever been stopped by the police for speeding or had been flashed by a speed camera within the past three years. Membership of the 'speeder' group varied with driver gender (M: 50%; F: 25%), age, reported annual mileage and current car engine size.

Respondents indicated whether they would drive faster, the same as usual, or slower compared to how they normally drove on their own for 18 journey scenarios. Half said they would speed up when late for a meeting or appointment, and a quarter when the traffic flow was faster than they normally drove. Almost all would slow down in fog and in heavy rain. Many indicated there were situations in which they would not vary their speed. Factor analysis identified three groupings: situations where the opportunities for fast driving were constrained by inclement weather conditions, darkness, unfamiliar roads or slow moving traffic; situations where obligations to present or distant others would bring speed change; and circumstances where situational arousal tended to increase the inclination to vary speed.

Those who had been detected speeding were significantly more likely to slow down for speed cameras and camera warning signs, on unfamiliar roads and with older passengers, but to drive faster when late for a meeting or appointment, when feeling stressed, or if the traffic around them was driving faster or slower than they normally drove.

More of those drivers who had been RTA involved within the previous 3 years indicated they would slow down for a speed camera and when driving with children in the car, and more of the RTA involved said they would drive faster when late, when feeling stressed, when listening to music and when the weather was hot.

Remediation: Combining Enforcement and Education to Change Speeding Behavior

Whilst there is no guarantee that changing these particular on-road responses to in-journey situational exigencies would reduce the future speeding behavior of drivers, or their future RTA involvement, there is a *prima facie* case for incorporating these research findings into the curricula of post-qualification retraining courses for speeding or for crash-involved drivers. Recently in some jurisdictions in the UK courses have been made available as diversions from prosecution for drivers charged with driving without due care and attention (Driver Improvement Courses) or detected exceeding the speed limit by up to 20 mph (Speed Awareness Courses).

Four Functions for a Speed Camera

Safety cameras on UK roads, whose deployment and operation are currently undertaken by partnerships between the police, local authorities and the courts, serve three main purposes but could serve a fourth.

1. Hazardous location indication. In the UK today most automatic safety cameras for detecting speeding motorists are located at crash hot spots. The deployment criteria being followed by the more than 40 Safety Camera Partnerships across the UK require speed cameras to be placed where there are elevated levels of recent, and speed-related, RTAs. The fixed-site cameras are also highly visible being painted yellow or, in Scotland, with yellow and red diagonal stripes. Their first function is thus to signal to the approaching driver 'Look out! Take extra care! This has lately proved to be a dangerous stretch of road.' They do not, however, provide any further site-specific hazard information ('What, exactly, should I be looking out for?') beyond this general alerting function.
2. General deterrence. Speed cameras slow down speeding drivers. In one study of newly installed speed cameras in built-up areas in Glasgow, Scotland (Campbell & Stradling, 2002a) baseline data showed 64% of passing motorists in excess of the speed limit. Installing speed camera housings reduced this to 37%. When the camera units became operational three months later the figure reduced further to 23%. Thus the number of speeders at the camera sites was reduced from two-thirds to one quarter in six months.
3. Specific deterrence. A study using the conviction and crash records of a large sample of drivers in Ontario, Canada (Redelmeier, Tibshirani, & Evans, 2003) concluded that "The risk of a fatal crash in the month after a [driving] conviction was about 35% lower than in a comparable month with no conviction for the same driver .. [but] The benefit lessened substantially by 2 months and was not significant by 3 – 4 months" (p. 2177), suggesting that conviction – detection and punishment – for a driving offence has only a brief and temporary effect on changing driver behavior.

   A study of 500 car drivers surveyed two months after receiving a speeding ticket in Glasgow, Scotland (Campbell & Stradling, 2002b) reported that speeding tickets changed the behavior of some, but not all, drivers, finding a mixture of speed sensitive drivers ('I now pay more attention to my speed while driving'), camera sensitive drivers ('I now keep more of a look out for speed cameras') and insensitive drivers, doing neither. Around half had become more sensitive to their speed and were driving more slowly, but one third reported only slowing down for speed cameras, and one sixth reported themselves unremediated, despite paying £60 and receiving 3 penalty points on their license (where 12 points brings temporary disqualification from driving), and not slowing down at all.

   Another recent study of Scottish drivers (Stradling et al., 2003) found that 23% of male and 15% of female Scottish car drivers had been flashed by a speed camera in the previous 3 years but when asked 'What happened the last time you were flashed by a speed camera?' 4 out of 5 of those (79%) replied 'Nothing' (typically because a smaller number of cameras are rotated amongst a larger number of housings, which still flash with no film inside). Detection without consequence is unlikely to be a powerful behavior change agent.
4. Detecting drivers in urgent need of help. We have long known that speed kills. The laws of physics inexorably dictate that the higher the speed at impact the more energy must be rapidly absorbed by hard metal, soft flesh and brittle bone. From the data reported here we have also seen that those drivers who had been stopped by the police for speeding or had been flashed by a speed camera had almost double the

incidence of recent crash involvement, with 22% of the detected speeders versus 12% of those who had not been detected speeding having been involved in an RTA as a driver in the previous 3 years.

The kinds of drivers who have been detected speeding are more likely to have been recently collision-involved. These people pose more risk to themselves and to other, usually more vulnerable, road users. They need help with changing their driving styles. There is support amongst the UK motoring public for such an approach. The 2002 RAC Report on Motoring (RAC Foundation, 2002) reported 57% of a large UK sample of drivers agreeing with the statement that 'All drivers should receive periodic refresher training'. Such driver refresher training could be duration-based (and be more frequent for young and old drivers) or incident-based following involvement in road traffic accidents or speeding infractions.

Driver retraining courses, where drivers pay for their own remediation, and pay more than the fixed penalty fine, combining classroom sessions ('Why to change') and on-road guided practice ('How to change') offer the possibility of undoing old habits and facilitating integrated, sustainable changes in driving style. Speed cameras spot 'crash magnets' in need of change. Changing KSA, addressing the knowledge, skills and attitudes of drivers, offers a potentially powerful route to changing KSI, and reducing the numbers killed and seriously injured on the roads.

References

Boyce, T. R. & Geller, E. S. (2002). An instrumented vehicle assessment of problem behavior and driving style: Do younger males really take more risks? *Accident Analysis and Prevention*, 34, 51-64.

Brook, L. (1987). *Attitudes to Road Traffic Law.* Contractor Report 59, Crowthorne: Transport and Road Research Laboratory.

Buchanan, C. (1996). *The Speeding Driver.* Edinburgh: Central Research Unit, Scottish Office.

Campbell, M., & Stradling, S. G. (2002a). The general deterrent effect of speed camera housings. In G. B. Grayson (Ed.), *Behavioural Research in Road Safety XI.* Crowthorne: Transport Research Laboratory.

Campbell, M., & Stradling, S. G. (2002b). The impact of speeding tickets on speeding behaviour. In G. B. Grayson (Ed.), *Behavioural Research in Road Safety XII.* Crowthorne: Transport Research Laboratory.

Campbell, M. & Stradling, S. G. (2003). *Speeding Behaviours And Attitudes Of Strathclyde Drivers.* Report to Strathclyde Safety Camera Partnership. Transport Research Institute, Napier University, Edinburgh.

Cooper, P. (1997). The relationship between speeding behavior (as measured by violation convictions) and crash involvement. *Journal of Safety Research*, 28, 83-95.

Elander, J., West, R., & French, D. (1993). Behavioural correlates of individual differences in road-traffic crash risk: An examination of methods and findings. *Psychological Bulletin*, 113, 279-294.

Evans, L., & Herman, R. (1976). Note on driver adaptation to modified vehicle starting acceleration. *Human Factors*, 18, 235-240.

French, D., West, R., Elander, J., & Wilding, J. (1993). Decision-making style, driving style and self-reported involvement in road traffic accidents. *Ergonomics*, 36, 627-644.

Gebers, M. A., & Peck, R. C. (2003). Using traffic conviction correlates to identify high accident-risk drivers. *Accident Analysis and Prevention,* 35, 903-912.

Horswill, M. S., & Coster, M. E. (2002). The effect of vehicle characteristics on drivers' risk-taking behaviour. *Ergonomics*, 45, 85-104

Ingram, D., Lancaster, B., & Hope, S. (2001). *Recreational Drugs and Driving: Prevalence Survey.* Development Department Research Programme: Research Findings No.102. Edinburgh: Central Research Unit, Scottish Executive.

Lajunen, T. (1997). *Personality Factors, Driving Style and Traffic Safety*. PhD Thesis. Faculty of Arts, University of Helsinki, Helsinki.

Lancaster, R., & Ward, R. (Entec UK Ltd). (2002). *The Contribution Of Individual Factors To Driving Behaviour: Implications For Managing Work-Related Road Safety*. Research Report 020. HSE Books.

Meadows, M., & Stradling, S. G. (2000). Are women better drivers than men? Tools for measuring driver behaviour. In J. Hartley, and A. Branthwaite, (Eds.), *The Applied Psychologist*, 2nd Edition. Open University Press.

Ormston, R., Dudleston, A., Pearson, S., & Stradling, S. G. (2003). *Evaluation of BikeSafe Scotland.* Edinburgh: Scottish Executive Social Research.

Parker, D., Manstead, A., Stradling, S., & Reason, J. (1992). Determinants of intention to commit driving violations. *Accident Analysis and Prevention*, 24, 117-131.

Quimby, A., Maycock, G., Palmer, C., & Grayson, G. B. (1999). *Drivers' Speed Choice: An In-Depth Study*. TRL Report 326. Crowthorne: Transport Research Laboratory.

Rajalin, S. (1994). The connection between risky driving and involvement in fatal accidents. *Accident Analysis and Prevention*, 26, 555-562.

Redelmeier, D. A., Tibshirani, R. J., & Evans, L. (2003). Traffic-law enforcement and risk of death from motor-vehicle crashes: Case-crossover study. *The Lancet*, 361, 2177-2182.

RAC Foundation. (2002). *RAC Report on Motoring 2002*. London: RAC Foundation.

Rietveld, P., & Shefer, D. (1998). Speed choice, speed variance and speed limits: A second-best instrument to correct for road transport externalities. *Journal of Transport Economics and Policy*, 32, 187-202.

Rimmo, P-A., & Aberg, L. (1999). On the distinction between violations and errors: Sensation seeking associations. *Transportation Research Part F*, 2, 151-166.

Shinar, D., Schechtman, E., & Compton, R. (2001). Self-reports of safe driving behaviors in relationship to sex, age, education and income in the US adult driving population. *Accident Analysis and Prevention*, 33, 111-116.

Silcock, D., Smith, K., Knox, D., & Beuret, K. (2000). *What Limits Speed? Factors That Affect How Fast We Drive*. Basingstoke: AA Foundation for Road Safety Research.

Stradling, S. G. (2002). Transport user needs and marketing public transport. *Municipal Engineer*, 151, 23-28. Special edition on Sustainable Transport Policy, March 2002.

_____ . (2003). Reducing car dependence. In J. Hine and J. Preston (Eds.), *Integrated Futures and Transport Choices.* Ashgate Publications.

Stradling, S. G., & Meadows, M. L. (2000). Highway code and aggressive violations in UK drivers. Global Web Conference on Aggressive Driving Issues, October 16 – November

30, 2000. Ontario Ministry of Transportation and the Traffic Safety Village. Available online at www.drivers.com

Stradling, S. G., Meadows, M. L., & Beatty, S. (2000). Characteristics of speeding, violating and thrill-seeking drivers. International Conference on Traffic and Transport Psychology, Bern, September 2000.

Stradling, S. G., & Ormston, R. (2003). Evaluating BikeSafe Scotland. Behavioural Studies in Road Safety XIII. Crowthorne: Transport Research Laboratory.

Stradling, S. G., Campbell, M., Allan, I. A., Gorrell, R .S. J., Hill, J. P., Winter, M. G., & Hope, S. (2003). *The Speeding Driver: Who, How and Why?* Edinburgh: Scottish Executive Social Research.

Sumer, N. (2003). Personality and behavioural predictors of traffic accidents: Testing a contextual mediated model. *Accident Analysis and Prevention,* 35, 949-964.

Ulleberg, P. (2002). Personality subtypes of young drivers. Relationship to risk-taking preferences, accident involvement, and response to a traffic safety campaign. *Transportation Research Part F*, 4, 279-297.

Wasielewski, P., & Evans, L. (1985). Do drivers of small cars take less risk in everyday driving? *Risk Analysis*, 5, 25-32.

Waterton, J. (1992). *Scottish Driver Attitudes to Speeding.* Edinburgh: Central Research Unit, Scottish Office.

In: Contemporary Issues in Road User Behavior and Traffic Safety ISBN:1-59454-268-6
Editors: D. Hennessy and D. Wiesenthal, pp. 125-133

Chapter 11

THE USE AND MISUSE OF VISUAL INFORMATION FOR "GO/NO-GO" DECISIONS IN DRIVING

Rob Gray
Arizona State University East

INTRODUCTION

In driving we are often required to make split-second decisions about whether or not there is sufficient time to complete a driving maneuver before colliding with a vehicle directly in front of us (the lead vehicle) and/or an oncoming vehicle. Two common situations where such judgments are required are when overtaking and passing another vehicle and when making a left-turn at an intersection. Accident analyses suggest that drivers frequently make critical judgment errors in these dangerous situations. Despite the fact that overtaking maneuvers are performed relatively infrequently, they account for roughly 10-15% of injury-causing automobile accidents in some areas (Clarke, Ward, & Jones, 1998; 1999; Jeffcoat, Skelton, & Smeed, 1973). Clarke et al. (1999) concluded that "the majority (of these accidents) arose from a decision to start the overtake in unsuitable circumstances ..and not a lack of vehicle control skills" (p. 849). Similarly, in an analysis of accidents involving left-turns (which accounted for 17% of all accidents), Larsen and Kines (2002) concluded that "the main problems for left-turning drivers lay in their attention errors or in misjudging the time they had to make a left-turn before the approaching traffic reached the intersection" (pg. 375). In this chapter, I will examine the visual information that can be used to perform these complex maneuvers with the goal of understanding some of the underlying causes of "go/no-go" judgment errors in driving.

VISUAL INFORMATION ABOUT TIME TO COLLISION AND DRIVING SPEED

When executing an overtaking maneuver or a left-turn, one of the primary things the driver must estimate is the time remaining until the oncoming vehicle arrives (the time to collision, TTC). Accurate information about the TTC with an oncoming car is available to the driver from the rate of change in size of the car's retinal image as:

$$TTC \approx \frac{\theta}{d\theta / dt} \quad [1]$$

where θ is the oncoming vehicle's instantaneous angular subtense and dθ/dt is its instantaneous rate of increase of angular subtense (Hoyle, 1957). It has been demonstrated that under ideal conditions drivers can accurately estimate TTC on the basis of equation [1] alone with estimation errors ranging from 2-12% of the actual TTC (Gray, 1998). In overtaking, it is crucial that the driver simultaneously monitor the TTC of the *lead* vehicle to avoid a rear-end collision; Clarke et al. (1999) reported that 37% of overtaking accidents involved a collision with the lead vehicle. The visual information expressed in equation [1] also provides an accurate estimate of the TTC with the lead vehicle.

As illustrated in Figure 1, to make a judgment about whether an overtake or left-turn can be completed safely the driver needs to compare the estimate of the TTC of the oncoming vehicle with an estimate of the time that will be required to complete the maneuver (the time required to overtake, TRO in Fig 1A or the time required to turn, TRT in Fig 1B). A safe overtaking maneuver requires that TTC>>TRO and a safe left-turn requires that TTC>>TRT. Very little is known about the information drivers use to estimate the time required to complete a particular driving maneuver. Estimation of TRO and TRT are presumably based on the driver's estimate of the current speed of self–motion and knowledge of the capabilities of one's own vehicle (e.g., the maximum acceleration). Visual information about the speed of self-motion is provided by the optical edge rate (i.e., the number of edges that pass the observer's eye in a given time period) (Larish & Flach, 1990), however, field studies have consistently demonstrated that drivers cannot accurately estimate their speed of travel: "errors in subjectively estimating speed are sufficiently great that drivers should consult speedometers (Evans, 1991) (pg. 128)". Studies using verbal estimates of speed (reviewed in Groeger (2000)) and studies using a procedure that requires drivers to adjust their speed to a specified level (e.g., halve their current speed) Denton (1976, 1977) have shown that speed estimates are highly inaccurate [errors range from 10-60 km/h (6-37 mph)] and are easily biased by factors such as the driving speed on the previous trial.

In summary, laboratory and field research has shown that under optimal conditions drivers can accurately estimate TTC but cannot reliably and accurately judge their own speed of travel. Therefore, we might expect that drivers will make frequent "go/no-go" judgment errors even under the best of circumstances. Evidence relevant to this point is provided in the next section.

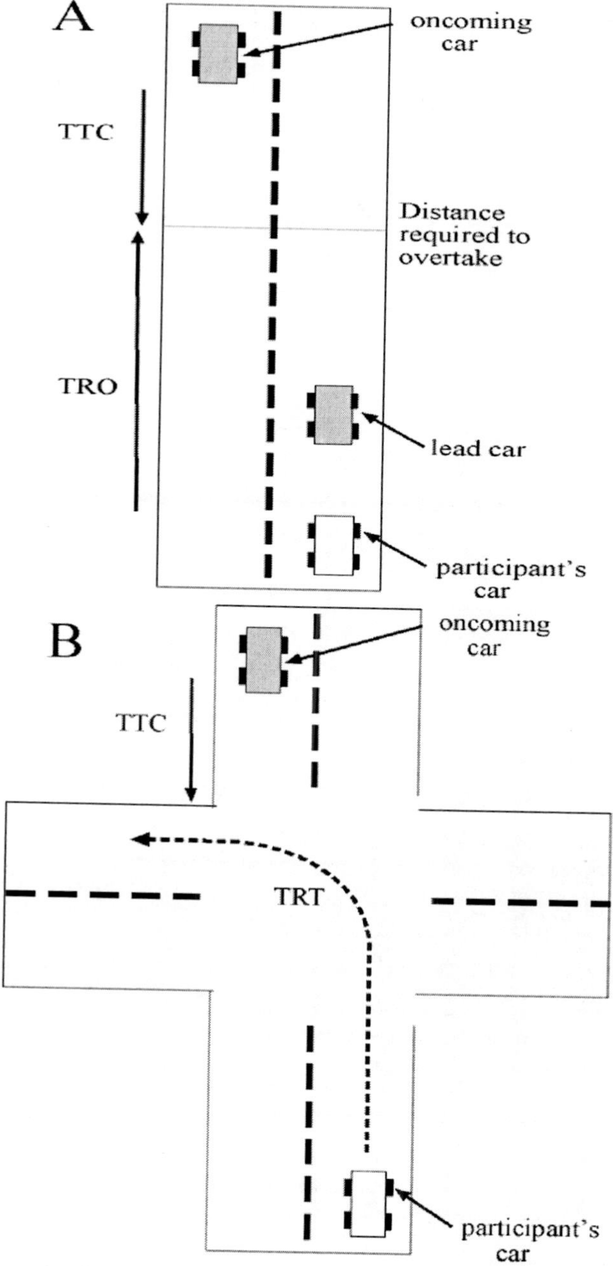

Figure 1. A: Calculation of safety margin for overtaking maneuvers. The distance required (shown by the dashed line) for the participant (white car) to overtake and pass the lead vehicle was calculated from real driving performance data (Gordon & Mast, 1970). The time required for the driver for the driver to overtake (TRO) was then compared to the time required for the oncoming car to reach this critical distance (TTC). B: Calculation of safety margin for left-turns. The time required for the driver to complete a left-turn (TRT) was calculated from pilot observations in the driving simulator. This value was then compared to the time required for the oncoming car to reach the middle of the intersection (TTC).

GO/NO-GO DECISIONS UNDER OPTIMAL DRIVING CONDITIONS

My colleagues and I have recently investigated go/no-go decisions in overtaking and left-turn execution using a fixed-base driving simulator (Gray & Regan, 2000, 2004). Let's first consider the overtaking maneuvers. On each trial, a lead car appeared on the road ahead and participants were instructed to adjust their speed so as to follow the lead car. After 5 sec. of car following, a brief auditory tone was played and an oncoming car appeared on the roadway. The speed and initial distance of the lead and oncoming vehicles was varied randomly from trial to trial to create different safety margins (i.e., different combinations of TTC and TRO). In the 'active' condition, participants were instructed that when they heard the tone they should overtake and pass the lead car as soon as it was safe to do so (i.e., either before or after the oncoming car had gone by). In the 'passive' condition, participants were instructed to press one of two response buttons on the steering wheel to indicate whether or not at the instant the tone was presented there would be sufficient time to pass safely. They were further instructed to respond as quickly as possible and response time was measured. The simulated roadway was flat so that participants had an unobstructed view of the lead and oncoming vehicles at all times.

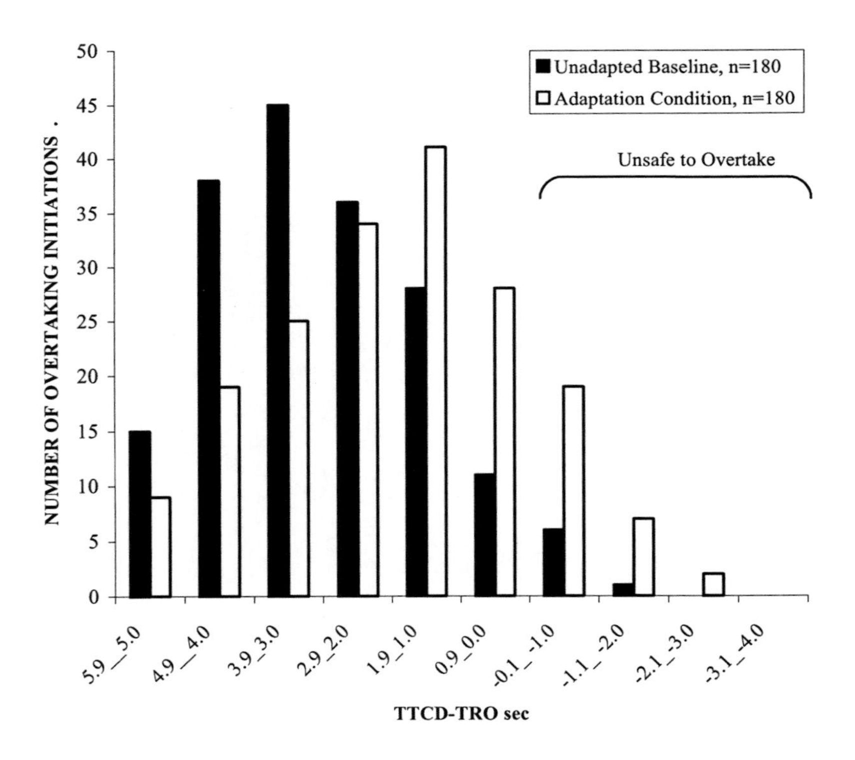

Figure 2. Number of overtaking maneuvers initiated for different ranges of the value of (TTC-TRO).

The solid bars in Figure 2 show the distribution of overtaking maneuvers initiated as a function of the value of TTC-TRO for 18 drivers. For all values less than 0.0 sec., an overtaking maneuver could not be completed without a collision unless the participant exceeded the speed limit or the oncoming vehicle decreased its speed. We defined this range to be unsafe. Drivers in our experiment initiated unsafe overtaking maneuvers (i.e., TTC<TRO) on 16% of the trials. Results from the judgment task were even worse as participants made unsafe judgments on 30% of the trials. I next consider results for left turn execution.

In left turn experiments, participants drove towards a simulated 4-way intersection with an oncoming car approaching the intersection (see Figure 1B). The speed and initial distance of the oncoming car was randomly varied from trial-to-trial to create different safety margins. A brief auditory tone was presented at a random time before the participant's car reached the intersection. In the 'active' condition, participants were instructed that when they heard the tone they should execute the left-turn as soon as it was safe to do so while in the 'passive' condition they judged whether or not it would be safe to turn by pressing a response button on the steering wheel. The solid bars in Figure 3 show the number of left-turns that were initiated for different ranges of (TTC-TRT). Again values of (TTC-TRT) less than zero were defined to be unsafe, because the oncoming car would have reached the intersection before the left-turn was completed. Drivers again made a substantial number of errors: unsafe left-turns were initiated on 13% of the trials and unsafe 'passive' judgments were made on 23% of the trials.

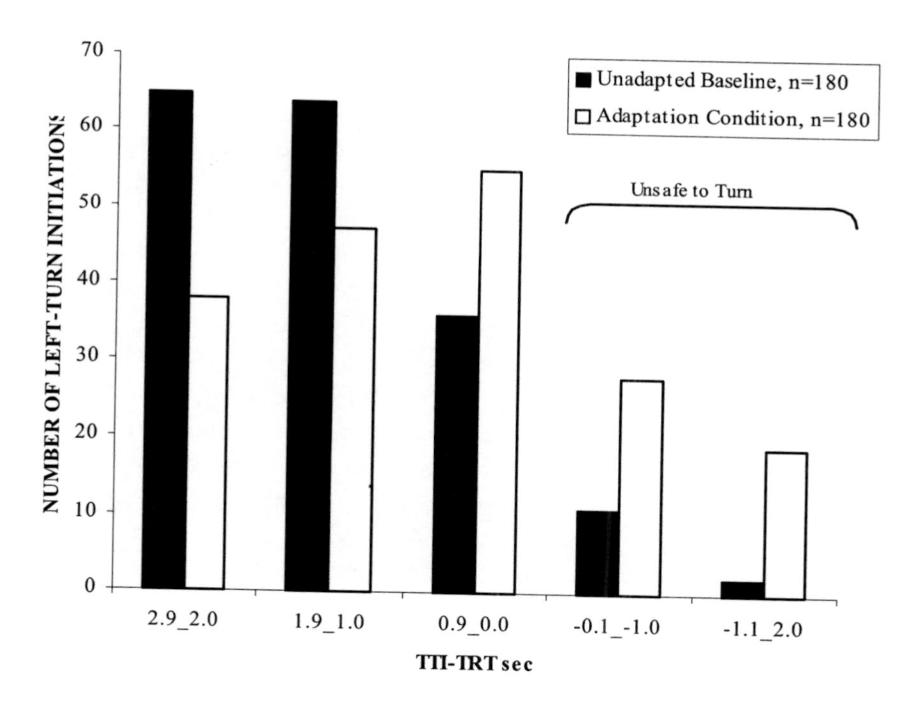

Figure 3. Number of left-turns initiated for different ranges of the value of (TTC-TRT). Open bars show data for the condition where observers adapted to motion by driving on a straight empty road prior to overtaking. Filled bars show data for the no-adapt baseline condition. (TTC-TRT) values less than zero were defined as unsafe. Open bars show data for the condition where observers adapted to motion by driving on a straight empty road prior to overtaking. Filled bars show data for the no-adapt baseline condition. (TTC-TRO) values less than zero were defined as unsafe.

Clearly participants in these experiments made a substantial number of errors in making the critical go/no-go decision for overtaking and left-turn execution. Why do drivers have such difficultly in these situations? One likely cause is the lack of accurate visual information about the time required to complete the maneuver (discussed in section 2 above). A second possible explanation for judgment errors was identified when we examined the driving maneuvers in more detail. For overtaking maneuvers, 11% of the participants used a strategy of initiating overtaking if the oncoming car's distance was greater than some critical value (regardless of its speed or TTC). The use of distance as a control variable is problematic for several reasons. First and foremost, this strategy is not robust across situations as it will be inaccurate for oncoming cars traveling at a very high speed. The second major problem associated with using the distance of the oncoming car as a control variable is that previous research has shown that drivers cannot accurately estimate the absolute distance of another vehicle on the roadway (Groeger, 2000; Teghtsoonian & Teghtsoonian, 1969). A final possible source of error in making go/no-go decisions is that in some driving situations the visual information expressed in equation [1] provides an erroneous estimate of the TTC of the oncoming vehicle. Two such situations are considered next in sections 4 and 5.

Go/No-Go Decisions When the Rate of Expansion of the Oncoming Car is Sub-Threshold

One of the major limitations with using an object's angular size to judge its TTC is that it requires that its rate of change of size ($d\theta/dt$) is large enough to be detectable. For objects with a small angular size (e.g., a motorcycle viewed from a distance of 300 m), observers cannot accurately estimate TTC from equation (1) because the object's rate of expansion is near the detection threshold (Gray & Regan, 1998). Hoffmann and Mortimer (1996) have estimated the threshold value of $d\theta/dt$ for driving to be roughly 0.003 rad/sec and have shown that $d\theta/dt$ can be well below this value in many driving situations.

In the overtaking experiments described above (Gray & Regan, 2004), there were several trials in which the rate of expansion of the oncoming car was below this threshold value. Analysis of these trials revealed that when the rate of expansion was sub-threshold many of our drivers reverted to using the dangerous strategy of initiating overtaking when the oncoming car's distance was above a critical value: 33% of our drivers used this strategy when the value of $d\theta/dt$ was below threshold vs. only 16% when it was above threshold.

The Effect of Closing Speed Adaptation on Go/No-Go Decisions

Recently, my colleagues I have shown that staring straight ahead during simulated driving on a straight open road can give the driver the illusion that the TTC with the lead and oncoming vehicles is longer than it really is (Gray & Regan, 2000). Following simulated highway driving on a straight empty road for 5 min, drivers initiated overtaking of a lead vehicle substantially later (220-510ms) than comparable maneuvers made following viewing a static scene. This *closing speed aftereffect* is quite distinct from the well–known adaptation of the perceived speed of self-motion that is caused by the expanding retinal flow pattern (Denton, 1976) (i.e., drivers underestimate their driving speed following adaptation) and is distinct from the classical motion aftereffect (Addams, 1834). Unlike these other phenomena, the *closing speed aftereffect* is produced by local adaptation of looming detectors that signal motion-in-depth for objects near the focus of expansion of the optical flow pattern (Regan & Beverley, 1979). How does this closing speed aftereffect influence go/no-go decisions in driving?

To address this question we used the same overtaking and left-turn conditions described above except that participants drove on a straight, empty simulated road for 10 min prior to completing the maneuvers. The open bars in Figure 2B plot the number of overtakes that were initiated following adaptation to closing speed. The results are striking: closing speed adaptation substantially increased the total number of unsafe overtaking maneuvers (from 16 to 29%). This increase in judgment errors following closing speed adaptation most likely occurred because drivers overestimated the value of TTC (i.e., they thought they had more time before the oncoming car would arrive). Analysis of the passive overtaking judgment revealed that this adaptation effect is even more dangerous. Reaction times for overtaking judgments were significantly slower and more variable following closing speed adaptation. The effect of closing speed adaptation on left-turn execution was similar to the effect on overtaking described above: drivers made more unsafe turns (47 vs. 13). Adaptation again had the effect of significantly reducing the driver's safety margin.

Conclusions

On the roads of the USA 41,821 individuals were killed and 3.2 million injured during the year 2000, and 2000 was a typical year (NHTSA, 2000). Accident analyses implicate errors in perception and decision-making as the probable cause of the majority of these accidents. Therefore, it is crucial that we understand what information driver's use to make critical decisions and identify situations in which this information is inaccurate and unreliable. This chapter examined a type a judgment that seems particularly prone to driver error and has been identified as a cause of a very high percentage of driving accidents: "go/no-go" decisions about whether there is sufficient time to complete a maneuver in the face of oncoming traffic. The experimental results presented in this chapter demonstrate that drivers are frequently inaccurate when making a "go/no-go" decision during overtaking and passing and left-turn execution. Analysis of these results identified two primary causes of these judgment errors: use of unreliable visual information and the use of visual information that is inaccurate under some driving conditions. I next consider these two errors in more detail and discuss some possible methods for reducing their frequency.

A substantial number of drivers appear to use the distance of the oncoming car (instead of its TTC) as the main source of information in deciding whether or not to initiate an overtaking maneuver i.e., "go" if the oncoming car is greater than 200m away. For the reasons discussed above this is a very dangerous control strategy. Why do some drivers use unreliable visual information to perform this maneuver? One problem could be the lack of driving training for overtaking maneuvers: since drivers are given very little practice in overtaking it is possible that many drivers do not know that the distance of the oncoming vehicle is not a reliable source of information. Clarke et al. (1998) have reported that faulty decisions to overtake are particularly prevalent in younger drivers. Because no training or practice with overtaking is provided it may be that the appropriate control strategy is only acquired through real-world driving experience. Another problem is that under some overtaking conditions the visual information that should be used by the driver (the TTC derived from the oncoming car's rate of expansion) is below threshold. Of course in this situation the driver should wait until the rate of expansion rises above threshold (i.e., the oncoming car gets closer) before making his/her decision. The results presented in this chapter suggest that the addition of instruction in overtaking maneuvers to driver training might reduce the prevalence of overtaking accidents. Driving simulations and/or filmed driving scenarios could be used to teach young drivers how to make a safe overtaking judgments based on visual information about TTC and to identify dangerous situations in which their visual perceptions can not be trusted (e.g., when the oncoming car's rate of expansion is below threshold).

When a driver is in a state of closing speed adaptation (Gray & Regan, 2000) visual information about the TTC of the oncoming car that can otherwise be used to make accurate "go/no-go" decisions now becomes dangerously distorted such that the driver overestimates the time available to perform an overtaking or left-turn maneuver. This adaptation effect results in decisions that are delayed, of higher risk, and more variable. Adaptation to closing speed can occur when a driver gazes fixedly at the road or at an oncoming vehicle rather than scanning the scene ahead (Regan & Beverley, 1979) (a state that is commonly referred to as 'highway hypnosis'). The ideal way for addressing this problem would again be to teach drivers that visual information cannot be trusted under these conditions, however our laboratory research suggests that drivers are almost never aware that they have adapted to closing speed (Gray & Regan, 2000). Another possible solution would be to avoid adaptation conditions by encouraging the driver to make more frequent eye movements. One possible method for achieving this would be to use an in-car eye tracker that would send a warning signal when there was a long period of fixation.

Acknowledgements

This work was supported by the National Science Foundation Faculty Early Career Development Program (Award # 0239657).

References

Addams, R. (1834). Mr. Addams on a peculiar optical phenomenon. *London and Edinburgh Philosophical Magazine and Journal of Science, 5*, 373-374.

Clarke, D. D., Ward, P. J., & Jones, J. (1998). Overtaking road-accidents: differences in manoeuvre as a function of driver age. *Accident Analysis and Prevention, 30*, 455-467.

Clarke, D. D., Ward, P. J., & Jones, J. (1999). Processes and countermeasures in overtaking road accidents. *Ergonomics, 42*, 846-867.

Denton, G. G. (1976). The influence of adaptation on subjective velocity for an observer in simulated rectilinear motion. *Ergonomics, 19*, 409-430.

_____ . (1977). Visual motion aftereffect induced by simulated rectilinear motion. *Perception, 6*, 711-718.

Evans, L. (1991). *Traffic safety and the driver*. New York: Van Nostrand Reinhold.

Gordon, D. A., & Mast, T. M. (1970). Driver's judgments in overtaking and passing. *Human Factors, 12*, 341-346.

Gray, R. (1998). *Estimating time to collision using binocular and monocular visual information*. Unpublished PhD, York University, Toronto.

Gray, R., & Regan, D. (1998). Accuracy of estimating time to collision using binocular and monocular information. *Vision Research, 38*, 499-512.

Gray, R., & Regan, D. (2000). Risky driving behavior: a consequence of motion adaptation for visually guided motor action. *Journal of Experimental Psychology: Human Perception & Performance, 26*, 1721-1732.

Gray, R., & Regan, D. (2005). Perceptual processes used by drivers during overtaking in a driving simulator. *Human Factors, In press.*

Groeger, J. A. (2000). *Understanding driving: Applying cognitive psychology to a complex everyday task*. Philadelphia, PA: Psychology Press.

Hoffmann, E. R., & Mortimer, R. G. (1996). Scaling of relative velocity between vehicles. *Accident Analysis and Prevention, 28*, 415-421.

Hoyle, F. (1957). *The black cloud.* Middlesex, England: Penguin.

Jeffcoat, G. O., Skelton, N., & Smeed, R. J. (1973). *Analysis of national statistics of overtaking accidents.* London: University of London, International Driver Behavior Research Association.

Larish, J. F., & Flach, J. M. (1990). Sources of optical information useful for perception of speed

of rectilinear self-motion. *Journal of Experimental Psychology: Human Perception and Performance, 16*, 295-302.

Larsen, L., & Kines, P. (2002). Multidisciplinary in-depth investigations of head-on and left-turn road collisions. *Accident Analysis and Prevention, 34*, 367-380.

NHTSA. (2000). *Traffic safety facts 2000.* Washington, D.C.: National Highway Traffic Safety Administration.

Regan, D., & Beverley, K. I. (1979). Visually guided locomotion: psychophysical evidence for a neural mechanism sensitive to flow patterns. *Science, 205*, 311-313.

Teghtsoonian, M., & Teghtsoonian, R. (1969). Scaling Apparent Distance in Natural Indoor Settings. *Psychonomic Science, 16*, 281-290.

In: Contemporary Issues in Road User Behavior and Traffic Safety ISBN:1-59454-268-6
Editors: D. Hennessy and D. Wiesenthal, pp. 135-149

Chapter 12

ROAD SAFETY IMPACT OF THE EXTENDED DRINKING HOURS POLICY IN ONTARIO

Evelyn Vingilis and Jane Seeley
Population & Community Health Unit, Family Medicine
University of Western Ontario
A. Ian McLeod
Statistical & Actuarial Sciences, University of Western Ontario,
Western Sciences Centre
Robert E. Mann
Centre for Addiction and Mental Health
Doug Beirness
Traffic Injury Research Foundation
Charles Compton
Transportation Research Institute, University of Michigan

INTRODUCTION

Over the past two decades, the progress to reduce the number of deaths and injuries related to impaired driving has decreased significantly in Canada and the United States as well as in other industrialized nations (Beirness, Simpson, Mayhew & Wilson, 1994; Sweedler, 2002) may have come to an end. This progress is being challenged by a political and economic mood against governmental control and regulation (Anglin, Kavanagh & Giesbrecht, 2002; Gliksman, Douglas, Rylett & Narbonne-Fortin, 1995). Moves to privatization, de-regulation, liberalization and fewer controls have been evident internationally (Vingilis, Lote & Seeley, 1998). One example of liberalization occurred in the province of Ontario with the extension of drinking hours in licensed establishments. On May 1, 1996, Ontario, Canada amended the *Liquor Licence Act* to extend the closing hours for alcohol sales and service in licensed establishments from 0100 to 0200 hours. This amendment provided an excellent natural experiment to evaluate an important alcohol policy.

BACKGROUND

The relationships among physical availability of alcohol, alcohol consumption, and alcohol-related problems are multi-faceted and complex (Ashley & Rankin, 1988; Skog, 2003). Availability theory posits that alcohol availability influences consumption levels, which influence alcohol problem levels in a population (Chikritzhs, Stockwell & Masters, 1997; Rush, Gliksman & Brook, 1986). One way to influence alcohol availability is through alcohol control policies, with the imposition of various "barriers", such as restricted drinking hours, that control consumer-product interaction (Ashley & Rankin, 1988). Although the theory is supported by literature, it is important to note that factors affecting availability are strongly related to aggregate alcohol consumption and problems only when other conditions remain unchanged (Room, Romelsjo & Makela, 2002; Skog, 1990, 2003). As Room et al., (2002) write, "effects of smaller changes in availability seem more variable, and often negligible in terms of the effects on total consumption" (p. 167). In addition to availability theory, "power drinking", (also called "last call" or "six o'clock swill") has been suggested as a competing hypothesis to explain what happens when closing hours are changed (Chikritzhs & Stockwell, 2002; Ragnarsdóttir, Kjartansdóttir & Daviosdóttir, 2002; Room, 1988). This hypothesis suggests that tight restrictions on closing times lead to great numbers of drinkers consuming as much alcohol as possible at "last call" for the service of alcohol, shortly before the licensed establishment closes. This means increased blood alcohol concentrations of patrons as they imbibe large amounts of alcohol (power drinking) over a short time period. These crowds of patrons leaving licensed establishments at closing times then become involved in increased levels of intentional and unintentional injuries and other types of problems. This hypothesis has often been cited as evidence that closing hours of licensed establishments should be less restricted as a way to reduce alcohol-related problems (Chikritzhs & Stockwell, 2002; Ragnarsdóttir et al., 2002). This was the rationale presented by the Ontario government for extending the drinking hours.

Research on the effects of changes to hours and days of sale on traffic safety measures are limited and have been conducted in Europe or Australia. A series of studies on the effects of increased hours of sale of alcoholic beverages in various cities and states of Australia have been reported by Smith. The increased days and hours were due to early openings (Smith, 1986), the introduction of Sunday alcohol sales in the cities of Perth, Brisbane and in the state of New South Wales (Smith, 1978, 1987, 1988a), the extension of hotel closings (Smith, 1988b, 1990), and the introduction of flexible trading hours (Smith, 1988c). In all these instances, significant increases in either fatal or injury-producing crashes were observed in the years in which alcohol became more available in comparison with previous years, control times periods or control areas where no changes were introduced. However, a number of methodological and statistical problems preclude firm conclusions being drawn from these studies.

More recently, evaluations of the public health and safety impact of extended trading permit hours were conducted in Perth, Australia (Chikritzhs et al., 1997). The permit allowed an additional hour of serving alcohol, typically at peak times, such as early Saturday or Sunday. Chikritzhs et al. (1997) conducted a study of 20 pairs of hotels matched on levels of assault prior to the introduction of late trading and wholesale purchase of alcohol in Perth between 1991-1995. Levels of monthly assaults more than doubled in hotels that had received extended hours permits compared to no changes in hotels with normal hours. However, no significant increases in road crashes were found related to the extended trading permits.

Only one preliminary assessment has been conducted of Ontario's new amendment. In August, 1996, four months after the law came into effect, the former Liquor Licence Board of Ontario (LLBO) (now amalgamated into the Alcohol and Gaming Commission of Ontario) surveyed their inspectors throughout Ontario to determine the extent and number of violations against the new regulations on cease sale of service at 0200 complaints regarding after-hours sale and service since extension, increases in police services (number of after-hours violations since extension), the operating hours in district, and perceptions of problems, significant differences in operating style, violations, etc. (Bolton, 1996). The results of the survey were consistent across the province. No major increases in problems with the extended hours were documented. In fact, in the period from May 1, 1995 to Aug 1, 1995, when the 0100 closing was still in effect, police services reported to the Board 15 offences of Sale and Service of Liquor Outside Prescribed Hours (SSLOPH) and 38 violations of Fail to Remove Signs of Sale and Service (FRSSS) by 0145. However, during this same period in 1996, only 3 breaches of SSLOPH and 5 offences of FRSSS by 0245 were reported (Bolton, 1996). Perceptions of inspectors of changes in drinking patterns were that some establishments had increased sales while other establishments stated sales were the same because patrons were arriving and leaving later. Inspectors also indicated that many licensed establishments in their local areas that had historically stayed open until 0100 had extended their drinking hours to 0200 on Thursday to Saturday nights, but still maintained their 0100 closing on weekday nights of Sunday to Wednesday. However, the research design and methodological problems of this study preclude conclusions being drawn about the impact of the new regulations. The purpose of this study was to conduct a comprehensive evaluation of the road safety impact of the extended drinking hours regulation in Ontario.

Hypotheses of Impact of Extended Drinking Hours in Ontario

Three competing hypotheses were tested: 1) Availability theory, 2) power drinking hypothesis and 3) temporal shift in drinking patterns hypothesis. Specifically, 1) Availability theory predicts that an extended drinking hour would increase alcohol consumption and lead to an overall increase in alcohol-related motor vehicle casualties. Evening road casualties should increase and shift by one hour; 2) The "power drinking" hypothesis posits that the former 0100 closing encouraged "loading up on the last call". This should lead to a large number of impaired patrons driving after the establishment closed. The government advocated the extension of drinking hours as a way of extending the same quantity of drinking over an extra hour, thereby reducing the 0100 exodus of patrons from licensed establishments. This hypothesis would thus predict an overall decrease in alcohol-related motor vehicle casualties; 3) The temporal shift in drinking pattern hypothesis arose from the preliminary survey conducted by the LLBO (Bolton, 1996). This hypothesis posits that the amount of consumption will stay the same because patrons will stay at licensed establishments the same length of time. Rather, patrons will shift their hours of patronage; for example, they will arrive one hour later and leave one hour later, as suggested by the LLBO survey. This should lead to a temporal shift in alcohol-related motor vehicle casualties, but no overall increase.

CONCEPTUAL FRAMEWORK

This study used a multi-methods, multiple-measures elaboration design, involving both process and outcome measures (Posavac & Carey, 1997; Rossi, Freeman & Lipsey, 1999; Vingilis & Pederson, 2002). The process component of the evaluation consisted of a survey of licensed establishments in Ontario in order to determine what percentage of establishments had taken advantage of the extended drinking hours and to obtain perceptions of proprietors of licensed establishments. Intermediate outcomes included overall trends in volume of sales of alcoholic beverages in Ontario to provide information on overall trends in purchase of alcohol in Ontario during the evaluation time period. The criterion outcome measure was motor vehicle fatalities. A quasi-experimental design using interrupted time series with a non-equivalent no-intervention control group assessed changes in total motor vehicle fatalities and motor vehicle fatalities with a positive blood alcohol concentration (BAC) with the introduction of the extended drinking hours regulations.

METHODS

Survey of Licensed Establishments

The LLBO provided their current list of all 16,466 licensed establishments in Ontario, and a 10% random sample of licensed establishments was computer generated. Using a modified Dillman method (Dillman, 1978) for questionnaire dissemination and follow-up procedures, an anonymous mail-in survey with covering letter and a return University of Western Ontario addressed and stamped envelope was sent to a sample of 1,647 licensed establishments on January 5, 2000. The questionnaire contained eight open- and closed-ended questions focussing on licensing information, type of establishment, current hours of business, whether and how the establishment changed hours since the LLBO amendment, and perceptions of changes to business, attendance, patrons' behaviors and operating costs. A total of 244 questionnaires were returned for a response rate of 16.0%.

Volume of Sales of Alcoholic Beverages

In Ontario, the purchase of any alcohol (wine, beer and spirits) used in licensed establishments or for "personal" consumption is through government regulated monopoly stores except for duty free and possible contraband purchases (Girling, 1994). Additionally, alcohol prices are government controlled so the prices of different brands remain the same throughout the province. Data were available on yearly sales in litres to licensees and to retail for domestic and imported wines, beers and spirits for Ontario for the years April 1, 1989 to March 31, 1999 (Statistics Canada, 2000).

Motor Vehicle Fatalities

The focus of this study was on total and BAC positive fatalities during the 2300 to 0400 time windows; specifically 2300-2359 (denoted as 11 p.m-12 a.m.), 2400-0059 (12-1 a.m.), 0100-0159 (1-2 a.m.), 0200-0259 (2-3 a.m.), and 0300-0359 (3-4 a.m.) for 4 years pre- and 3 years post policy change in Ontario compared to New York and Michigan, two neighbouring American states. Additionally, these data were disaggregated into two groups by day of week: 1) Sunday - Wednesday nights (Sunday 2300 to Thursday 0400), and 2) Thursday - Saturday nights (Thursday 2300 to Sunday 0400).[1] The neighbouring U.S. states of New York and Michigan were chosen as comparator regions to control for climatic and history effects (Cook & Campbell, 1979). Adjacent Canadian provinces either did not have sufficient fatalities (Manitoba) or had introduced a variety of drinking-driving prevention policies that may have confounded the data (Québec). Measures and trends of alcohol consumption, relative importance of alcohol by sector and per capita consumption for Canada and the United States indicate very similar patterns and trends (Anglin, Mann & Smart, 1995; Produktschap voor Gedistilleerde Dranken, 1997). Similarly, motor vehicle fatality and drinking-driving fatality data and trends are comparable (Mayhew, Brown & Simpson, 1996; Ministry of Transportation of Ontario, 1996; NHTSA, 1996). Licensed establishments close at 0200 in Michigan while in New York they range between 0100 to 0400, depending on the county.

Two datasets used were: 1) Traffic Injury Research Foundation (TIRF)[2] and 2) the U.S. Fatal Analysis Reporting System (FARS) databases. The TIRF Fatality Database is a comprehensive source of objective data on alcohol use among persons fatally injured in motor vehicle crashes occurring on and off the highways (e.g. snowmobiles, boats, pedestrians). Objective information on the presence and quantity of alcohol (concentrations detected by chemical tests on blood, urine or other body fluids) as well as information needed to interpret the results of chemical tests - such as time of death – are included in the TIRF database. Because of the high BAC testing rates, (e.g. 82.2% in 1996) for Ontario drivers fatally injured within six hours of collision, this database provided the most sensitive measure of changes in alcohol-related fatalities. In 1996, 37% of fatally injured drivers were BAC positive and of those 30% were over Canada's legal limit of .08% (Mayhew et al., 1996).

The FARS database on police-reported fatal motor vehicle collisions is gathered from states' source documents. In order to be included in FARS, a crash must involve a motor vehicle travelling on a road or highway customarily open to the public and result in the death of a person (either vehicle occupant or non-motorist) within 30 days of the crash. Nationwide in 1996, a total of 16,689 fatally injured drivers had BAC test results out of 24,456, or 68.2%. Testing rates for Michigan and New York, the two comparison states, were 71.2% and 47.8%, respectively. In Michigan for 1996, 1505 drivers died of whom 41% were BAC positive and 32% were over .10%, while in New York, 1564 drivers died of whom 34% were BAC positive and 24% were over .10%.

---

1 A post policy survey by the LLBO found that licensed establishments in smaller communities often maintained their 0100 am closing time for Sunday through Wednesday nights because of lack of sufficient business, but kept open until 0200 am on Thursday through Saturday nights (Bolton, 1996).

2 The TIRF Fatality Database is funded by Transport Canada and the Canadian Council of Motor Transport.

Statistical Time Series Analyses

The data from TIRF and FARS were aggregated into monthly counts according to hour (five groupings between 2300 – 0359) and "weekgroup" (Sunday-Wednesday and Thursday-Saturday), generating ten time series for each data set. Following exploratory analyses, the simple step intervention model (Box & Tiao, 1976) was fitted to test statistically for the presence of shifts in the level of the time series. This model may be written,

$$z_t = c + g(x_t) + I_t + N_t$$

where z_t denotes the observed time series value corresponding to the t-th observation number, c is the constant term, $g(x_t)$ is a term involving a linear function or a transfer-function of one or more covariate time series, I_t represents the intervention term and N_t represents the error or disturbance term assumed to follow an ARIMA or SARIMA time series model. For the simple step intervention model, we may write

$$I_t = \omega \cdot S_t^{(T)}$$

where,

$$S_t^{(T)} = \begin{matrix} 0 & t \leq T \\ 1 & t \geq T \end{matrix}$$

and T is the time the effect of the intervention starts which in this case is the observation number corresponding to May 1996, ω is the parameter which determines the effect of the intervention. In fitting this model, a suitable model for the noise term can be identified by examining the pre-intervention series or by following a two-stage identification process (Hipel & McLeod, 1994). In the two-stage process, the model is initially assumed to be white noise, or in other words, N_t, is assumed to be normal distributed and statistically independent. After the model is fit, plots of the residual autocorrelation function are examined to check for possible autocorrelation. If autocorrelation is found, then a suitable model is determined for N_t, and the model is refit using maximum likelihood estimation.

In most cases, our analysis identified the term N_t as white noise, so linear regression analysis could be used. For the TIRF datasets it was found that the time series of monthly fatalities, when disaggregated into hour and week groups, consisted almost entirely of very small integer values, usually zeros, ones and twos. In this case, the normality assumption is not satisfied and a generalized linear model (McCullagh & Nelder, 1989) was used. Since it was found that there was no significant autocorrelation present in the TIRF and FARS time series, models assuming independence can be used.

The generalized linear model, formulated for a simple step intervention, is comprised of three components:

a) The probability function, $f(z_t, \mu_t, \theta)$, where μ_t is the expected value of z_t and θ represents the distributional parameter or parameters,
b) the linear predictor, $\eta_t = c + \omega \cdot S_t^{(T)}$
c) the link function, $\eta_t = l(\mu_t)$

In some cases, the data were adequately represented by the Poisson distribution, while in other cases it was necessary to use a negative binomial regression (Venables & Ripley, 2002) due to over-dispersion in the data. The log link function was used. Poisson and negative binomial regression were fit using exact maximum likelihood estimation (Currie, 1995). In general, we found that these models agreed very well with the results obtained by fitting the normal linear regression model as might be expected from the robustness result of Hjort (1994).

RESULTS

Survey of Licensed Establishments

As noted earlier, the response rate for the survey was low. Of the licensed establishment respondents, 78.4% of the businesses were first opened before 1995, 60.7% listed themselves as bar/taverns or restaurant/bar and grill, (17% classified themselves as only bars or taverns), while the remainder included sports clubs, private/country clubs/legion, hotel/resorts, banquet hall/service clubs, nursing homes, race tracks, curling arenas and convention centres. Table 1 presents reported closing times of surveyed licensed establishments. Licensed establishments are most likely to close at 0200 on Thursday through Saturday nights, while 0100 is more common on Sunday through Wednesday nights.

Table 1. Percentage Reported Closing Times of Licensed Establishments from 2000 p.m. to 0200 a.m.

Weekday	Percentage Reported Closing by Time of Day								
	2000	2100	2200	2300	2400	0100	0200	Total	N
Monday	4.0	7.5	11.1	11.1	10.6	24.1	15.6	84.0	199
Tuesday	3.3	7.1	10.0	11.4	11.0	24.8	16.2	83.8	210
Wednesday	2.8	8.9	10.7	11.7	9.8	25.2	16.4	85.5	214
Thursday	4.6	7.9	10.2	11.6	9.3	20.4	22.7	86.7	216
Friday	3.2	5.5	6.8	10.5	6.4	21.5	32.0	85.9	219
Saturday	3.7	6.0	6.0	10.2	6.5	21.4	32.6	86.4	215
Sunday	6.8	10.8	11.9	17.6	9.1	11.4	14.2	81.8	176

Among respondents, 49.6% indicated that they had changed their hours since the LLBO amendment to the extended hour. Of those who changed their hours, 31.7% indicated that their closing hours shifted from 0100 to 0200 for all days of the week, 26.6% indicated that their closing hours shifted from 0100 to 0200 for Thursday through Saturday only, and

41.59% changed their hours in other ways, such as shifting of hours, shorter hours, and various combinations of openings throughout the whole week.

Establishments were asked how they have changed as a result of the change in hours of business. Specifically, 67.7% reported no change in liquor sales, 15.2% reported increases/great increases and 17.2% reported decreases/great decreases; 71.7% reported no change in food sales, while 9.8% reported increases/great increases and 18.4% reported decreases/great decreases; 64.6% reported no changes in attendance while 14.6% reported increased/greatly increased attendance and 20.9% reported decreased/greatly decreased attendance; 12.0% reported that patrons' unruly behavior had increased/greatly increased, 66.3% indicated that it had remained the same and 21.8% said it decreased/greatly decreased; 51.0% stated that operating costs had increased/greatly increased, 43.9% that it stayed the same and 5.1% that it had decreased/greatly decreased.

Volume of Sales of Alcoholic Beverages

The trends indicate that the volume of sales in thousands of litres for beer and per capita 15 years and over for Ontario between 1989 and 1999 has decreased between 1994 and 1998, while the consumption of wine and spirits decreased in the early 1990's and increased in the late 1990's (Figure 1).

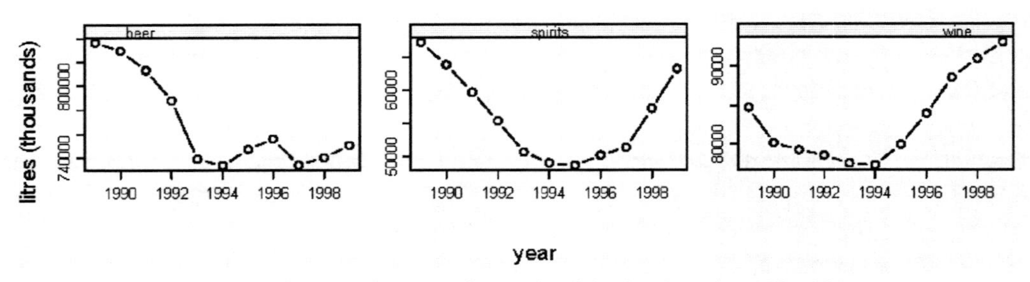

Figure 1. Annual Volume of Sales of Beer, Wine and Spirits for Ontario 1989-1999.

Motor Vehicle Fatalities

Figure 2 shows Ontario motor vehicle BAC positive driver fatalities. Ontario total motor vehicl driver fatalities and New York/Michigan total drive fatalities time series anlayses for the extended drinking hour transition time periods of Thursday to Saturday (actually referring to Friday-Sunday morning 0100-0159 (T1) and 0200-0259 (T2)), and Sunday-Wednesday (actually referring to Monday-Thursday morning 0100-0159 (S1) and 0200-0259 (S2)). For BAC positive driver fatalities, TIRF data showed significant downward trands for the (S1) Sunday-Wednesday (Monday-Thursday morning 0100-0200 (p = 014) time period, but no other trends were significant. For the Ontario Total Driver Fatalities, TIRF data showed significant reductions for the (S1) Sunday-Wednesday (Monday-Thursday morning) 0100-0200 ($p = .006$) time period, for the Sunday-Wednesday (Monday-Thursday morning) 2400-0100 time period ($p = .044$) (not shown in graph) and one almost significant upward trend for the Thursday-Saturday (Friday-Sunday morning) 0300-0400 ($p = .088$) time period (not shown in graph). No other trends were significant for the hourly time periods between 2300 and 0400 for Sunday through Wednesday and Thursday through Saturday evenings. In New York and Michigan (FARS data), no significant trends were identified coincident with the Ontario extended drinking hours intervention.

The total and BAC positive TIRF and FARS driver fatality data were aggregated over the 2300-0400 time period to determine whether there had been overall increases in BAC positive driver fatalities over the evening drinking hours. The results of the intervention analysis indicated no significant changes for Sunday-Wednesday and Thursday-Saturday groups for total driver fatalities for both TIRF and FARS data. For BAC positive driver fatalities, downward trends were observed for Ontario TIRF data for both Sunday-Wednesday ($p = .07$) and Thursday-Saturday ($p = .06$) groups, while a significant downward trend was observed for the control group FARS data for the Thursday-Saturday ($p = .00$) group.

To control for possible overall downward trends in Ontario collision rates, the BAC positive TIRF driver fatality data were re-analyzed using the TIRF BAC negative data as a covariate. The results did not differ from the analyses without the covariate.

To determine whether there was a temporal shift in BAC positive driver fatalities, BAC positive TIRF data were collapsed over the pre- and post- amendment time periods for the different hours and Sunday-Wednesday and Thursday-Saturday week groups. Figure 3 indicates that for the Sunday-Wednesday time period, the peaks for alcohol-related driver fatalities occurred between 2400 and 0200 pre-amendment while the peaks occurred between 0200 and 0400 post-amendment. However, the distribution is different for the Thursday-Saturday time periods. Prior to the amendment the peaks of alcohol-related driver fatalities occur at 2300-2400 and 0100-0200 while following the amendment the distribution has flattened over the different time periods.

DISCUSSION

These findings appear to support the contention of Room et al. (2002) and Skog (1990, 2003) that the effects of smaller changes in availability seem variable, and may be negligible. In Ontario the drinking in licensed establishments was extended for only one hour and thus possible effects on motor vehicle fatalities were expected to be small. Multiple measures were gathered to enhance validity by seeking convergence of findings, thereby enhancing the interpretability of findings.

The multiple datasets converge in support of the power drinking or temporal shift hypotheses. Availability theory was not supported because no increases in BAC positive driver fatalities were observed in Ontario after the amendment to extend drinking hours. Even controlling for overall trends in BAC negative driver fatalities, the trends for BAC positive driver fatalities were not in the direction predicted by availability theory. However the data obtained from the survey of licensed establishments indicated that many licensed establishments did not implement the extended drinking hours and indeed the hours of closing were quite variable among licensed establishments across Ontario.

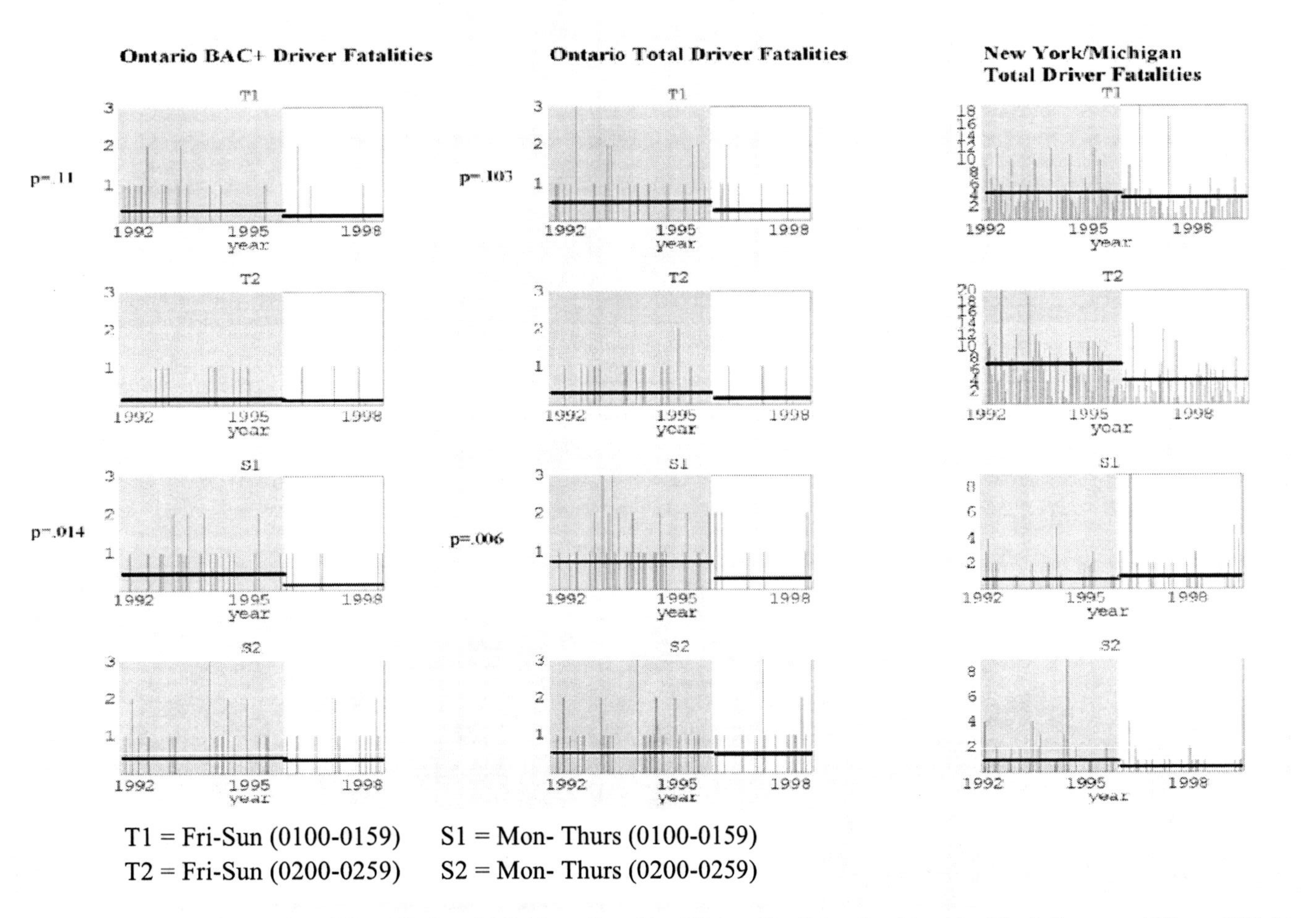

Figure 2. Ontario BAC+, Ontario Total, and New York and Michigan State Total Driver Fatalities Combined by Week Group and Time of Night.

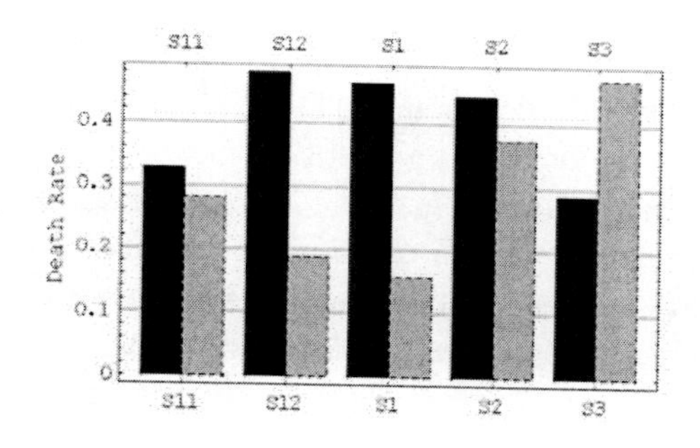

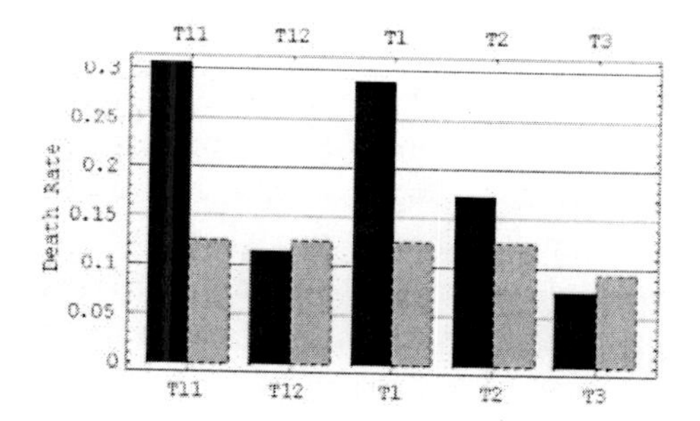

Figure 3. Average BAC+ Driver Fatality Rate (#deaths/month) by Time of Night and Week Group. The black bar corresponds to the pre-amendment death rate and the gray bar corresponds to the post-amendment death rate. (S11-Sun-Wed 2300-2359, etc, T11=Thurs-Sat 2300-2359, etc.)

Of the 17% of licensed establishments most likely to stay open late, bars and taverns, two thirds reported remaining open until 0200 on Thursday and somewhat over four fifths reported staying open until 0200 on Friday and Saturday, although the low response rate limits the generalizability of these findings. Other types of licensed establishments were much less likely to stay open until 0200. This would suggest that alcohol availability might not have substantively increased in spite of the regulation change.

BAC positive driver fatality trends reflected downward trends for Sunday-Wednesday 2400-0200 and Thursday-Saturday 0100-0200 for Ontario, and downward trends for Thursday-Saturday 2300-0100 and 0200-0400 for New York and Michigan, suggesting diverging patterns between the extended drinking hours jurisdiction of Ontario and the control jurisdictions of New York and Michigan. However, the trends in total driver fatalities and BAC positive driver fatalities for Ontario were not divergent. If New York and Michigan are to be considered the expected trend, the lack of concomitant significant reductions in Ontario could suggest that the extended drinking hours moderated the expected downward trend. However, there are lower BAC testing rates for New York and Michigan. This issue makes it difficult to assume that the New York and Michigan data are an ideal control series. Ontario total driver fatalities may be a more appropriate comparison measure.

An inspection of the pre- and post-amendment distribution curves for BAC positive driver fatalities for Ontario in Figure 3 suggests that two different phenomena may be occurring for Sunday-Wednesday and Thursday-Saturday nights. For Sunday-Wednesday nights, the pre-amendment 2400-0200 peaks for BAC positive driver fatalities seems to have shifted to 0200-0400 post-amendment, while for Thursday-Saturday the 2300-2400 and 0100-0200 pre-amendment peaks seem to have decreased and flattened out over the 2300-0400 time periods. These different distributions could suggest different patterns of drinking for weekdays and weekends by patrons of licensed establishments. It could well be that problem drinkers are more likely to engage in weekday evening drinking, while social drinkers are more likely to go out on weekends.

The Sunday-Wednesday shifting of peak BAC positive driver fatalities could support the temporal shift hypothesis, while the Thursday-Saturday reduction and flattening out of the BAC positive driver fatality distribution could support the power drinking hypothesis that would predict an overall decrease in alcohol-related motor vehicle casualties that seems to have occurred. Furthermore, the alcohol sales data reveal decreases in beer sales, the most commonly sold beverage in taverns and bars (Gruenewald & Ponicki 1995). Moreover, the survey of licensed establishments found that almost two thirds of respondents observed no increases in alcohol sales following the increase in drinking hours. Thus, although it is not possible to choose between the power drinking or temporal shift hypothesis, the different data sets converge to suggest that the road safety impact of these smaller changes in availability was minimal in Ontario as a whole.

It is also possible that drinking driving fatality rates were changing in response to several factors including a number of road safety initiatives that occurred within a two-year interval before and after the change in hours of sale. Ontario introduced a Graduated Licensing System in the spring of 1994 and a 90-day Administrative Driver's Licence Suspension for those charged with a drinking driving offence, on November 28, 1996. Both these initiatives have been associated with reductions in drinking-driving behavior, and collisions and alcohol-related driver fatalities (Boase & Tasca, 1998; Mann et al., 2000, 2002). These initiatives may have created a declining trend in the drinking driving problem that masked the effects of a small increase in alcohol availability.

Limitations of the study include the fact that the impact of extended drinking hours would be limited to those who drink in licensed establishments during early morning hours. Although these establishments are a substantial source of the proportion of drinking drivers, those who typically drink at home or at parties would not necessarily be affected by the longer hours. Moreover, the inability to disaggregate Ontario alcohol sales by licensed establishments and by hour of sales meant that we could not obtain verification of the survey findings that the majority of respondents perceived no changes in alcohol sales. Additionally, the limited number of BAC positive driver fatalities by hour of day and the rarity of motor vehicle fatalities may have made this data set a less sensitive measure to detect changes. For example, Chikritzhs and Stockwell (1997) found no significant increases in road crashes related to the extended trading permits, although they did find differences in levels of violent assaults on or near licensed establishments.

To summarize, these findings support international research that increases in alcohol availability can be seen to affect aggregate alcohol consumption and alcohol-related problems only when other conditions remain unchanged (Room et al., 2002; Skog, 1990, 2003). The small change in policy, the limited implementation and other societal factors such as economic conditions and road safety countermeasures may have mitigated any affect on alcohol-related motor vehicle fatalities in Ontario.

Acknowledgements

The study received support from the Donner Canadian Foundation.

References

Anglin, L., Mann, R. E., & Smart, R. G. (1995). Changes in cancer mortality rates and per capita consumption in Ontario, 1963-1983. *International Journal of Addiction,* 41, 124-131.

Anglin, L., Kavanagh, L., & Giesbrecht, N. (2002). Public Opinion analysis suggesting demographic characteristics of persons trying to favour internal versus external control or drinking behavior. *Journal of Substance Abuse*, 7, 214-220.

Ashley, M. J., & Rankin, J. G. (1988). A public health approach to the prevention of alcohol-related health problems. *Annual Review of Public Health*, 9, 233-271.

Beirness, D. J., Simpson, H. M., Mayhew, D. R., & Wilson, R. J. (1994). Trends in drinking driver fatalities in Canada. *Canadian Journal of Public Health*, 85, 19-22.

Boase, P. & Tasca, L. (1998). *Graduated Licensing System Evaluation: Interim Report '98.* Toronto, Canada: Ministry of Transportation Ontario.

Bolton, T. (1996). *Inspector Survey*. [Unpublished data]. Toronto, Canada: Liquor Licensing Board of Ontario (LLBO).

Box, G. E. P., & Tiao, G. C. (1976). Intervention analysis with applications to economic and environmental problems. *Journal of the American Statistical Association*, 70, 70-79.

Chikritzhs, T., Stockwell, T., & Masters, L. (1997). *Evaluation of the public health and safety impact of extended trading permits for Perth hotels and night clubs*. Curtin University of Technology, Australia: National Centre for Research into the Prevention of Drug Abuse.

Chikritzhs, T., & Stockwell, T. (2002). The impact of later trading hours for Australian public houses (Hotels) on levels of violence. *Journal of Studies on Alcohol*, 63, 591-599.

Cook, T. D., & Campbell, D. T. (1979). Quasi-experimentation Design and Analysis Issues For Field Settings. Skokie, Il.: Rand McNally.

Currie, I. D. (1995). Maximum likelihood estimation and Mathematica. *Applied Statistics*, 44, 379-394.

Dillman, D. A. (1978). *Mail and Telephone Surveys: The Total Design Method.* New York, NY: Wiley-Interscience.

Girling, S. (1994). *Preventing The Use Of Illegal Liquor In Establishments Licensed By The LLBO: A Study In General Deterrence.* (May 3) Toronto: Liquor License Board of Ontario.

Gliksman, L., Douglas, R. R., Rylett, M., & Narbonne-Fortin, C. (1995). Reducing problems through municipal alcohol policies: The Canadian experiment in Ontario. *Drugs Education Prevention Policy, 2,* 105-118.

Gruenewald, P. J., & Ponicki, W. R. (1995). The relationship of the retail availability of alcohol sales to alcohol-related traffic crashes. *Accident, Analysis and Prevention, 27,* 249-259.

Hipel, K. W., & McLeod, A. I. (1994). *Time Series Modelling of Water Resources and Environmental Systems.* Amsterdam: Elsevier.

Hjort, N. L. (1994). The exact amount of t-ness that the normal model can tolerate. *Journal of the American Statistical Association, 89*, 665-675.

Mann, R. E., Smart, R. G., Stoduto, G., Adlaf, E. M., Vingilis, E., Beirness, D., & Lamble, R. (2000). Changing drinking-driving behavior: The effects of Ontario's administrative driver's licence suspension law. *Canadian Medical Association Journal*, 162, 1141-1142.

Mann, R. E., Smart, R. G., Stoduto, G., Vingilis, E., Beirness, D., & Lamble, R. (2002). The early effects of Ontario's administrative driver's licence suspension law on driver fatalities with a BAC > 80 mg%. *Canadian Journal of Public Health*, 93, 176-180.

Mayhew, D. R., Brown, S. W., & Simpson, H. M. (1996). *Alcohol Use Among Drivers and Pedestrians Fatally Injured in Motor Vehicle Accidents.* (TP-11759-96 E) Ottawa, Canada: Transport Canada Publication.

McCullagh, P., & Nelder, J. A. (1989). *Generalized Linear Models*. Boca Raton: Chapman and Hall/CRC.

Ministry of Transportation of Ontario (MTO) (1996). *'96 Ontario Road Safety Annual Report*. Downsview, Ontario: Safety Research Office.

National Highway Traffic Safety Administration (NHTSA) (1996). *Traffic Safety Facts 1996 Annual Report*. Washington, D.C.: U.S.: Department of Transportation, Accessed on October 23, 2003 at http://www-nrd.nhtsa.dot.gov/departments/nrd-30/ncsa/AvailInf.html

Posavac, E. J., & Carey, R. G. (1997). *Program Evaluation Methods and Case Studies* (5th ed). New Jersey: Prentice Hall Publishing.

Produktschap voor Gedistilleerde Dranken. (1997). *World Drink Trends 1997*. Netherlands: NTC Publications Ltd, Accessed on October 23, 2003 at *http://www.pgd.nl/img/wrldcon96.pdf*

Ragnarsdóttir, P., Kjartansdóttir, A., & Daviosdóttir, S. (2002). Effect of extended alcohol serving-hours in Reykjavik. In R. Room (Ed.). *The Effects of Nordic Alcohol Policies. What Happens To Drinking And Harm When Alcohol Controls Change?* (No.42, pp. 145-154). Netherlands: NAD Publications.

Room, R. (1988). The dialectic of drinking in Australian life: From the rum corps to the wine column. *Australian Drug and Alcohol Review*, 7, 413-437.

Room, R., Romelsjo, A., & Makela, P. (2002). Impacts of alcohol policy: The Nordic experience. In R. Room (Ed.). *The Effects of Nordic Alcohol Policies. What Happens To Drinking And Harm When Alcohol Controls Change?* (No.42, pp. 167-174). Netherlands: NAD Publications.

Rossi, P. H., Freeman, H. E., & Lipsey, M. W. (1999). *Evaluation: A Systematic Approach*. California: SAGE Publications.

Rush, B. R., Gliksman, L., & Brook, R. (1986). Alcohol availability, alcohol consumption and alcohol-related damage. The distribution of consumption model. *Journal of Studies on Alcohol*, 47, 1-10.

Skog, O-J. (1990). Future trends in alcohol consumption and alcohol-related problems: Anticipation in light of the efforts at harmonization in the European Community. *Contemporary Drug Problems*, 17, 575-593.

_____ . (2003). Alcohol consumption and fatal accidents in Canada, 1950-98. *Addiction*, 98, 883-893.

Smith, D. I. (1978). Impact on traffic safety of the introduction of Sunday alcohol sales in Perth, *Western Australia Journal of Studies on Alcohol*, 39, 1302-1304.

_____ . (1986). Comparison of patrons of hotels with early opening and standard hours. *International Journal of Addictions*, 21, 155-163.o

_____ . (1987). Effect on traffic accidents of introducing Sunday hotel sales in New South Wales, Australia. *Contemporary Drug Problems*, Summer, 279-294.

_____ . (1988a). Effect on traffic accidents of introducing flexible hotel trading hours in Tasmania, Australia. *British Journal of Addictions*, 83, 219-222.

Smith, D. I. (1988b). Effect on casualty traffic accidents of the introduction of 10 pm Monday to Saturday hotel closing in Victoria. *Australian Drug and Alcohol Review*, 7, 163-166.

_____ . (1988c). Effect of traffic accidents of introducing flexible hotel trading hours in Tasmania, Australia. *British Journal of Addictions*, 83, 219-222.

_____ . (1990). Effectiveness of legislature and fiscal restrictions in reducing alcohol related crime and traffic accidents. In J. Vernon (Ed.). *Australian Institute of Criminology (AIC)*

Conference Proceedings. Alcohol and Crime (No. 1, April 4-6, 1989). Canberra, Australia: Australian Institute of Criminology.

Statistics Canada. (2000). *The Control and Sale of Alcoholic Beverages in Canada* Catalogue No. 63-202-X1B.

Sweedler, B. M. (2002). Worldwide trends in drinking and driving: Has the progress continued? In D. Mayhew and C. Dussault (Eds.). *Proceedings of the International Conference of Alcohol, Drugs and Traffic Safety*. (CD-ROM). Montréal, Québec: International Committee on Alcohol, Drugs and Traffic Safety.

Venables, W. N., & Ripley, B. D. (2002). *Modern Applied Statistics with S.* (4th Ed.). New York: Wiley.

Vingilis, E., Lote, R., & Seeley, J. (1998). Are trade agreements and economic co-operatives compatible with alcohol control policies and injury prevention? *Contemporary Drug Problems*, 25, 579-620.

Vingilis, E., & Pederson, L. (2002). Using the right tools to answer the right questions: The importance of evaluative research techniques for health services evaluation research in the 21st century. *Canadian Journal of Program Evaluation*, 16, 1-26.

Part 4
Alcohol and Driving

In: Contemporary Issues in Road User Behavior and Traffic Safety ISBN:1-59454-268-6
Editors: D. Hennessy and D. Wiesenthal, pp. 153-166

Chapter 13

Characteristics of Persistent Drinking Drivers: Comparisons of First, Second, and Multiple Offenders

William F. Wieczorek
Center for Health and Social Research
State University of New York College at Buffalo
Thomas H. Nochajski
School of Social Work
University at Buffalo, The State University of New York

Introduction

Despite years of educational programs, enhanced enforcement, and other interventions, alcohol-impaired driving (DWI) remains a major traffic safety and public health problem (Hingson & Winter, 2003). Sharp decreases in alcohol-related crash fatalities seen in the late 1980s and early to middle 1990s have leveled off. Although the number of alcohol-related crash fatalities was about 33% lower in 2002 compared to 1982, the absolute number of deaths in 2002 (17,419) shows a continuing massive societal impact (National Highway Traffic Safety Administration [NHTSA], 2003). Even more important is the fact that alcohol-related crash fatalities have actually increased since 1999 (NHTSA, 2003).

Other measures such as general population and roadside surveys show that drinking and driving is still a relatively common behavior. A national survey of 5,733 persons age 16 and older conducted for the National Highway Traffic Safety Administration (Royal, 2000) found that 21% reported driving within two hours of drinking. The national roadside survey conducted in 1996 (Voas, Wells, Lestina, Williams, & Greene, 1998) found that 16.9% of weekend nighttime drivers had positive blood alcohol concentrations (BAC). The total prevalence in 1996 showed a substantial decrease from 25.9% in 1986, but the percentage of drivers at higher BACs (i.e., above .05 percent ethanol by volume) did not significantly change. In addition to these research-based examples, arrests for impaired driving in the U.S. continue to be one of the most common crimes, with about 1.4 million arrests each year (Maguire & Pastore, 2003).

The identification of a subgroup of especially risky and highly impaired drivers has been offered as one potential explanation for the lack of continued success in reducing the toll from impaired driving. Simpson and colleagues (Simpson & Mayhew, 1991; Simpson, Mayhew, & Bierness, 1996) are the trailblazers in recognizing this subgroup of drinking drivers, which they call "hard core drinking drivers." The working definition of hard core drinking drivers is "those individuals who repeatedly drive after drinking, especially with high BACs (.15 percent ethanol by volume or greater) and who seem relatively resistant to changing this behavior" (Simpson et al. 1996, p. 9). Sweedler (1995) called these highly problematic individuals persistent drinking drivers, a term that can be used interchangeably with hard core drinking drivers. By analyzing fatal crash and other data, Simpson and Mayhew (1991) showed that persistent drinking drivers are greatly over-represented in fatal crashes. They estimated that persistent drinking drivers caused a majority of alcohol-related fatalities.

The task of operationalizing the definition of persistent drinking drivers is non-trivial. The issue is that it may be an easier task to identify persistent drinking drivers after a fatal crash rather than before one. Persons convicted of more than one drinking and driving offense are by definition persistent drinking drivers. However, the repeat offender is relatively under-studied from a descriptive viewpoint because they are only a limited proportion of the total drinking driver offender population. Even obtaining accurate estimates of the rate of recidivism for drinking and driving convictions is a difficult task. Problems in record keeping, expunging of records after a period of time (e.g., three or five years), and offenses in different states that do not share records result in underestimates of the actual size of the recidivist population (Simpson, 1995). Just by lengthening the follow-up period for first offenders, the recidivism rate can increase from less than one-quarter to over one-third (Hedlund, 1995). There is no currently available procedure for identifying persistent drinking drivers among the pool of first-time offenders, although Rauch et al. (2002) suggest that any alcohol-related driving event is a significant predictor of future recidivism.

There are a notable proportion of identified drinking drivers in fatal crash data, despite the limitations of the official arrest/conviction records. These drivers are considered to be persistent drinking drivers because they continued to drink and drive (as evidenced by the fatality) despite the earlier offense. The initial research in this area was conducted by Fell (1993) using a national database of fatal crashes and driving record checks for the past three years to examine the proportion of fatality subgroups with a previous drinking-driving offense. He found that 4.5 % of the fatalities had a previous impaired driving offense, and among those with highly elevated BAC (greater than .10 percent ethanol by volume), 11.8% had a previous conviction. The patterns found by Fell are similar to the current (2002) proportion of fatalities with previous impaired driving convictions (NHTSA, 2003). In the current data, the elevated risk is clearest among fatalities with a BAC above .15 in which 10% had a previous offense compared to only 1% of the fatalities with no alcohol involvement (NHTSA, 2003). In studies of fatalities in specific jurisdictions, even higher rates of previous offenses have been found (Bailey, 1993, 1995; Simon, 1992).

Almost by definition, general alcohol and other problem severity is expected to be greater among persistent drinking drivers compared to others. There is, however, a relative dearth of detailed assessments of persistent drinking drivers based on individual interviews. Some of the difficulties include obtaining a sufficiently large sample and utilizing appropriate measures of the multiple domains associated with drinking and driving and criminal offending in general (Nochajski & Wieczorek, 1998; Welte, Zhang, & Wieczorek, 2001). There are a few examples available. McMillen, Adams, Wells-Parker, Pang, and Anderson (1992) examined personality and other predictors between first and repeat offenders, although the focus was not on persistent drinking drivers. A cluster analysis of persistent drinking drivers found at least two groups of these offenders when differentiated along scales of alcohol-related behaviors and problem severity (Wieczorek, Callahan, & Nochajski, 2000). A study of social bonds among persistent drinking drivers found that contrary to expectations, higher levels of bonding were associated with increased risk of recidivism (Ahlin, Rauch, Zador, Baum, & Duncan, 2002). Siegal, Falck, Carlson, Rapp, Wang, and Cole (2002) conducted a qualitative study of a small number of incarcerated persistent drinking drivers and found extreme levels of alcohol problems.

Although the general characteristics of drinking drivers are well documented , those of persistent drinking drivers are not. A brief overview of characteristics of drinking and driving offenders provides an important foundation for interpreting research on persistent drinking drivers. In general drinking drivers are predominantly male, white, working class, high school educated, and drink more heavily than the general population (Jones & Lacey, 2001). Drinking and driving does not decrease substantially until after about age 45, although the youngest drinking drivers are at the highest risk of fatal crashes (Hingson & Winter, 2003). The majority of alcohol-related fatalities and arrests are in the 21 – 45-year age range (Jones & Lacey, 2001; Maguire & Pastore, 2003). Psychiatric problems, risk taking personality, and risky driving histories are associated with impaired drivers (Vingilis, 2000). These factors and others are examined in this chapter because they are also likely to differentiate specific characteristics or subtypes of persistent drinking drivers.

Rationale for the Current Study

One purpose for the current study is to expand the body of extant research on specific characteristics of persistent drinking drivers, which is relatively limited. The analyses presented in this chapter examined characteristics from multiple domains to differentiate first, second, and multiple (three or more) drinking and driving offenders. The second and multiple offenders are defined as persistent drinking drivers. We hypothesize that differences will exist between first and second, and second and multiple offenders, with each group showing greater problem severity. We also hypothesize that the multiple offender group will have the greatest severity of alcohol dependence and anti-social behavior.

Methods

Sample and Procedures

The sample was recruited through Buffalo City Court records, the Erie County Probation Department, and the Drinking Driver Program (DDP). The project utilized a protocol that was reviewed and approved by the IRBs at the Research Institute on Addictions and Buffalo State College. Subjects were recruited using pamphlets distributed to the drinking-driving offenders by probation officers, the DDP staff, or mailed directly to them (court sample). Recruitment rates for the subsamples were: 56 % for probationers (311 out of a possible 560); 43 % for DDP participants (145 out of a possible 335); and 22% for the court cases (200 out of a possible 911). Trained research staff conducted confidential face-to-face interviews with each participant in a private interview room. The categorization into first, second, and multiple offenders was based on self-report and official records (available for 446 subjects). The final categorization was based on the highest total number of drinking-driving offenses, regardless of the source. The offenders had no incentive to either minimize or maximize their count of offenses for these research interviews because they provided consent to obtain their official records and none of their responses were shared with the courts or probation. For the analyses in this chapter, comparisons were made between first offenders (n = 286), second offenders (n = 160), and multiple offenders (n = 204).

MEASURES

A broad assessment of demographics was conducted including date of birth, marital status, education level, ethnicity, occupation, employment status and household income. Quantity-frequency items developed by Armor and Polich (1982) were used to measure alcohol consumption. The Alcohol Use Inventory (AUI) was used to measure drinking style, consequences, perceived benefits, concerns, and a general alcoholism score (Horn, Wanberg, & Foster, 1987). DSM-III-R alcohol dependence was assessed using a modified version of the Diagnostic Interview Schedule (Robins, Helzer, Cottler, & Goldring, 1989). The family history items from the Research Diagnostic Criteria were used to assess familial alcoholism (Andreasen, Rice, Endicott, Reich, & Coryell, 1986). Drug use was assessed using a items derived from the National Household Survey on Drug Abuse (now known as the National Survey on Health and Drugs) (Substance Abuse and Mental Health Services Administration, 2003). A history of victimization experiences (including sexual and physical assaults) during adulthood was obtained.

A series of questions pertaining to arrests and convictions for charges other than DWI were used to assess criminal history. Official Department of Motor Vehicles records and self-reports of driving incidents were available. Donovan's driving-related attitude measures developed for DWI populations were used (Donovan, 1980; Donovan & Marlatt, 1982; Donovan, Marlatt, & Salzberg, 1983). The attitudes assessed included a 12-item driving-related aggression scale, a 6-item competitive speed scale, a 7-item driving for tension reduction scale, and a 3-item driving inhibition scale. Knowledge of drinking-driving laws and the amount of alcohol consumption necessary to reach legal BAC levels were assessed through a series of questions asking (1) the blood alcohol levels for driving while impaired and driving while intoxicated (using New York State statutes); (2) the number of drinks required in one hour to reach .10 BAC for a 160-pound man; and (3) the number of bottles of beer that equal a 1-ounce shot of liquor.

The Symptom Checklist-90 Revised (SCL-90-R) was used to measure psychiatric severity (Derogatis, 1983). The SCL-90-R is composed of nine symptom dimensions and includes symptom counts and an overall severity rating. A 46-item Socialization Scale derived from the California Psychological Inventory (Megargee, 1972) was used to measure sociopathy. The Sensation Seeking Scale (SSS) Form V was used to assess the respondents' risk taking propensities (Zuckerman, 1979). Locus of control was measured using Levenson's (1981) Internal, Powerful Others, and Chance scales. A five-item measure, derived from the Marlowe-Crowne scale, developed by Hays, Hayashi, and Stewart (1989) was used to assess socially desirable responding. A modified version of Rosenberg's (1965) measure of self-esteem was used to assess self-concept. The motivation to change drinking-driving behavior was assess using the stages of change scale (DiClemente & Prochaska, 1982; McConnaughy, Prochaska, & Velicer, 1983). Each stage (i.e., Pre-Contemplation, Contemplation, Action, and Maintenance) is measured on an 8-item scale. Information on previous utilization of various alcohol-related treatments was obtained.

RESULTS

The purpose of the data analyses was to describe the variation associated with three categories of DWI offenders: first offenders (only one offense), second offenders (two offenses), and multiple offenders (three or more offenses). ANOVA and chi-square statistics were used to compare the three categories across individual variables. There is a risk of chance findings due to multiple comparisons, however, highly conservative adjustments (e.g., Bonferroni) raise the chances of missing true differences (Perneger, 1998). Scheffé tests were used to supplement the ANOVAs to perform contrasts on non-categorical variables. Logistic regression and multinomial regression were used to develop multivariate models. Space limitations preclude tables for all analyses discussed in the text; all differences and associations discussed in the text are significant at the $p \leq .05$ level.

For the majority of demographic characteristics there appears to be a linear trend, such that the first offenders were more likely than the second and multiple offenders to be female, never married, and have an education beyond high school. In contrast, the second and repeat offenders were more likely than the first offenders to be older, unemployed, divorced, have less than a high school education, and a lower income. The likelihood of having a criminal history increases with the number of prior drinking-driving offenses. The mean total number of arrests, especially for the repeat (3.57) and multiple offenders (5.35), shows the high level of involvement these individuals have with the criminal justice system. The two family history characteristics that are most strongly associated with number of DWI offenses were whether the father had a drinking problem or if a relative was arrested for a DWI. In both instances, family problems increased with the number of drinking-driving convictions.

Table 1 shows comparisons across drinking-related characteristics for the different groups of drinking-driving offenders. In the month (i.e., the thirty-day period) prior to the most recent arrest, there is a linear trend for the number of days the individual drank, number of days intoxicated; number of days drove after 10 drinks; and number of days drove drunk. As one progresses from one to two to three or more offenses, the likelihood of drinking heavily in the month prior to the arrest increases. Both second and multiple offenders spent substantially more money on alcohol during the month prior to their most recent drinking-driving arrest. For drinking-related behaviors associated with the most recent drinking-driving arrest, repeat and multiple offenders were more likely to be drinking alone, in a multiple vehicle crash, to have refused a breath/chemical test, and reported higher BACs than first offenders. Drinking with female friends was more common among the first-time offenders than in the other groups.

Table 1. Comparison of First, Second, and Multiple DWI Offenders on Drinking-related Measures

	First (n=286)	Second (n=160)	Multiple (n=204)	Significance
Money spent on alcohol per week for month prior to arrest	\$46.02[a] (54.22)	\$79.07[b] (74.83)	\$77.43[b] (69.52)	F(2,303)=6.93 p = .001
Days drinking out of month prior to the arrest	12.72[a] (9.40)	14.57[ab] (9.89)	15.67[a] (10.55)	F(2,647)=5.52 p = .004
# Days high from alcohol in month prior to arrest	6.91[a] (8.03)	9.01[ab] (9.25)	10.66[b] (10.01)	F(2,647)=10.61 p < .001
# Days drove after 10 drinks in month prior to arrest	2.81[a] (6.54)	4.18[ab] (7.01)	5.33[b] (8.24)	F(2,648)=7.33 p = .001
# Days drove drunk in month prior to arrest	6.98[a] (8.83)	8.90[ab] (9.11)	10.99[b] (10.41)	F(2,647)=10.83 p < .001
Drinking alone before DWI arrest	5.2%	11.9%	14.2%	X^2 = 12.10 df = 2 n = 650 p = .002
Drinking with female friend(s) before DWI	42.7%	28.1%	31.8%	X^2 = 11.34 df = 2 n = 650 p = .003
Multiple vehicle Accident at DWI arrest	39.3%	40.0%	60.5%	X^2 = 5.36 df = 2 n = 149 p = .069
Refused Chemical Test after DWI arrest	16.0%	29.2%	42.3%	X^2 = 39.65 df = 2 n = 650 p < .001
Self-Report BAC at arrest	0.159[a] (0.053)	0.182[b] (0.061)	0.173[ab] (0.059)	F(2,442)=5.94 p = .003
Total DSM-III-R Criteria (Lifetime)	4.08[a] (2.46)	5.52[b] (2.37)	6.39[c] (2.34)	F(2,644)=54.81 p < .001
Alc Diagnosis (lifetime) No Diagnosis	3.9%	0.6%	1.0%	X^2 = 47.43 df = 4
Abuse	27.7%	13.2%	6.4%	n = 647 p < .001
Dependent	68.4%	86.2%	92.6%	
Alc Diagnosis (current) No Diagnosis	12.9%	11.3%	7.8%	X^2 = 12.77 df = 2
Abuse	40.9%	33.8%	32.8%	n = 649 p = .002
Dependent	46.2%	55.0%	59.3%	

Note: different superscripts indicate Scheffe p <.05 differences between groups.

Alcohol dependence was also more common in the second and multiple offender groups compared to the first offenders (see Table 1). This finding held for both lifetime and current (i.e., in the past 12 months) diagnoses. Alcohol abuse was more common among the first offenders, suggesting lower problem severity in this group. The high severity of alcohol problems among the persistent drinking drivers is shown by the strong linear increase in total number of dependence criteria. Among multiple offenders, severe alcohol dependence is the norm, as shown by an average of over six dependence criteria. These findings are supported by the results for the 15 subscales derived from the AUI (not shown in Table 1). In 14 of the 15 subscales (Gregarious Drinking being the only exception with no significant differences), there were linear trends with the first offenders always reflecting the lowest scores. This continues to reinforce the previous findings that showed the link between heavy drinking, drinking problems, and persistent drinking-driving behavior.

Table 2 shows lifetime drug use for the three categories of offenders. These results indicate that many drinking-drivers have substantial drug use histories in addition to heavy drinking. The rates of use across all three groups are notably high and not limited to "soft" drugs like marijuana. Over 40% of all offenders have used cocaine, hallucinogens, and amphetamines. Even narcotic use (i.e., opiate-based drugs) was reported by about 10% of the first offenders.

Table 2. Comparison of First, Second, and Multiple DWI Offenders Across Drug Use Characteristics (Lifetime)

	First (n=286)	Second (n=160)	Multiple (n=204)	Significance
Ever used marijuana/hashish	84.1%	92.5%	87.2%	$X^2 = 6.44$ df = 2 n = 646 p = .040
Used hallucinogens	43.7%	55.0%	57.6%	$X^2 = 10.71$ df = 2 n = 647 p = .005
Used amphetamines/speed	44.2%	61.3%	60.3%	$X^2 = 17.53$ df = 2 n = 649 p < .001
Used tranquilizer not prescribed	25.6%	41.3%	50.0%	$X^2 = 31.83$ df = 2 n = 649 p < .001
Used barbiturates not prescribed	13.4%	30.0%	35.3%	$X^2 = 34.56$ df = 2 n = 648 p < .001
Used codeine not prescribed	23.5%	38.8%	39.2%	$X^2 = 17.60$ df = 2 n = 649 p < .001
Used narcotics	10.9%	21.9%	25.5%	$X^2 = 19.01$ df = 2 n = 649 p < .001
Used cocaine	44.9%	66.3%	66.5%	$X^2 = 30.04$ df = 2 n = 648 p < .001
Used inhalants	26.7%	36.3%	30.4%	$X^2 = 4.47$ df = 2 n =649 p = .107
Used angel dust	17.3%	20.8%	30.2%	$X^2 = 11.71$ df = 2 n = 645 p = .003

In addition, Table 2 shows many significant differences in drug use by the number of drinking-driving offenses. The persistent drinking drivers show a strong trend for higher levels of drug use than the first offenders. The trend tends to show a linear increase with the number of offenses. There are a few exceptions to this trend (e.g., amphetamines, marijuana, inhalants) in which the highest rates are found in the second offender category. Current drug use (not shown in a table) is substantially lower and shows fewer significant differences; however, many in the sample (especially the second and multiple offenders) are currently participating in substance abuse treatment, which is likely to influence drug consumption patterns.

Table 3 shows treatment history and stages of change for the drinking-drivers. A history of alcohol treatment is the norm for persistent drinking drivers. Over two-thirds of the second offenders and almost 90% of the multiple offenders have been in alcohol treatment. There is a linear trend with number of drinking-driving offenses for total number of treatments, ever attended Alcoholic's Anonymous (AA), current attendance at AA, prior inpatient, and outpatient treatment. Although the first offenders had lower rates of treatment, there was a substantial proportion (almost one-third) with a treatment history.

The effect of treatment may be questionable because so many of the persistent drinking drivers have substantial treatment histories. One of the key issues in treatment success is the stage of change, or readiness for change, of those in treatment. All three groups scored highly on the precontemplation stage (see Table 3). This result indicates that many offenders have not as yet developed the motivation to change their behavior, regardless of the number of prior drinking-driving convictions. They have not begun to think about making changes or the need for any action regarding their behavior. The trend for contemplation, action, and maintenance show linear increases with the number of offenses. This suggests that more persistent drinking drivers, as compared to first offenders, are starting to recognize the need to alter their drinking-and-driving behaviors. Perceptions toward the need to change may be influenced by multiple treatment experiences and the accumulation of negative consequences associated with multiple offenses. These findings also highlight the necessity for treatment providers to address motivation toward change in the offenders.

Table 3. Comparison of First, Second, and Multiple DWI Offenders Across Alcohol Treatment Histories and Motivation for Change

	First (n=286)	Second (n=160)	Multiple (n=204)	Significance
Percent with prior alcohol treatment	31.8%	69.4%	87.7%	$X^2 = 163.66$ df = 2 n = 650 p < .001
Ever attended an AA meeting	34.7%	76.7%	91.1%	$X^2 = 178.27$ df = 2 n = 646 p < .001
Currently attend AA or similar program	19.8%	46.4%	64.2%	$X^2 = 80.66$ df = 2 n = 535 p < .001
Ever in inpatient treatment	9.6%	31.8%	41.1%	$X^2 = 66.96$ df = 2 n = 639 p < .001
Ever in halfway house program	0.7%	3.8%	9.0%	$X^2 = 20.36$ df = 2 n = 638 p = .001
Ever in outpatient program	29.0%	67.5%	84.2%	$X^2 = 157.88$ df = 2 n = 643 p < .001
Pre-contemplation (T-Score)	56.40[a] (10.06)	51.32[b] (9.31)	48.46[c] (9.23)	F(2,647)=42.63 p < .001
Contemplation (T-Score)	29.41[a] (17.86)	38.84[b] (15.45)	45.56[c] (17.05)	F(2,647)=66.02 p < .001
Action (T-Score)	38.58[a] (13.64)	45.53[b] (10.93)	50.86[c] (8.82)	F(2,647)=67.59 p < .001
Maintenance (T-Score)	39.42[a] (10.61)	45.63[b] (10.06)	49.22[c] (8.53)	F(2,647)=61.49 p < .001

Note: different superscripts indicate Scheffe p <.05 differences between groups.

Results for driving-related measures and trauma history are shown in Table 4. For driving characteristics, linear trends for number of lifetime tickets and number of lifetime crashes were apparent. The multiple offenders had the highest number of violations and crashes while the first offenders had the least. The repeat offenders appeared to exhibit the most driving inhibition. Differences between repeat offenders (second and multiple) and first offenders were found for the tension reduction driving scale. Further, the crashes and violations suggest that the repeat offenders are bad drivers in addition to their propensity to drink-and-drive, which compounds safety problems.

Table 4 also shows that traumatic injuries are common among drinking-drivers. The trends show that first offenders were less likely than the persistent drinking drivers to have had emergency treatment, broken bones, or have been an assault victim. Although the rates for traumas in the first offenders are lower, the absolute value is quite high even among this group. These results, which are consistent with Cherpital's (1995, 1997) association between emergency room visits and problem drinking, suggest that emergency treatment settings were commonly visited by all drinking-drivers, and that multiple adult-age traumas are the norm for persistent drinking drivers.

Table 4. Comparison of First, Second, and Multiple DWI Offenders Across Driving and Trauma Measures

	First (n=286)	Second (n=160)	Multiple (n=204)	Test of Sig.
Number of tickets for violations (lifetime)	4.30^{a} (6.23)	5.92^{a} (7.06)	9.91^{b} (12.50)	$F(2,647)=24.28$ $p < .001$
Number of accidents (ever)	0.59^{a} (1.13)	1.22^{b} (1.74)	1.86^{c} (3.30)	$F(2,646)=20.48$ $p < .001$
Driving Inhibition Score	1.02^{a} (1.14)	1.32^{b} (1.15)	1.15^{ab} (1.14)	$F(2,647)=3.48$ $p = .031$
Tension Reduction Score	2.83^{a} (1.78)	3.27^{ab} (1.80)	3.29^{b} (1.84)	$F(2,647)=5.00$ $p = .007$
Needed emergency medical treatment	51.7%	72.5.%	76.5%	$X^2 = 37.60$ df = 2 n = 650 $p < .001$
Number of times needed emergency treatment	1.39^{a} (3.03)	2.78^{b} (6.23)	2.43^{b} (3.37)	$F(2,646)=6.99$ $p = .001$
Hurt badly enough to leave scar	47.6%	56.9%	57.4%	$X^2 = 5.92$ df = 2 n = 650 $p = .052$
Had any broken bones since age 18	32.2%	49.4%	53.4%	$X^2 = 25.39$ df = 2 n = 650 $p < .001$
Victim of Violent Assault	44.0%	58.9%	66.2%	$X^2 = 25.03$ df = 2 n = 646 $p < .001$

Note: different superscripts indicate Scheffe $p < .05$ differences between groups.

There were a large number of significant differences between the three groups on psychological and psychiatric measures (not shown in tables). There was a linear trend for the CPI socialization scale, reflecting poorest socialization for the multiple offenders. Results for the sensation seeking scales showed curvilinear relationships, with the second offenders scoring highest and the first and multiple offenders scoring similar to one another. For the locus of control scales, only the chance subscale showed significant differences, reflecting a linear trend with the multiple offenders scoring the highest. There was also a linear trend for self-esteem and social desirability, with the multiple offenders scoring lowest. For the attitudes towards DWI, there was a threshold effect, with the first offenders differing from the second and multiple offenders.

In almost every instance, the repeat offenders (second and multiple) were significantly higher than the first offenders on the SCL-90-R assessment of psychiatric symptoms. The hostility scale was the only one to not show significant differences. These results point to the utility of using such measures for identifying potential recidivists among the first offender group. In addition, the high level of psychiatric symptoms highlights the importance of taking these factors into account when developing an individualized treatment plan.

A multivariate multinomial regression analyses was conducted to better understand the transitions between first, second, and multiple drinking-driving offenders. The multinomial regression compared first versus second (repeat) offenders, and second versus multiple offenders across a broad selection of variables. The multinomial regression allows linear and threshold/plateau effects to be identified.

The multinomial regression analysis identified twelve measures that significantly ($p < .05$) discriminated between the first and second offenders. Of the significant and marginal differences, five were related to driving or drinking-driving behaviors and perceptions. Compared to the first offenders, second offenders were more likely to have refused the breath test, expressed more emotions regarding driving inhibitions, reported more crashes in their lifetime, drank at fewer locations, and indicated being able to drive safely after more drinks than the first offenders. The second offenders tended to take more health risks and score lower on the powerful others scale of locus of control than the first offenders. Finally, the second offenders were less involved with alcohol at the current time, more likely to be taking steps to change their drinking behavior, have sought treatment previously for their drinking problem, and be aware of the problem and receptive to changing their behavior.

The multinomial contrast between the second offenders and multiple (three or more) offenders also identified twelve significant ($p < .05$) differences. Second offenders tended to be younger than multiple offenders and had higher incomes than the multiple offenders. In contrast to multiple offenders, second offenders expressed more emotions regarding driving inhibitions, were less likely to refuse the breath test, less likely to have a family member who was arrested for DWI, and had fewer crashes and traffic violations. Relative to the multiple offenders, the second offenders scored lower on the powerful others and chance subscales of the locus of control measure and higher on the sensation seeking scale, the SCL-90R phobic anxiety subscale, and the additional items from the SCL-90R. The multiple offenders were also more likely than the second offenders to be correct about the alcohol content for beer and liquor. Finally, the multiple offenders scored higher than the second offenders on the action subscale of the stages of change and the AUI awareness subscale.

Discussion

There appear to be a number of factors that are capable of distinguishing between first, second, and multiple drinking-driving offenders. Some of these are indicators of social stability (education, income, unemployment, criminal history, traffic accidents and infractions). These signify that the persistent drinking driver is an individual with potentially a low skill level and with perhaps few resources available to improve on the situation. The implication is that interventions and sanctions should involve more than just a simple focus on the drinking or drug problems of these individuals. Other indicators are related to the period prior to the current arrest (heavy drinking and drinking-and-driving). In these variables we find that the individuals with three or more offenses were more likely to have drank more heavily and to have driven drunk more frequently in the period prior to the arrest. This result suggests that there may be a recognizable threshold level that could be used as a potential indicator of risk of becoming a persistent drinking driver. Other characteristics that appear to distinguish between first, second, and multiple offenders are related to drinking problems or drinking variables (alcohol problems and diagnoses, specific criteria).

The implication is that we have identified factors that could be used in some form to help recognize those individuals at higher risk of becoming a repeat offender. In addition, some of these factors might identify specific life function and psychological domains where intervention is needed. For instance, the social instability factors suggest that there may be a need for a global type of intervention, with a focus on building self-concept and efficacy. Another insight from these results is the need for treatment interventions to focus on individual characteristics. Part of this focus should address the motivation to change. Furthermore, the association of other psychological problems with heavy drinking indicates a need to include these issues in providing or planning for the treatment of drinking-drivers.

The multinomial regression results indicate that while some relationships are in fact linear (refusal, crashes), others are quadratic in nature (driving inhibition, sensation seeking, powerful others), and still others may have a threshold level. These findings suggest that using a strictly linear approach may miss measures that could potentially identify persistent drinking drivers. It is also encouraging to note that there were a number of indicators that discriminated between the second and multiple offenders, indicating that these two groups are distinct. This finding also suggests that it may be possible to intervene before a second offender becomes a multiple offender.

Conclusion

The initial hypotheses were generally supported by the results of the study. The multiple offenders had the greatest level of alcohol problems and anti-social tendencies. The level of general problem severity tended to increase with the number of drinking-driving offenses. Persistent drinking drivers clearly have substantially greater levels of almost all problems than do first offenders. However, not all variables followed a linear trend, with some showing a quadratic form with higher levels among the second offenders. These findings have major relevance for developing treatment interventions for persistent drinking drivers. In addition, the findings provide guidance for efforts to develop screening and assessment tools to identify potential second and multiple offenders at the time of the first drinking-driving arrest. This study is only a starting point, much additional research is necessary to develop and validate screening instruments to identify potential persistent drinking drivers. Furthermore, additional samples of persistent drinking drivers need to be studied in detail to confirm the current findings.

Acknowledgements

This chapter is a greatly expanded and updated version of a paper presented by the authors at the T-2000 International Conference on Alcohol, Drugs and Traffic Safety held in Stockholm, Sweden. The categorization of the drinking-driving offenders was updated since then by using additional official records. All of the results reported in this current chapter were from analyses conducted specifically for this publication.

We wish to thank Marla Fulton and Jeannine Dudziak for their editorial assistance. We are also grateful for grants to support this research from the Alcoholic Beverage Research Foundation and the National Institute on Alcohol Abuse and Alcoholism.

REFERENCES

Andreasen, N.C., Rice, J., Endicott, J., Reich, T., & Coryell, W. (1986). The family history approach to diagnosis. *Archives of General Psychiatry,* 43, 421-429.

Ahlin, E.M., Rauch, P.L., Zador, H.M., Baum, H.M., & Duncan, D. (2002). Social bonds as predictors of recidivism among multiple alcohol-related traffic offenders participating in an ignition interlock license restriction program in Maryland. In D. R. Mayhew & C. Dussault (Eds.), *Proceedings of the 16th international conference on alcohol, drugs and traffic safety- T2002* (pp. 177-184). Montreal, Quebec, Canada: Societe de l' assurance automobile du Quebec.

Armor, D. J., & Polich, J. M. (1982). Measurement of alcohol consumption. In C. M. Pattison & E. Kaufman (Eds.), *Encyclopedic Handbook of Alcoholism.* (pp. 72-79). New York: Gardner Press.

Bailey, J.P.M. (1993). Criminal and traffic histories, blood alcohol and accident characteristics of drivers in fatal road accidents in New Zealand. In H.D. Utzelmann, G. Berghaus, & G. Kroj (Eds.), *Alcohol, drugs, and traffic safety-T 1992* (pp. 883-846). Cologne, Germany: Verlag TUV Rheinland.

Bailey, J.P.M. (1995). Hard core offenders among drinking drivers in fatal accidents. *Alcohol, drugs, and traffic safety- T 1995* (pp. 605-609). Adelaide, Australia: NHMRC Road Accident Research Unit.

Cherpital, C.J. (1995). Alcohol and injury in the general population: data from two household samples. *Journal of Studies on Alcohol,* 56, 83-89.

Cherpital, C.J. (1997). Alcohol and injury: a comparison of three emergency room samples in two regions. *Journal of Studies on Alcohol,* 58, 323-331.

Derogatis, L.R. (1983). *SCL-90-R: Administration, scoring & procedures manual-II.* Towson, MD: Clinical Psychometric Research.

DiClemente C.C., & Prochaska, J.O (1982). Self-change and therapy change of smoking behavior: a comparison of processes of change in cessation and maintenance. *Addictive Behaviors,* 7, 133-142.

Donovan, D. (1980). Drinking behavior, personality factors and high-risk driving (Doctoral dissertation, University of Washington, 1980). *Dissertation Abstracts International,* 41, 4256.

Donovan, D.M., & Marlatt, G.A. (1982). Personality subtypes among driving-while-intoxicated offenders: relationship to drinking behavior and driving risk. *Journal of Consulting and Clinical Psychology,* 50, 242-249.

Donovan, D.M., Marlatt, G.A., & Salzberg, P.M. (1983). Drinking behavior, personality factors and high-risk driving. *Journal of Studies on Alcohol,* 44, 395-428.

Fell, J. C. (1993). Repeat DWI offenders: their involvement in fatal crashes. In H.D Utzelmann, G. Berghaus, & G. Kroj (Eds.), *Alcohol, drugs and traffic safety T-1992* (pp. 1044-1049). Cologne, Germany: Verlag TUV Rheinland.

Hays, R. D., Hayashi, T., & Stewart, A. L. (1989). A five-item measure of socially desirable response set. *Educational and Psychological Measurement,* 49, 629-636.

Hedlund, J. (1995). Who is the persistent drinking driver? Part I: USA. In B.M. Sweedler (Ed.), *Transportation research circular 437: strategies for dealing with the persistent drinking driver* (pp. 16-19). Washington, DC: Transportation Research Board.

Hingson, R., & Winter, M. (2003). Epidemiology and consequences of drinking and driving. *Alcohol Research and Health,* 27, 63-78.

Horn, J. L., Wanberg, K.W., & Foster, F.M. (1987). *Guide to the alcohol use inventory (AUI).* Minneapolis, MN: National Computer systems, Inc.

Jones, R.K., & Lacey, J.H. (2001). *Alcohol and highway safety 2001: a review of the state of knowledge.* Washington, DC: National Highway Traffic Safety Administration.

Levenson, H. (1981). Differentiating among internality, powerful others, and chance. In H.M. Lefcourt (Ed.), *Research and the locus of control construct* (pp.16-63). New York, NY: Academic Press.

Maguire, K. & Pastore, A.L. (Eds.). (2003). *Sourcebook of criminal justice statistics 2002. Washington D.C.:* U.S. Department of Justice, Bureau of Justice Statistics.

McConnaughy, E.A., Prochaska, J.O., & Velicer, W.F. (1983). Stages of change in psychotheraphy: Measurement and sample profile. *Psychotheraphy: Theory, Research and Practice,* 20, 368-375.

McMillen, D.L., Adams, M.S., Wells-Parker, E., Pang, M.G., & Anderson, B.J. (1992). Personality traits and behaviors of alcohol-impaired drivers: a comparison of first and multiple offenders. *Addictive Behaviors,* 17, 407-414.

Megargee, E.I. (1972). *The California psychological inventory handbook.* London: Jossey Bass.

National Highway Traffic Safety Administration (NHTSA). (2003). *National center for statistics and analysis. 2002 annual assessment of motor vehicle crashes based on the fatality analysis reporting system, the national accident sampling system and the general estimates system.* Washington, D.C.: U.S. Department of Transportation.

Nochajski, T.H., & Wieczorek, W. F. (1998). Identifying potential drinking-driver recidivists: do non-obvious indicators help? *Journal of Prevention and Intervention in the Community,* 17, 69-84.

Perneger, T.V. (1998). What is wrong with Bonferroni adjustments. *British Medical Journal,* 136, 1236-1238.

Rauch, W.J., Zador, P.L., Ahlin, H.M., Baum, H.M., Duncan, D., Raleigh, R., Joyce, J., & Gretsinger, N. (2002). Any first alcohol-impaired driving event is a significant and substantial predictor of future recidivism. In D. R. Mayhew & C. Dussault (Eds.), *Proceedings of the 16th international conference on alcohol, drugs and traffic safety- T 2002,* (pp. 161-168). Montreal, Quebec, Canada: Societe de l' assurance automobile du Quebec.

Robins, L., Helzer, J., Cottler, L., & Goldring, E. (1989). *NIMH Diagnostic Schedule: Version III Revised (DIS-III-R).* St. Louis, Missouri: Washington University.

Rosenberg, M. (1965). *Society and the adolescent self-image.* Princeton, NJ: Princeton University Press.

Royal, D. (2000). *Racial and ethnic group comparisons: national surveys of drinking and driving. Attitudes and behavior: 1993, 1995, and 1997. Volume 1: Findings; Volume 2: Methods.* Washington, D.C.: U.S. Department of Traffic Safety, National Highway Traffic Safety Administration.

Siegel, H.A., Falck, R.S., Carlson. R.G., Rapp, R.C., Wang, J., & Cole, P.A. (2000). The hardcore drunk driving offender. In H. Laurell & F. Schlyter (Eds.), *Proceedings of the 15th international conference on alcohol, drugs, and traffic safety-T 2000.* Retrieved May 19, 2004, from *http://www.vv.se/traf_sak/t2000/ 128.pdf*

Simon, S.M. (1992). Incapacitation alternative for repeat DWI offenders. *Alcohol, Drugs and Driving,* 8, 51-60.

Simpson, H.M. (1995). Who is the persistent drinking driver? Part II: Canada and elsewhere. In B.M. Sweedler (Ed.), *Transportation research circular* 437: *strategies for dealing with the persistent drinking driver (*pp. 21-25). Washington D.C.: Transportation Research Board.

Simpson, H.M., & Mayhew, D.R. (1991). *The hard core drinker driver.* Ottawa, Canada: Traffic Injury Research Foundation of Canada.

Simpson, H.M., Mayhew, D.R., & Bierness, D.J. (1996). *Dealing with the hard core drinking driver.* Ottawa, Canada: Traffic Injury Research Foundation of Canada.

Substance Abuse and Mental Health Services Administration. (2003). *National Survey on Drug Use and Health (NSDUH).* Washoington, D.C.: Department of Health and Human Services.

Sweedler, B.M. (Ed.). (1995). *Transportation research circular 437: strategies for dealing with the persistent drinking driver.* Washington, D.C.: Transportation Research Board.

Vingilis, E. (2000). Driver characteristics: what have we learnt and what do we still need to know? In H. Laurell & F. Schlyter (Eds.), *Proceedings of the 15th international conference on alcohol, drugs and traffic safety- T 2000.* Retrieved May 19, 2004, from *http://www.vv.se/traf_sak/t2000/PLENARY6.pdf*

Voas, R.B., Wells, J., Lestina, D., Williams, A., & Greene, M. (1998). Drinking and driving in the United States: the 1996 national roadside survey. *Accident Analysis and Prevention,* 30, 267-275.

Welte, J.W., Zhang, L., & Wiecorek, W.F. (2001). The effects of substance use on specific types of criminal offending in young men. *Journal of Research in Crime and Delinquency,* 38, 416-438.

Wieczorek, W.F., Callahan, C. P., & Nochajski, T.H. (2000). An empirical typology of persistent drinking drivers. In H. Laurell & F. Schlyter (Eds.), *Proceedings of the 15th international conference on alcohol, drugs, and traffic safety-T 2000.* Retrieved May 19, 2004, from *http://www.vv.se/traf_sak/t2000/132.pdf*

Zuckerman, M. (1979). *Sensation seeking: beyond the optimal level of arousal.* New York: Lawrence Erlbaum.

In: Contemporary Issues in Road User Behavior and Traffic Safety ISBN:1-59454-268-6
Editors: D. Hennessy and D. Wiesenthal, pp. 167-182

Chapter 14

PERSONAL DRINKING AND DRIVING INTERVENTION: A GRITTY PERFORMANCE

J. Peter Rothe
Alberta Centre for Injury Control and Research
Public Health Sciences, Faculty of Medicine and Dentistry, University of Alberta

INTRODUCTION

There is a chorus of voices that resonate around the theme of drinking and driving. Few traffic-safety-related issues are more thoroughly studied, and with good reason. According to Transport Canada's report, 'The Alcohol-Crash Problem in Canada: 2000," 981 people died in alcohol-related crashes in 2000. This includes off-road vehicles as well as pedestrians with alcohol in their blood. The Insurance Institute for Highway Safety recently reported statistics in the United States for the same year. Forty-one percent of traffic fatalities involved drinking drivers (17,732 deaths to be exact). There has been little appreciable change over the past few years. Yet interventions at many different levels of life are touted as candidate for reducing carnage and destruction. Hence a closer look at the intricacies of intervention is warranted to reveal their complexities, limitations and risks.

The most often espoused drinking and driving interventions are personal; people engaged in different one-on-one drinking situations. They need to be further explored to better appreciate their chance for success. Goffman's (1979) thesis on performance serves as the basis for that exploration. In short, interventions for Goffman are everyday dramas consisting of social performances that are based on individuals' beliefs in the impression of reality that they attempt to engender in those with whom they find themselves at different times. Drinking drivers and interveners project definitions of a situation by virtue of their responses to one another and by virtue of the actions they take with respect to one another. Rationalizations and justifications are used to account for their activities, which in turn foster reactions, grist for further analysis.

INSTITUTIONAL INTERVENTION

On the agenda is the question: what is being done about drinking and driving? Different organizations, agencies, government departments, corporations and faculties have developed interventions to manage drinking and driving, many of which do little more than serve the self-interests of the sponsoring groups. Institutional determinism serves as their guiding light. The interventions are institutionalized in the principles of rules, interests, and meanings that underscore politics (e.g., government administration), law and order, business procedure, medicine, education and the media. For example, from a local political perspective, interventions may reflect zoning and municipal regulations, alcohol sales outlets hours of service, and retail licensing. Research evidence suggests that the greater the number of alcohol outlets per resident in a community, the greater the rate of drinking driving (Watts & Rabow, 1983). Increased sales of beer, and to a lesser extent, spirits and wine have been positively associated with crashes (Gruenewald & Ponicki, 1995).

Law and order intervention may focus on strategies of control and enforcement like the implementation of sobriety checkpoints, level of punishment for those who are caught drinking and driving and other enforcement/deterrence strategies. Stuster and Blower (1995) and Lacey, Jones and Fell (1997), established that sobriety checkpoints help reduce jurisdictional alcohol-related traffic crashes. Publicity about sobriety checkpoints also appears to heighten their effectiveness.

An example of a business-based intervention, is the design and implementation of drinking establishment server programs to change the behavior of those selling and serving alcohol. The key objective for responsible beverage server programs is to prevent intoxicated patrons from driving. The intent is not necessarily to reduce the volume of alcohol a patron consumes, but to discourage the drunk customer from driving. In some cases they have had success. Holder and Wagenaar (1994) showed an 11 per cent decrease of drinking and driving charges after the first year of Oregon's state-mandated server-training program.

High schools develop safe graduation programs to help reduce drinking and driving in conjunction with graduation ceremonies. Parents and community agencies collaborate with students to provide safe alternative transportation. Furthermore, some school systems sponsor alcohol education programs that help youths gain awareness of: the dangers of alcohol, societal values related to alcohol consumption by minors; lifestyle influence of alcohol advertising, physical, social and psychological effects of alcohol, alcohol and driving use and the enforcement of drinking driving laws.

Intervention also entails the use of mass media. A popular approach is to warn people of inevitable police enforcement, drinking and driving doomsday scenarios like death and maiming, social misplacement, and severe financial losses. Six basic media-based strategies are routinely practiced: (a) mass media campaigns (using social marketing and entertainment education), (b) embedded messages, (c) media advocacy, (d) media literacy, (e) small media (community newsletter), and more recently (f) Internet interventions. Organizations such as Mothers Against Drunk Driving (MADD) and Students Against Destructive Decisions (SADD) have actively used the media and public forums to convey their message. Both the direct and indirect effects media has on changing behavior must be considered (Yanovitzky and Bennett, 1999; Worden, Flynn, Solomon, & Secker-Walker, 1996). Direct effects refer to the impact of mass media messages on beliefs, intentions, attitudes, social norms, and efficacy, which in turn affect behavior. Indirect effects refer to the impact of mass media messages on mental frameworks, social institutions, interpersonal communication, and other processes that can affect targeted behavior. Capella (2001) reported that national media attention to problems such as drinking driving in the United States has affected state legislative activity in the sense that new legislation on drinking and driving was debated and later introduced in the government house.

Comprehensive community-based programs have been designed to increase the individual's perception of being arrested for drunk driving: publicize the health and safety features of drinking and driving, create a greater awareness of consequences resulting from drinking and driving, decrease the opportunities for drinking and driving, and motivate individuals not to drink and drive.

PERSONAL INTERVENTIONS

Whereas institutional involvement in drunk driving interventions dominates, significant emphasis has fallen on personal intervention. It is an approach punctuated by moments of fulfillment, doubt, and tension. Personal intervention is defined as a citizen trying to prevent a potential drinking driver from driving, to alter the behavior of a potential drinking driver, or redesign the drinking and driving circumstances to stop the person from driving. Common sense is often used as face-to-face interaction. It is performed whenever a drinking person is convinced not to drive, a friend refuses a ride with a drinking driver and chooses alternative modes of transportation, someone takes a drinking driver's keys, a friend gives or finds the drinking driver a ride or makes it easier for the latter to sleep it off, and whenever a designated driver is selected.

Personal intervention can succeed, but it also has a good chance of failing. Individuals engage their friends, colleagues or family members, exemplifying the popular slogans "Friends don't let friends drink and drive," or "if you drink, let others drive." While personal interventions appear simple and straightforward, their interactive scenarios are complex encounters. Individual interventions occur as a result of any number of factors including strong personal beliefs, relationships, culture, self-image, propensity to take risks, mood, self-interest, and personal power/autonomy. Because of the complex nature of personal interventions, their outcomes are unpredictable, making consistent measurement of impact difficult (Aspler, Harding, & Goldfein, 1987; Dejong & Wallack, 1992; Stewart, 1992). A citizen connects with someone at-risk and in the process, mobilizes a great deal of energy, tact and poise to achieve a goal of safety. In a way, the engagement has a far-reaching impact upon the common good. By stopping one person from drinking and driving, the intervener potentially saves other road users from harm. In this way interveners embrace fiduciary responsibility, helping another person. They disavow themselves of the popular life slogan, "Every person for him or herself" (Rothe & Elgert, 2003).

Personal intervention is a dominant highly socially marketed collection of countermeasures. It is an interlocking series of actions people take to deter an alcohol consumer from becoming a drinking driver. Personal interventions reflect a ray of hope (Stotland, 1969), but may yield a dark shadow. Personal drinking and driving interventions are often on the edge – turbulent and willful. In the text that follows they are featured as highly differentiated forms of social interactions featuring successes and failures in accordance to principles of everyday life. The description draws on findings from *Precedent Rules: An Inquiry into Drinking and Driving Among Alberta's Youth* (Rothe & Elgert, 2003), a study examining ways in which people live, and reconstruct personal interventions as live experience.

Querying the Drunk

A typical rule exists. People do not blindly and unconsciously accept rides with friends who have been drinking. They engage in sympathetic introspection, analyzing where the potential drinking driver's thinking is at, trying to understand his or her meanings and motives. They try to understand the sensual, ideational and practical modalities, to help them better comprehend the situation at hand.

When an exchange about driving occurs between friends who are drinking alcohol, the potential driver maintains face by describing himself or herself as fit to drive, a performance designed to "open up closure", or to stop the discussion (Goffman, 1959). Often the query ends. The intervener accepts the drinking driver's response or verbal performance and foregoes any further action. Based on recent research by Rothe and Elgert (2003), a chart could readily be developed to script typical responses provided by drinking drivers. Often used are appeals to superior driving abilities, having the natural characteristic to limit the effect of alcohol on those innate driving abilities, having ingested a limited quantity of alcohol and/or possessing phenomenal insight to recognize limitations. In spite of some skepticism, interveners often accept the overall impression as compatible and consistent with that offered by drinking drivers. Although there may be some discordance, interveners may overlook it and acquiesce from further questioning. It is recursive! A drinking driver wants to drive. To get his friends to approve, he rationalizes that he is able to drive, regardless of his state of mind at the time. His friends listen to his rationalizations. They may agree! Because his friends agree the drinking driver now believes that he really is able to drive. His friends' agreement justifies his original predisposition. In the Rothe and Elgert (2003) study, a young man knowingly accepted a ride from a drinking driver who self-assessed that he was not drunk, conveying the message that he did not have much to drink. He indirectly empowered the potential drinking driver to continue the high-risk game of drinking and driving:

> "I think last week end or the week end before or some weekend a bunch of us got in a car with a guy that I kind of thought he was drunk and I asked him if he was drunk and he said no, so I didn't push it. But I kind of think he was drunk... . So, yeah I think he was. But he didn't have a lot to drink" (Rothe & Elgert, 2003).

Although interveners may suspect a potential drinking driver as misrepresenting his or her impression, they accept the performance cues as sufficient to stop further questioning. The intervener halts the inquiry and concedes to the impression offered by the drinking driver, blinding the former to the weakness or dissonance in the performance. Interveners may even accept a ride with the drinking driver. Witness a young woman speaking about her experiences when she questioned a potential drinking driver about his state of mind:

> "I would ask first of all if he felt that he was okay to drive because usually you get different responses from people. But, I don't know. If I find that they're overly drunk and I know they can't drive, I will say: "you shouldn't drive," you know "let me take your keys," like let's just stay here... . They have their half-wits about them still, you know there's not much you can say or do to, you know, take their keys or take anything away from them, you know. They're still gonna drive. So, there is only so much you can say to a person and have them actually listen. So I let them be..." (Rothe & Elgert, 2003).

Being questioned about sobriety, state of being, driving skills, and rationality can be uncomfortable, even disturbing. But as the conversation continues, there is a shift to conciliation. The research participant's concern transforms into a concession.

The want-to-be intervener may recognize the drinking driver's fallibility and seek to re-establish order and stability, a sense or credibility. The "you" and "me" become a "we". The intervener and drinking driver reach a fragile, if not tenuous agreement whereby the drinking driver will drive his vehicle with the intervener riding along. The path brinks on soft determinism, whereby persons with different perspectives join loosely together in the process of mutual doubt yet support (Matza, 1969).

Others step out. Interveners are convinced that the drinking driver lacks the wherewithal to drive safely. They refuse to expose themselves to danger. A young man provided a short yet meaningful message:

> "Hum, if it continued I probably would not get into any sort of physical (3) tussle. I'd find an alternative way home" (Rothe & Elgert, 2003).

The logical extreme pointed out by the speaker motivates her to seek an alternative course of action. She would calculate the threat of violence and break away accordingly by seeking out transportation options.

Staying Out of a Drinking Driver's Vehicle

Personal intervention suggests emerging action. In some cases the solution to the problem lies not in continued discourse, but in the changing dynamic. The intervener refuses to ride with a suspected drinking driver. Mental anguishes, like suspicion, fear, disruption, caution or hopelessness enter into the drinking driver/intervener discussion. The drinking driver's excuses and justifications are discredited and a practical judgment is made – no ride! The intervener walks. But, walking may not be a routine, safe or practical solution. It depends on factors like the distance between drinking site and home, time of night and fear of safety, season, weather conditions, and the ethos of the neighborhood.

The refusal-to-ride-with-a-drinking-driver rule can be invoked any number of times in the intervention cycle. Whereas the intervener may have originally decided to be conciliatory and drive with the drinking driver, she can change her mind midway. Turbulent events happen on the trip. The passenger feels at risk. She gets out of the car midstream and walks to her destination. Distance is no longer a factor. Survival is on her mind. A young woman from Edmonton captured the kinesis of walking without boundaries:

"It was bad! I got out of the vehicle with him when he was drunk. And I'm like: "You're drunk. I don't want to be driving with you." After he ran like three lights. I'm like: "Pull the fucking car over", I'm like: "I'm getting out of here", I yelled at him… and then I got out of the truck and I walked back to Edmonton. It's like what can I do you know… . I thought he was okay" (Rothe & Elgert, 2003).

Whereas the speaker above approved the driver's earlier improprieties, his actual driving performance shocked and startled her as the passenger and forced her to drastically change her course of action. When the driver ran three red lights she decided that the risks were unmanageable. She was fortunate to gain her leave.

The Potential Drinking Driver and Intervener

Personal interventions are unpredictable, essentially unstable, fragile and often ambiguous. To establish a sense of order and predictability, to enhance the chance of changing a potential drinking driver's mind about driving, the intervener is best armed with alternatives. And sometimes they don't work. One option is to help find the potential drinking driver a ride, whether that be with the intervener, someone else on location, a parent, friend or mate outside of the drinking site or a professional driver like a cabby or representative from a designated driver service. The new driver is expected to be sober or to have consumed less liquor than the potential drinking driver. Consider the following three scenarios common to personal drinking driving interventions. The first one is a young woman who reflects upon the peace of mind from choosing to remain sober so she could drive:

> "I used to be the designated driver for them. Like: "Oh, she doesn't drink. We could ask her to drive us all over the place so, I used to always stay sober in the bar because I could drive them home and it was good for me because I knew they'd get home safe" (Rothe & Elgert, 2003).

Staying sober for the above speaker was an act of conscience, an expression of safety. The flowing scenario describes a situation when the intervening driver who had consumed alcohol takes a hard-nosed line on least-at-risk:

> "Well actually two years ago, I was with one of my ex-boyfriends in his hometown and we were at a bar and he had been drinking a lot more than I had. I had one drink and he wanted to drive home and I wouldn't let him. And he was kind of mad about it or whatever but I said like: 'Either I'm going to stay here like or I'm going to drive.' So he let me drive home" (Rothe & Elgert, 2003).

Certain facts are clear. The speaker, although having consumed alcohol, still judged herself to be a more capable driver than her ex-boyfriend who had consumed more alcohol. Although still at-risk, she did lower the risk of a drinking driver-caused collision. And she maintained expressive control of the situation. Third is a scenario where the intervening driver consumed significant numbers of drinks – enough to be impaired – a scene that pervades the drinking and driving landscape:

"I know the driver didn't drink all that much. Like he only had a couple still its like he was impaired but not that bad. He shouldn't have been for the big guy" (Rothe & Elgert, 2003).

The speaker allowed the new driver to drive, although he was impaired. The speaker drove home with him. The three scenarios form a continuum of intervention driving. At the far end of the spectrum is the new driver as sober, in the middle is the new driver with a few drinks, while at the other end of the spectrum is the new driver as impaired.

A fourth scenario can be introduced. It is chance determined. A drinking passenger and drinking driver leave in a car. A chaotic situation may arise whereby the passenger becomes unsure riding with the drinking driver and demands to drive. A sample quote taken from Rothe and Elgert (2003) shows how the drinking passenger turned driver met an unfortunate fate:

"Well, my friend she was drinking and driving and I was on the passenger side I was drinking with her... picked up a few friends we were all drinking and driving. I don't know, she started kind of fishtailing and so I noticed she was going too fast and I was like you know, you better slow down. Like she started driving all wild and so I told her to stop the vehicle. And I was like I'll drive, I'm fine to drive. So, I started driving that night and it turns out that I was the one to hit the ditch. Anyways, but who knows it could have been worse because I have seen her drive before. Like every time she drinks and drive something like gets knocked off her car like a window. A mirror, you know, a door sometimes. Pretty nasty drunk!"

A drunken passenger does not translate into a safe driver. It only replaces one problem with another; potentially a greater risk as the passenger never had intentions of driving in the first place and is freshly entering into an emotionally charged situation.

Often interventions that appear obvious are problematic. For example, someone at the bar can provide a potential drinking driver a lift home. But the option is often ruled out because of the potential drinking driver's concern about leaving his or her vehicle on site for the night. There is a worry that the vehicles may be stolen or vandalized. A young man captured the essence in his statement:

"There's no way I'm gonna leave my truck at the bar for the night. I worked too hard to have it broke into and vandalized. No way..." (Rothe & Elgert, 2003).

A similar case can be made about taking a taxi. At first glance it is simple, safe and expedient. But it is problematic. For many, cost is a major obstacle. With a bit of pre-planning, to round out the evening of revelry, partygoers chip in for the cab before they spend freely at the bars. A beneficial outcome is organized.

Taking the Keys and...

There is power in the ability to influence the actions of others. It can be realized in many ways. In personal interventions, power can be practiced through removal of a potential drinking driver's keys, an intrusive act that is qualitatively different from other personal intervention strategies discussed to this point. It is one of the most emotionally charged actions someone can take. For many, keys symbolize ownership, who they are through what they possess. Taking them can lead to "fight and defend" response. Although keys are not stolen per se, they are taken, forcibly or secretively, out of an owner's grasp for a short period of time. The keys may be hidden before drinking begins or any time during/after alcohol consumption. By taking a person's keys, the intervener risks an immediate or short-term conflict, but succeeds in the long term. A drinking driver is not on the road! Witness several people who spoke about taking a friend's car keys:

> "We usually just steal the keys and hide them. They (potential drinking drivers) usually get pretty mad at us and the next morning they're just over it."
>
> "I take the keys right out of their hands. I wouldn't let them in the vehicle because it's not worth it for anyone involved or for anyone out there for that matter. I mean it's a very potentially dangerous situation that I would feel the need to prevent" (Rothe & Elgert, 2003).

Taking someone's car or truck keys without consent carries risks. It creates anger and defensiveness and may lead to severed friendships. Two young drivers (Rothe & Elgert, 2003) capture this sentiment in the following comments:

> "Friendships have been absolutely severed as a result of that type of stuff you know. People are getting very, very angry. They get so defensive about it right. They don't want to be called assholes by their so-called friends. And yet the individual who asked for the keys oftentimes is not unhappy that he was asked for the keys. He thought that was the responsible thing to do."
>
> "The situation is that the person who's drunk is very hostile and they won't listen and they say: 'I'm not drunk, I can drive, I'm fine.' And so the person who's trying to intervene becomes the victim. Then the person you know they're going looking for their keys, to find their keys. So it's not a good situation."

To re-state a previous comment: keys are personal possessions that carry meaning beyond practicality and utility. They reflect personal possessions, a way of life. Keys are physical artifacts or symbolic representations of responsibility, and personal belonging. They are meaningful creations of people's behavior. Taking them can and does create turmoil. The following account underscores the sense of loss:

> "The keys are your life, you know, my keys are my life. It's not just the key to my vehicle, the keys to my house, I supposed I could give them my car keys, but then when I come back to get it in the morning, I don't have it and where is he? You lose control of your life in a way .... We start out with one. We start out with the latch key that mom gave us you know, and then we get the key to our car and then we get the key to our office and it multiplies. And you don't want to let go of those. Keys mean access, independence and freedom, and rights too. Once you have those, you have access to my stuff" (Rothe & Elgert, 2003).

Picture the key-taking paradox. "A" takes the car keys from friend "B", to prevent "B" from hurting herself. "A" is acting responsible, possibly saving "B" and others from harm. "A" becomes the focus of "B's" anger, scorn and possible verbal/physical attack. "A" chances the risk of a ruptured or scarred friendship, or the guilt of feeling not enough was done to prevent injury or tragedy. Herein lies the unpredictability and turbulence of personal drinking and driving intervention.

The Designated Driver

The designated driver is perhaps the most often discussed personal intervention strategy in traffic safety. Drivers know of it, use it, and many come to depend on it. Anti-drinking and driving lobby groups propose it, beer companies promote it, and community agencies implement it. Parents of young drivers and young people themselves are instructed to believe, submit to and trust in this socially blessed and narrowly analyzed intervention strategy – the designated driver. In some respect, the designated driver is the choice of the decade – the panacea of risk-free driving, a mythology of traffic safety.

Designated Driver Selection

For many people, the decision to drink alcohol or the amount they consume is factored by the availability of a designated driver. Having a designated driver in place is like a security blanket that allows group members to drink and behave in carefree ways. Rules are in place.

A successful designated driver plan is contingent on the availability of a vehicle, the selection of a person who has a license, who is prepared to drive and who is willing to stay sober. The latter is especially relevant in the glaring pressure of bars and beer parties where drivers are challenged to remain drink free in a coercive drinking environment.

Assigning the role of designated driver is typically done by self-selection. In some cases, that decision is based on a turn-taking routine in which group members are expected to participate. Typically there is an order to the selection, as a young person in a small town expressed: "*Usually... everybody has to take a turn at it.*" Some groups are more fortunate than others because one or more of its members prefer not to drink or "*nurse a drink for the night.*" This makes them natural targets for designated drivers. Other groups have no volunteers, so someone is assigned. The strategy has limited success. Those assigned as designated drivers often do consume alcohol in order to maintain a consistent performance with their friends (Rothe & Elgert, 2003).

At first blush, the featured rule for becoming a designated driver is *thou shall not drink.* But it is not that simple! In everyday life, the blanket rule is negotiated or scaled down to a measured drinking rule, a more practical choice in life. Rather than zero tolerance, the designated driver is granted leeway - several drinks, proportioned over time. Two well-positioned perspectives illustrate the relevance of this theme:

> "Because it takes say one hour to get rid of one drink. If you don't have anything for like an hour and a half. Then its fine with me but once they start drinking like steady."
>
> "First thing is if I have, have a designated driver like I'd make sure they don't have drink or anything. If anything, like they have a drink maybe every hour and half (Rothe & Elgert, 2003).

The turn-taking selection process is most often an offering made by a person on behalf of the group. "*Someone is good about it and offers.*" And it does not have to be a special offering. It may be in total rhythm with what the candidate was planning to do for personal safety, consistency with personal beliefs or sense of responsibility.

Sometimes selecting a designated driver becomes an economic decision. Group members who can't afford to buy alcoholic drinks, but wish to be with friends nominate themselves to drive. Consider the two examples:

> "With our group there is always somebody poor enough that they couldn't drink or you know It was usually X or Y. They never had a lot of money. And X never really drank because he couldn't handle it. Like he always got sick really easy. And so he'd always be designated drive and Y just got pressured into it. We'd tell him: 'We need a designated driver.' 'Well I guess I could drive' (Rothe & Elgert, 2003).
>
> "Someone will do it because they can't// they don't have money. So, they just don't bother. They don't bother drinking because they can't afford it at the bar. We were taking turns because we found like say a person was doing all the time. Then we would offer to [drive], someone else would offer or we'd ask. Hum, if it's a birthday or something like obviously we like to take the person. I think its basically most on people who couldn't afford to drink would offer to drive" (Rothe & Elgert, 2003).

Because some persons lack the finances to buy alcohol, there is less pressure on them to breach the sobriety rule. And in many cases there is no stigma attached to being temporarily broke.

Several friends meet for a couple of drinks and nachos after work. A few more friends join and more beer is consumed as the hours pass by. More and more pressure is placed on the designated driver to participate. Whereas originally the designated driver follows a *measured or no drinking* rule pressure mounts to have a few more drinks. And he or she may do that!

Choosing a designated driver while partying can be perilous. According to informants, the process is effective, or as one participant offered, "*It just seems to work out that way.*" (Rothe & Elgert, 2003). Most often the group member who is the "least drunk" or s/he who is most in control, despite the number of drinks consumed becomes the designated driver.

Designated Driver Problems

Success is achieved if the designated driving intervention worked, that is the designated person was successfully chosen, did not drink alcohol, and brought the group home problem free. But, major problems arise, placing the success of designated driver intervention in doubt. Consider the following examples in which two persons in their late twenties talk about their exposure to a designated driver:

> "Before we used to go out drinking all the time and I mean we would designate a driver but it would cause fights. That would be the big problem. It would cause, like well, that person wouldn't want to drive sober you know. They would want to party with us you know. And so they'd feel left out being a designated driver. And so hum that was a lot of the reason

> that contributed to us not going out and drinking as much because it was always a problem. There was always an issue about who's gonna drive" (Rothe & Elgert, 2003).
>
> "Well, actually that's what happened to us one time. Our DD (designated driver) ditched us. And it was me and (my friend). It was like I don't care and she was going to get in the car and drive. I mean that's pretty psycho, because we were both totally drunk that night. Like really, we don't even remember leaving the bar, but we're lucky her cousin stopped us and noticed that we were just drunk, trying to drive away. So he hopped in and drove for us" (Rothe & Elgert, 2003).

The designated driver may become intoxicated or drink more than is considered to be a safe amount, leaving friends without warning or options. This sentiment was detailed further in interviews with police officers and community counselors:

> "I see that a lot of times when people are going to be driving I notice they do try to not drink as much. I've seen that. But they do drink. And eventually it gets worse. But I don't think they go to extremes like the others – get really drunk. They do try to control it a bit" (Rothe & Elgert, 2003).
>
> "I think sometimes they plan to have a designated driver but the designated driver gets drunk too" (Rothe & Elgert, 2003).
>
> "In some cases it's a situation where there's an apparent loss of control, loss of decision making because of the alcohol. In some cases it could be a situation where I went out with my friends, we agreed that so and so was going to be a designated driver, they all got drinking, they drank more than me. I had no choice but to drive" (Rothe & Elgert, 2003).

Discussion on designated driver practices include the unforeseen, the unpredictable. A young woman volunteered to be a designated driver and experienced major problems because there were no insurance papers for the person's car she was driving. She was engaging in safe behavior, but the car she was driving had no insurance. The police stopped her. She spent five days in jail and was fined $2,800 (Rothe & Elgert, 2003).

A Designated Driver Moral Dilemma

There is a conflict of meaning that impinges on the moral side one takes. For example, it is illegal for teenagers under 18 to drink alcohol. Zero tolerance is the logical reality. Teenagers grow up hearing that, "friends don't let friends drive drunk," suggesting that drinking alcohol under the age of 18 is transformed from illegality to begrudging acceptance. Moral significance has shifted. It is an issue that triggers debate in public schools. The superintendent of schools for Portland, Oregon, argued that the evil feature of designated driver is when students, despite their age, make plans ahead of time to drive drinkers to or from school events. From the perspective of the school official, pre-planned drinking will take place; an illegal and immoral act. But, according to the educator, the designated driver rule is worthwhile when friends give rides to students who are drinking at a school function.

Knight, Glascoff, and Rikard (1993) found that what constitutes as a designated driver differs from person to person. They concluded that although many participants defined a designated driver as one who abstained from alcohol, in reality, they learnt that many designated drivers experience varying degrees of alcohol consumption. They were not sober.

Personal Intervention Risks

Personal drinking and driving interventions may be commonplace events, but they are not simple direct and straightforward. They are complex social acts that are embedded in a conglomerate of psychological, social, economic and legal designations.

Within the universal belief that something should be done about drinking drivers there are individual footnotes of doubt. Interventions include interveners and recipients, the latter of which may not share the goals of the former. Hence the opportunity for anger and conflict become played out through physical violence, verbal abuse or isolation.

Risk of Violence

Potential for confrontation can arise whenever personal interventions take place Emotions run high and episodes of anger and aggression are possible, even probable. In some cases the potential of violence serves as the foundation for how interveners pattern or style their personal intervention. The following descriptions show how personal interventions can have potential violent consequences:

> "I've just taken keys away. Like, we just stuck them somewhere so that they couldn't ah, get a hold of them or whatever. But other times I'm guilty because some of them fight or whatever and then I usually just end up letting them go and… you feel like a jerk" (Rothe & Elgert, 2003).
>
> "I had this happen once. I took a friend's keys, but then he started to fight and for all intents and purposes, let just say, I'm just a black belt, so I'm not allowed to fight people. Unless it's a life or death situation for myself. So when I fight a drunk, even with just a push, I could go to jail for a while, so when I take their keys I got to maybe fight…" (Rothe & Elgert, 2003).

When keys are taken the intervener feels it necessary to prepare for conflict. They appreciate that their actions can lead to risky outcomes. The threat of violence or actual occurrence of violence can occur during the course of encounters involving considerable emotion and image saving. As detailed in the quotations, encounters can be hazardous because of the possibility that the drinking driver's status is disconfirmed or damaged by the intervener's actions (Lyman & Scott, 1970).

Most of the personal interventions begin with talking – asking a friend about his state of sobriety and ability to drive. The drinking person offers a response – a presentation that all is well. When the counter-point does not sway the intervener, the drinking person may become angry elevating the query to indignation (intense anger awakened by anything unworthy, cruel or mean), rage (vehement expression of anger) and fury (excess of rage). A respondent in the Rothe and Elgert (2003) study explained:

> "An argument between two people that have been drinking is not going to be the most rational argument. People aren't always going to make sense, because as soon as someone decides to get angry or physically angry, then most people want to stop the argument and just appease that person. And end up saying: 'Oh, they're just drinking and driving. Let them go.' They can justify it by saying: 'Well, I'm not on the road, so he's not going to hit me. All my

family is at home so they are not going to hit them.' So, if something does happen then no one I know is going to get hurt. So, why push it" (Rothe & Elgert, 2003).

According to the speaker, an instinctual, natural way to respond to anger and aggression is to appease the angry person. It is an adaptive response to threats posed by the drinking person. The production of anger and aggression due to the influence of alcohol usually depends on the individual. Some persons are conciliatory and apathetic while others become angry:

"Depends on the individual, some people are very much 'like that's a good idea.' Some people, actually get violent at times" (Rothe & Elgert, 2003).

"From my experience, nah I don't care. As long as I get home somehow. But, yeah I'm a different kinda guy you know…" (Rothe & Elgert, 2003).

"Well, from my past experience, if the person you're dealing with, if it's explained to him in the proper way usually there isn't a problem. You will always get the person that alcohol fuels anger. And there is nothing you can do about that. You've heard the story of the happy drunk. Right? And you've heard the story of: 'Gees, don't fool around with him, when he's drinking, he'll kill you.' Right? Well, that's depends on who you're talking to" (Rothe& Elgert, 2003).

The third speaker touches on the theme that logic defeats anger, because anger becomes irrational. So, by explaining things to the drinking driver, she will get a better-balanced perspective. Sometimes it worked often it did not. The dynamics are broad and can include hostile words to actions (e.g, grabbing keys), which propel a potential drinking driver into physical violence. Some vignettes:

"Well, a friend of mine was at birthday party for two people. My friend, and his friend that he's known for quite a while, have the same birthday. We went and partied. We got a cab back to his friend's apartment, the boyfriend of the birthday girl wanted to drive home and he was very, very drunk. So my friend grabbed his keys and ran up to the apartment and it became a violent altercation after that as the person wanted his keys and wanted to drive home. But, we dissuaded him from that and we smartened him up. But it took some physical manhandling to do" (Rothe & Elgert, 2003).

"So we just took his keys and well, he started to fight like for his keys.

I had this happen once: took a friend's keys but then he started to fight… .

…And then he got mad at us. He's like, where the hell are my fucking keys blah, blah, blah. And so you we were just like okay you know this is exactly where they are. Go look for them fine, you know like just don't hit us" (Rothe & Elgert, 2003).

The fear of violence is a restrictive conception of what are normal or appropriate personal interventions. The potential of violence, and desire to drink and drive are intimately entwined.

Risks of Verbal Abuse

One of the main ways of expressing anger is through expression. Some call it verbal abuse. It can include the use of vulgarity, shouting threats and insults, and other degrading expressions. Being highly emotional and trying to "save face", the potential drinking driver may use profanity and yells at the intervener(s). The intervener may find it hard to respond in a supportive or agreeable way. She fights back! Hence a clash of head-on shouting and swearing! The verbal attacks may lead to physical abuse initiated by one or the other actors. As can be seen in the following narrative produced by a young man, the risk of verbal abuse can be a factor in personal drinking and driving strategies:

> "A friend of mine wanted to drive, but me and my other friend told him there's no way you're going to drive. So we put him in the back of the car and the designated driver drove him home instead. Because he started arguing and getting loud and insulting, we'd figured we better do it now (put him in the car) because if we wait it's going to get even worse. So we dealt with it right there and then" (Rothe & Elgert, 2003).

The speaker recognized the warning signs and took action before the verbal abuse had a chance to escalate. It was a calculated response to a highly charged situation. And it proved to be successful!

Risk of Social Rupture

Taken to the extreme acts of physical violence, verbal abuse and guilt trips can lead to severed relationships. Friends, lovers or colleagues break their bond or doubt the authenticity of their shared lives. Drinking and driving-related events become personal differences: aggressive assertions for which there appear to be no socially approved ways of dealing. According to a campus security officer:

> "Taking someone's keys, that's a tough one. It's really tough to take your friend's keys if he's been drinking and wants to drive. Yeah. You might save his life but you might ruin a friendship trying to take someone's keys…" (Rothe & Elgert, 2003).

A potential drinking driver may impose wounding criticisms or guilt trips on his friend that extends beyond the moment. Whereas the intervention originally carried a meaning related to the intervener and potential drinking driver, it now shifts to include a boyfriend or girlfriend or amongst friends and family member relationships. The event is re-framed, candidate for social disruption or character assassination.

Conclusion

Traffic safety professionals have borrowed the cachet of personal drinking and driving interventions and endorsed them without clarifying the *seen-but-unnoticed* undercurrents supporting them. Many of the difficulties faced in personal interventions remain unresolved. People have conflicting attitudes, commitments and priorities that influence their behavior in face-to-face interactions. They do not magically disappear when suggestions are made that persons take personal responsibility to act on other persons, or when a pronouncement is made for citizens to plan for personal interventions.

The concept of personal intervention covers a wide range of phenomena in ordinary discourse. While it underscores a transcendent theme – drinking and driving is unacceptable, improper, dangerous and illegal, it incorporates social compromises and situational factors to allow drinking and driving to continue. Personal interventions are more than first-level strategies designed solely for safety. They are mechanisms by which people interact on the basis of socially approved or disapproved performances, personal interests, and other social/personal risks intervening may bring. These factors need to be understood for more intense sense-making of personal intervention research, strategies, implementation, and evaluation.

Personal drinking and driving interventions are governed by potentially individualistic, volatile, irregular and, to some extent, unpredictable situations. They reflect competition, individualism, concern for private property, maintenance of defensible self-image and core values such as taking responsibility for self and others. By looking at personal interventions through the lens of social interaction, a picture is drawn that captures opportunities for making significant changes. People are in a better position to discuss risks and limits of personal drinking and driving interventions and offer alternative opportunities. It is within the ideological repertoire of realism and common sense that this text is offered.

REFERENCES

Aspler, R., Harding, W., & Goldfein, J. (1987). The review and assessmentof designated driver programs as an alcohol countermeasure approach:*DOT HS 807 108 final report.* Washington, DC: U.S. Department of Transportation.

Capella, J. (2001, August 9-10). How the Media Can Change Behavior. *2nd National Conference on Drug Abuse Prevention Research,* Washington D.C.

Dejong, W., & Wallack, L. (1992). The role of designated driver programs in the prevention of alcohol-impaired driving: a critical reassessment. *Health Education Quarterly, 19,* 429–442.

Goffman, I. (1959). The Presentation of Self in Everyday Life. New York: Anchor Press.

Gruenewald, P., & Ponicki, W. (1995). The relationship of retail availability and alcohol sales to alcohol related traffic crashes. *Accident Analysis and Prevention, 27(2),* 249-260.

Holder, H. D., & Wagenaar, A. C. (1994). Mandated server training and reduced alcohol-involved traffic crashes: a time series analysis of the Oregon experience. *Accident Analysis and Prevention, 1,* 89-97.

Insurance Institute for Highway Safety (PDF) *Status Report, 38*, No. 2, February 8th, 2003.

Knight, S. M., Glascoff, M. A., & Rikard, G. (1993). A view from behind the wheel: College students as designated drivers. *Health Values, 17,* 21 – 27.

Lacey, J., Jones, R., & Fell, J. (1997, September 1-6). *The Effectiveness of the Checkpoint Tennessee Program.* 40th Proceedings, 14th International conference on Alcohol, Drugs, and Traffic Safety.

Lyman, S. M., & Scott, M. B. (1970). *A sociology of the absurd.* New York: Appleton-Century-Crofts.

Matza, D. (1969). *Becoming deviant.* Chicago: University of Chicago Press.

Rothe, J. P., & Elgert, L. (2003). *Precedent rules: An inquiry into drinking and driving among Alberta's youth*. Edmonton, AB: Alberta Centre for Injury Control and Research.

Stewart, K. (1992). Designated drivers: The life of the party? *Prevention Pipeline, 5,* 4–5.

Stotland, E. (1969). *The psychology of hope*. San Francisco: Jossey-Bass.

Stuster, J. W., & Blowers, *P. A.* (1995). Experimental evaluation of sobriety checkpoint programs. Washington DC: National Highway Traffic Safety Administration.

Transport Canada (2000). The Alcohol Crash Problem in Canada.

Watts, R. K., & Rabow, J. (1983). Alcohol availability and alcohol related problems in 213 California cities. *Alcoholism: Clinical and Experimental Research, 7,* 45-58.

Worden, J. K., Flynn, B. S., Solomon, L. J., & Secker-Walker, R. H. (1996) Using mass media to prevent cigarette smoking among adolescent girls. *Health Education Quarterly, 23,* 453-468.

Yanovitzky, I., & Bennett, C. (1999). Media attention, institutional response, and health behavior change: The case of drunk driving, 1978-1996. *Communication Research, 26,* 429-453.

PART 5
TREATMENT/DRIVER CHARACTERISTICS

In: Contemporary Issues in Road User Behavior and Traffic Safety ISBN:1-59454-268-6
Editors: D. Hennessy and D. Wiesenthal, pp. 185-197

Chapter 15

THE EFFECTS OF MULTIPLE VARIABLE PROMPT MESSAGES ON STOPPING AND SIGNALLING BEHAVIORS IN MOTORISTS

David L. Wiesenthal
York University
Dwight A. Hennessy
State University of New York College at Buffalo

INTRODUCTION

Prompts are events that help initiate a response (Kazdin, 1994) and have their roots in the operant conditioning and behavior modification literatures. Preceding a response, prompts are intended to facilitate or encourage the desired response. In other words, prompts are instructions, gestures, signs or other cues that generate a specific behavior. A common example would be anti-littering messages on product wrappers or the health warnings on cigarette packets and bottles of wine or liquors intended to discourage or limit the consumption of these products. As Kazdin (1989) has indicated, the effectiveness of prompts may depend upon how clearly specified the desired behavior is, the proximity of the prompt to where the behavior may be performed, how convenient the requested behavior is, along with polite phrasing of the message and its noticeablity.

Prompts have been found to be a highly effective behavioral change tools in a wide variety of situations (Hopper & McCarl-Nielsen, 1991). For example, written signs, written handbills, and verbal reminders have been found to reduce littering in a variety of diverse environments such as a high school cafeteria (Houghton, 1993), university cafeteria (Craig & Leland, 1983; Dixon, Knott, Rowsell, & Sheldon, 1992), football stadium (Baltes & Hayward, 1976), campsite (Wagstaff & Wilson, 1988), and grocery store (Geller, Witmer, & Orenbaugh, 1976; Geller, Witmer, & Tuso, 1977). Prompting strategies have also proven useful in motivating an increase in recycling in an apartment building (Reid, Luyben, Rawers, & Bailey, 1976), university offices (Austin, Hatfield, Grindle, & Bailey, 1993), and grocery store (Geller, Farris, & Post, 1973); to increase consumption of both french fries (Martinko, White, & Hassell, 1989) and salads (Wagner & Winett, 1988) in national restaurant chains; to increase the prevalence of cervical examinations (Duer, 1982), breast self-exams (Solomon et. al., 1998), blood donations (Ferrari, Barone, Jason, & Rose, 1985), and physical exercise (Lombard, Lombard, & Winett, 1995) in health care settings; and to decrease lawn walking (Hayes & Cone, 1977; Smith & Bennett, 1992) in a number of neighbourhoods.

Factors Affecting Prompt Efficiency

One of the simplest means of increasing behavior change is to increase the rate or frequency of prompt presentations. Repeated prompts have been found to successfully increase return rates of questionnaires (Winett, Stewart, & Majors, 1978) and frequency of physical exercise (Lombard et al., 1995). Geller, Winett, and Everett (1982) found that the most successful prompt messages were polite, referred to a specific behavior which was convenient to emit, and were placed in close proximity to the area in which the behavior was to be emitted. Others have verified the success of these characteristics in reducing litter and increasing recycling (Craig & Leland, 1983; Dixon, et al., 1992; Reid, et al., 1976). However, research utilizing younger participants and target behaviors other than recycling and littering have yielded conflicting results. Younger cafeteria patrons were not influenced by the specificity or the politeness of prompt signs in decreasing littering behavior compared to older patrons (Durdan, Reeder, & Hecht, 1985). Similarly, politeness was not influential in reducing littering at a football game (Baltes & Hayward, 1976) or recycling in a university dormitory (Witmer & Geller, 1976). Finally, polite and specific prompts were ineffective in reducing litter in a grocery store when researchers placed a great deal of litter on the floor prior to introducing the prompt (Geller et al., 1977). When participants had seen that others had littered in the face of the prompt, it had little effect on their behavior.

Some have argued that the effects of prompt programs tend to be modest and typically dissipate when the prompts are removed (Smith & Bennett, 1992; Stern & Oskamp, 1987). In contrast, a number of researchers have found behavior change to be quite resilient following prompt removal (Dixon et al., 1992; Houghton, 1976; Thyer, Geller, Williams, & Purcell, 1987). Wiesenthal and Ford (1992) suggest a similar concern in that behavior change may not persist when prompts are used for prolonged periods. Prompts do have a clear effect on behavior, but most previous research has employed an operant design in which prompts are introduced and removed in cycles so as to continually maintain a level of novelty toward the prompt. These ABAB designs (where A is the time period assessing the baseline or operant level and B is the period where a reward or other stimulus is introduced) are useful for demonstrating that behavior is under the control of a reward or specific stimulus, but have the effect of reducing the likelihood of respondents habituating to the changed stimulus or reward as the introduction of these changes is usually for a brief duration with A and B frequently rotated. If one is going to base public policy decisions on research derived these operant designs, it is vital that the research provide assurance that the behaviour changes are enduring rather than transitory. Wiesenthal and Ford (1990) found that although a prompt sign was effective in increasing the number of complete stops at a stop sign, behavior change and the length of presentation of their prompt were negatively correlated.

Prompting Safe Driving Behaviors

Previous research has successfully applied prompt programs to increase safe driving behavior in the traffic environment. Specifically, verbal and written prompts have been found to increase seat belt use by up to 30% (Austin, Alvero, & Olson, 1998; Cox, Cox, & Cox, 2000; Geller, Bruff, & Nimmer, 1985; Gras, Cunill, Planes, Sullman, & Oliveras, 2003; Thyer et al., 1987). This is a significant increase in safety when it is considered that the leading cause of fatalities in the United States among those aged 5-34 is automobile crashes, and that seat belt use has been estimated to reduce injuries by 65% (Thyer et al., 1987). Simple prompt signs have also been used to increase the use of designated drivers in a college pub (Brigham, Meier, & Goodner, 1995). Given that approximately 40% of drivers in the United States will at some point be involved in an alcohol related traffic accident (Brigham et al., 1995), reducing the potential number of individuals who might drive while intoxicated can only serve to reduce such a hazard. Another driving habit that has been linked with accident involvement is speeding (Rothengatter, 1987). Van Houten and Nau (1983) found that speeding could be reduced on city streets by handing out written prompts containing positive feedback regarding safe driving speeds to individuals prior to driving.

Traffic signs are used to provide drivers with legal driving directions such as when and where to stop, the location of crosswalks, the legal speed limit, and appropriate lane usage. The purpose of such signs is to ensure maximum efficiency and minimum danger on the roads. Unfortunately, this information is often overlooked by drivers. Hughes and Cole (1986) had individuals report the sources of their attention while driving and found that over half the information was irrelevant to the driving situation, such as pedestrians or product advertisements. Naatanen and Summala (1976) interviewed drivers at a roadside stop and found that most were unable to identify traffic signs that they had just passed. Overlooking such vital traffic information could potentially lead to dangerous driving behaviors such as speeding or failing to stop at traffic lights, stop signs, or pedestrian crosswalks. Van Houten (1988) noted that 15% of all traffic deaths in Canada are the result of pedestrians being struck by motorists. Prompt signs have been used successfully to increase the number of vehicles that stop at pedestrian crosswalks, especially when used in combination with clearly marked stop lines on the road (Van Houten, 1988; Van Houten & Retting, 2001; Van Houten, Malenfant, & Rolider, 1985). Further, Van Houten and Malenfant (1992) reported the maintenance of these effects one year following the withdrawal of the prompt sign, although the stop lines may have served as a continued prompt to drivers.

Habituation and Decreased Prompt Effectiveness

Wiesenthal and Ford (1990) also successfully prompted drivers to make complete stops at a stop sign, but, the behavior change was negatively related to the duration of presentation of the prompt. Complete stops peaked during the initial introduction of the prompt, but decreased steadily over the next six weeks (although the effect was still greater at six weeks than during the baseline). The authors proposed that habituation to the prompt had occurred, which decreased the long term effects. Over time, the novelty of the prompt likely decreased, leading to decreased attention to the message. As a result, the prompt may simply have become just another traffic sign, which often go unnoticed. Most previous research has employed an operant design in which prompts are introduced and removed in cycles, which may actually maintain a level of novelty toward the prompt.

PREDICTIONS

If habituation to a single prompt occurs when the novelty of the message wanes, it may be possible to delay habituation by varying the prompt messages. Also, a greater number of target behaviors may be influenced in a single setting if the variable prompt messages refer to different behaviors. It was predicted that "complete stops" would increase and that "rolling stops" and "failed stops" would decrease when drivers were prompted to stop at a stop sign. Similarly, it was predicted that signalling of turns would increase when drivers were prompted to signal. Finally, by rotating the type of message displayed, it was predicted that the prompt effects would resist habituation and endure over the course of the prompting procedure.

METHOD

Subjects

A total of 23,690 vehicles were unobtrusively observed over a nine week period in suburban North York, Ontario, Canada: 7,645 during the baseline and 16,045 during the prompting periods. Although located on the York University campus, these roads are used extensively by North York residents, as well as university patrons, as they connect to major city streets. Baseline measurements were taken for three weeks in late May to early June and prompt condition measurements were taken for six weeks in early July to mid-August. During the baseline 6,348 vehicles travelled straight and 1,297 vehicles turned at the chosen intersection, while 12,942 and 3,103 vehicles travelled straight and turned, respectively, during the prompting period.

Materials

Three signs were professionally constructed using white corrugated plastic boards 0.5 cm thick, 58.2 cm wide, and 43.0 cm high. Messages were displayed in black block letters 5.1cm high and 7.6 cm wide, centred on the message boards. The three messages used were "*Make Complete Stops*", "*Signal Turns*", and a blank sign with no message.

A black painted wood sign frame was constructed in such a way as to allow the sign messages to slide in and out from the top. The top of the frame stood 152.4 cm from the ground and the bottom of the sign segment stood 101.6 cm from the ground. The sign opening was 58.4 cm wide and 43.2 cm high.

Procedure

For both the baseline and prompting periods, measurements were recorded Monday to Friday, from 0800 to 1000 hours and from 1600 to 1800 hours. During the baseline period, no sign frame or message signs were used. The observer was located approximately 30 m perpendicular from the proposed sign frame location, hidden from traffic by a set of trees. Over a six week period, vehicles were observed as they approached the target stop line. Each vehicle was ranked according to their level of stopping and whether or not they signalled prior to turning (when applicable). The level of stopping (or stop type) was defined as either a *complete stop*, in which the vehicle ceased all forward movement/momentum before proceeding through the intersection, a *rolling stop*, in which vehicles slowed down, but did not come to a complete stop, or a *failed stop*, in which vehicles did not slow down prior to proceeding through the intersection. The vehicles were further categorized based on the type of vehicle (heavy truck, light truck and vans, or cars) and their direction (turning or non turning) for more detailed analysis. Vehicles that were forced to stop by another vehicle stopping or turning in their path were excluded from the study as it would be impossible to determine if the sign had any influence on their behavior.

For the prompting period, the sign frame was anchored to the ground at a 65° angle from the road, and 30 m from the stop line. As with the baseline session, the investigator was located approximately 30 m perpendicular to the sign frame location. Each day, a single prompt sign was used in both the morning and afternoon. The signs were changed each day in an ABC order for the first three weeks and a CBA order for the next three weeks. Within a three week period, each sign was represented on each day of the week once. The same measurements and categorization techniques were used as in the baseline period.

RESULTS

Stop Type

Three separate ANOVA analyses were conducted using the three stop types as the dependent variables and using the four sign conditions (baseline, stop, signal, or blank message), two times of day (morning or afternoon), five days of the week (Monday-Friday), seven week numbers (baseline or prompt Weeks 1-6), three vehicle types (heavy truck, light truck, or car), and two vehicle directions (turn or non turn), as the levels of the independent variables. For each day, all data was aggregated into an average daily total of complete stops, failed stops, and rolling stops.

Complete Stops

Although not predicted, an interaction of sign condition X direction was found ($F(3,212 = 5.84$, $p<.05$). Non-turning vehicles displayed more complete stops for the "Stop" prompt sign than the baseline, the "Signal" prompt sign, or the "Blank" prompt sign (see Figure 1). However, both the "Signal" and "Blank" prompt signs also led to a greater number of stops than the baseline condition. For turning vehicles, the "Stop", "Signal", and "Blank" prompt signs, all generated more complete stops than the baseline condition, but there was no difference between the three prompt signs. Within each condition, fewer stops were committed by turning than non-turning vehicles. As predicted, the "Stop" prompt led to increased stopping behavior over the baseline. Surprisingly, the other two prompts had equally beneficial effects on the level of stopping for turning vehicles.

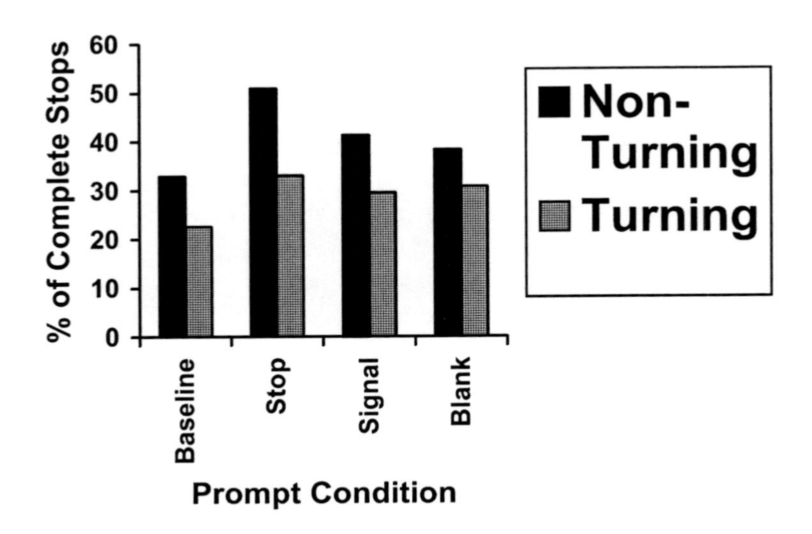

Figure 1. Proportion of Complete Stops as a Function of Prompt Condition X Vehicle Direction.

Table 1. Average Stopping Behavior Across Baseline and Experimental Week Conditions

	Baseline	Week 1	Week 2	Week 3	Week 4	Week 5	Week 6
Complete Stops	28.57	32.26	40.75	39.07	41.25	39.24	36.50
Rolling Stops	24.14	17.96	12.80	9.35	9.74	7.17	8.96
Non-Stops	47.27	49.76	46.44	51.57	48.99	53.58	54.53

Finally, as expected, a significant main effect for week number was found ($F(6,212) = 22.17$, $p<.05$) indicating that habituation was delayed as complete stops were greater in all six prompt sign weeks than in the baseline (see Table 1). Within the prompt condition, fewer stops were found in Week One than all other weeks except Week Six, but complete stops were lower in Week Six than Weeks Two and Four.

Failed Stops

As with complete stops, an interaction of sign condition X direction was found ($F(3,212) = 3.56$, $p<.05$). For non turning vehicles, fewer vehicles failed to stop during the prompt conditions of the "Stop", "Signal", and "Blank" signs than the baseline condition, but there were no differences between the three prompt types (see Figure 2). Similarly, for turning vehicles, fewer vehicles failed to stop during the prompt conditions of the "Stop", "Signal", and "Blank" signs than during the baseline condition. Within each condition, a greater number of turning vehicles failed to stop compared to non-turning vehicles. As predicted, the

"Stop" prompt led to fewer failed stops compared to the baseline condition, but unexpectedly, the "Stop" prompt was no more effective than the other two signs. A significant interaction between time of day and direction was also found ($F(1,212) = 3.74, p<.05$). In general, fewer turning vehicles failed to stop during the afternoon ($M = 17.26, n = 135, SD = 18.01$) than the morning ($M = 23.16, n = 130, SD = 19.58$) conditions. Finally, a significant main effect for week number was found ($F(6,212)=39.20, p<.05$). As predicted, habituation was not evident, since fewer failed stops were found in all six prompt sign weeks than in the baseline (see Table 1). Week One also represented greater failed stops than the other five prompt weeks.

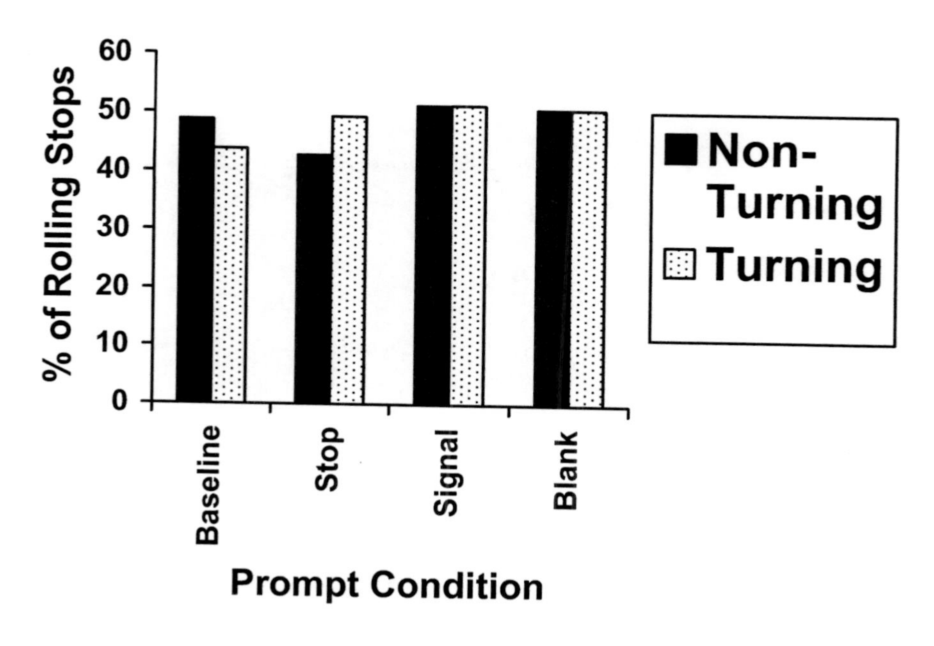

Figure 2. Proportion of Rolling Stops as a Function of Prompt Condition X Vehicle Direction.

Rolling Stops

A significant interaction of sign condition X direction was found ($F(3,212) = 10.03, p<.05$). Contrary to prediction, the "Stop" prompt had no effect on rolling stops compared with the baseline. For non turning vehicles, there was no difference between the number of rolling stops in the baseline compared to the "Stop", "Signal", or the "Blank" prompt conditions. However, the "Stop" prompt exhibited superiority in decreasing rolling stops over the other two prompt messages. Also contrary to expectations, rolling stops *increased* for the "Signal" and the "Blank" prompts compared with the baseline. No difference was found between the "Stop" prompt condition and any other condition. Turning vehicles exhibited greater rolling stops than non turning vehicles in only the baseline and "Stop" conditions. Finally, a significant main effect of week number was found ($F(6,212 = 4.03, p<.05$). Unexpectedly, rolling stops were greater only in Week Six than in the baseline (see Table 1). Within the prompt conditions, fewer rolling stops were found in Week Two than Weeks Five and Six.

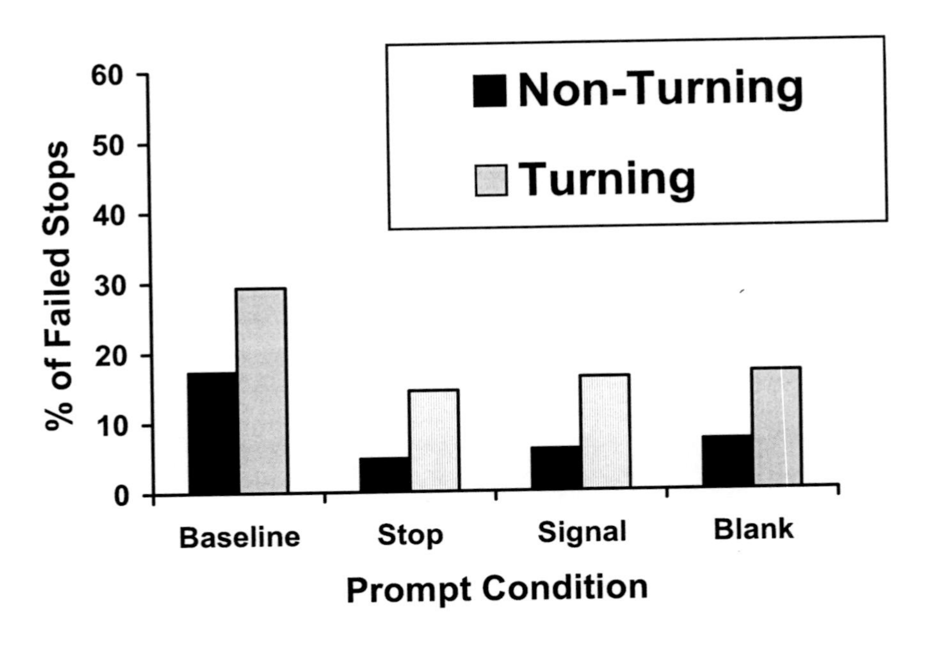

Figure 3. Proportion of Failed Stops as a Function of Prompt Condition X Vehicle Direction.

Signalling

A significant sign condition main effect was found ($F(3,100) = 18.99$, $p<.05$). As predicted, signalling was greater than the "Signal" ($M = 76.22$, $n = 60$, $SD = 17.36$) prompt than the baseline ($M = 54.68$, $n = 90$, $SD = 23.36$). Unexpectedly, the "Stop" ($M = 66.96$, $n = 60$, $SD = 20.65$), and "Blank" ($M = 64.94$, $n = 60$, $SD = 22.66$) prompts also resulted in greater signalling than the baseline, although the "Signal" prompt had a greater effect than the other two prompts. The main effect of week number was also found significant ($F(6,100) = 11.22$, $p<.05$). As predicted, habituation to the prompt message was not evident in that signalling was greater in all six prompt weeks ($M_{\text{Week 1}} = 62.00$, $M_{\text{Week 2}} = 63.19$, $M_{\text{Week 3}} = 66.13$, $M_{\text{Week 4}} = 70.39$, $M_{\text{Week 5}} = 70.36$, and $M_{\text{Week 6}} = 69.00$ respectively) compared to the baseline ($M = 54.68$). No differences in signalling were found between the six prompt weeks.

Discussion

The present findings supported the hypotheses that prompts would increase complete stops and decrease failed stops. This is consistent with previous research that prompts can influence stopping behaviors (Van Houten & Malenfant, 1992; Wiesenthal & Ford, 1990). Contrary to prediction, rolling stops were not influenced by the "Stop" prompt. A possible explanation could be that, if complete stops, roll stops, and failed stops are considered ordered levels of stopping behavior, some individuals may have simply changed a "level" of stopping when prompted rather than proceeding to completely stopping. In other words, some who would normally not stop would roll stop and some who would normally roll stop would completely stop. In such cases, there is still an improvement in stopping behaviour, but the roll stop group would not have appeared to change. This would seem the most logical explanation as complete stops increased and failed stops did decrease significantly with the inclusion of the "Stop" prompt.

Also as predicted, signalling increased when drivers were prompted. The present findings show that two separate behaviors can be influenced in a single program simply by varying more than one prompt message. More importantly was the finding that generalization occurred between the three prompt messages. The "Signal" and "Blank" prompts also had an impact on complete stops and failed stops, while the "Stop" and "Blank" signs also increased signalling behavior. This is consistent with Ludwig and Geller (1999) who found that pizza delivery drivers prompted to wear their seat belts also increased their frequency of turn signalling following the prompt implementation. This represents an important contribution to the applied use of prompt research in that a single prompt may potentially change multiple behaviors that have been previously linked within a single implementation program. If a single reminder can lead drivers to engage in multiple safe driving behaviors, overall traffic safety should increase exponentially.

Finally, the prediction that habituation could be delayed by using changeable messages was also confirmed. The behavior changes that were found did not vary significantly from the baseline over the six week prompt period. Previously, Wiesenthal and Ford (1990) found that a single prompt message was effective in increasing complete stops at a stop sign, but over time the effect decreased steadily from the first presentation of the prompt. They felt that habituation to the prompt led to its decreased effectiveness. By varying the prompts in the present study, a level of novelty may have prevented habituation from occurring. The overall effects appeared to be as strong following the sixth week as it did during the first week of presentation. The only effect that was not predicted was that turning vehicles displayed fewer complete stops and more failed stops than non-turning vehicles in both the baseline and prompt conditions. One possible explanation for this difference could be that drivers must naturally slow their vehicle to turn any corner and may not consider it necessary to slow down any further to ensure safe driving. As a result, drivers may develop "rules of thumb" that hold that stop signs apply mainly to non-turning vehicles.

The present research has very important implications in driving settings. Driving rules, such as making complete stops at stop signs and signalling turns, are created to maximize traffic safety for drivers, passenger, and pedestrians. Proceeding through a stop sign can cause conflicts and collisions with other vehicles or pedestrians. Signalling turns provides information to other drivers and pedestrians valuable to the preparation of appropriate responses on their part. Turning without signalling can lead to uncertainty and unpredictable behavior on the part of others, which can in turn lead to collisions. However, people often bend or cheat driving rules due to laziness, inattention, forgetfulness, or outright defiance. Any steps that can be implemented to increase traffic rule compliance can only serve to increase traffic safety. The variable prompt technique used in the present study increased rule compliance to *two* distinct driving behaviors, even in the absence of a specific prompt. The effects also showed relative consistency in two distinct driving settings, turning and non turning, which have inherently different "rules of thumb" or driving cultures regarding how much deviation from the law is deemed acceptable.

Electronic changeable message signs are presently used in some areas to provide traffic information such as distance and location of construction or lane closures. The use of this technology may be used to conveniently present variable safety prompts and, thus, improve traffic safety, especially if used in combination with other techniques such as feedback, personal appeals, education, and incentives as has been the case with single prompt techniques. The current findings lend confidence that frequently changing prompts on these signs reduces the likelihood of drivers habituating to an unchanging message. One potential limitation with the present research was the fact that safety could not be measured directly, but could only be implied from the fact that stopping and signalling were influenced by the prompts. Perhaps closed road or simulator studies could investigate driving performance as well as prompting effects. Another limitation was that only a single intersection in a suburban area was investigated. Responses to the prompts may have been different in a more congested area such as the downtown core where there are a greater number of demands on a driver's attention. Given the recent popularity of cellular telephones and GPS navigational systems used by drivers, these devices impose a cognitive workload on drivers which may reduce attention to prompts and other roadway stimuli (Singhal & Wiesenthal, 2003). However, previous research has shown that single prompt messages can increases stopping behavior in multiple locations (Van Houten & Malenfant, 1992; Wiesenthal & Ford, 1990). Also, it may be possible that the multiple prompt effect was particular to stopping and signalling behaviors. Future research may investigate the generalizability of the multiple prompt technique by using different prompts and multiple driving settings. Finally, it may be argued that the behavior change observed in the prompt conditions may have been due to a concern that police may be monitoring the intersection to ticket violators. According to Lonero et al (1994), threat of punishment can lead to safe driving behavior when the threat is real and imminent, such as under potential police presence. As such, the change would simply be due to fear of repercussions rather than a reminder to stop or signal turns. The fact that some change occurred even under the blank sign condition lacking a prompt, suggests that threat of punishment cannot be the only explanation. In addition, Nasar (2003) found that prompted stopping behaviors persisted "downstream" at a later unprompted intersection, indicating that a true prompted change had occurred. Nonetheless, future research may be needed to examine the long term impact of prompted behaviors across conditions and over time.

REFERENCES

Austin, J., Alvero, A. M., & Olson, R. (1998). Prompting patron safety belt use at a restaurant. *Journal of Applied Behavior Analysis, 31*, 655-657.

Austin, J. Hatfield, D. B., Grindle, A. C., & Bailey, J. S. (1993). Increasing recycling in office environments: The effects of specific, informative cues. *Journal of Applied Behavior Analysis, 26*, 247-253.

Baltes, M. M., & Hayward, S. C. (1976). Application and evaluation of strategies to reduce pollution: Behavioral control of littering in a football stadium. *Journal of Applied Psychology, 61*, 501-506.

Brigham, T. A., Meier, S. M., & Goodner, V. (1995). Increasing designated driving with a program of prompts and incentives. *Journal of Applied Behavior Analysis, 28*, 83-84.

Cox, B. S., Cox, A. B., Cox, D. J. (2000). Motivating signage prompts safety belt use among drivers exiting senior communities. *Journal of Applied Behavior Analysis, 33*, 635-638.

Craig, H. B., & Leland, L. S. (1983). Improving cafeteria patrons' waste disposal. *Journal of Organizational Behavior Management, 5*, 79-88.

Dixon, R. S., Knott, T., Rowsell, H., & Sheldon, L. (1992). Prompts and posted feedback: In search of an effective method of litter control. *Behavior Change, 9*, 2-7.

Duer, J. D. (1982). Prompting women to seek cervical cytology. *Behavior Therapy, 13*, 248-253.

Durdan, C. A., Reeder, G. D., & Hecht, P. R. (1985). Litter in a university cafeteria: Demographic data and the use of prompts as an intervention strategy. *Environment and Behavior, 17*, 387-404.

Ferrari, J. R., Barone, R. C., Jason, L. A., & Rose, T. (1985). The effects of a personal phone call prompt on blood donor commitment. *Journal of Community Psychology, 13*, 295-298.

Geller, E. S., Bruff, C. D., & Nimmer, J. G. (1985). "Flash for life": Community-based prompting for safety belt promotion. *Journal of Applied Behavior Analysis, 18*, 309-314.

Geller, E. S., Farris, J. C., & Post, D. S. (1973). Prompting a consumer behavior for pollution control. *Journal of Applied Behavior Analysis, 6*, 367-376.

Geller, E. S., Witmer, J. F., & Orenbaugh, A. L. (1976). Instructions as a determinant of paper-disposal behaviors. *Environment and Behavior, 8*, 417-439.

Geller, E. S., Witmer, J. F., & Tuso, M. A. (1977). Environmental interventions for litter control. *Journal of Applied Psychology, 62*, 344-351.

Geller, E. S., Winett, R. A., & Everett, P. B. (1982). *Preserving the environment: New strategies for behavior change*. Elmsford, NY: Pergamon.

Gras, M. E., Cunill, M., Planes, M., Sullman, M. J. M., & Oliveras, C. (2003). Increasing safety belt use in Spanish drivers: A field test of personal prompts. *Journal of Applied Behavior Analysis, 36*, 249-251.

Hayes, S. C., & Cone, J. D. (1977). Decelerating environmentally destructive lawn-walking. *Environment and Behavior, 9*, 511-534.

Hopper, J. R., & McCarl-Nielsen, J. (1991). Recycling as altruistic behavior: Normative and behavioral strategies to expand participation in a community recycling program. *Environment and Behavior, 23*, 195-220.

Houghton, S. (1976). Littering in high schools. *Educational Studies, 19*, 247-254.

Hughes, P. K., & Cole, B. L. (1986). What attracts attention while driving? *Ergonomics, 29*, 377-391.

Kazdin, A. E. (1989). *Behavior modification in applied settings* (4th ed.). Pacific Grove, CA: Brooks\Cole.

Kazdin, A. E. (1994). *Behavior modification in applied settings* (5th ed.). Pacific Grove, CA: Brooks/Cole Publishing Co.

Lombard, D. N., Lombard, T. N., & Winett, R. A. (1995). Walking to meet health guidelines: The effects of prompting frequency and prompt structure. *Health Psychology, 14*, 164-170.

Lonero, L. P., Clinton, K., Wilde, G. J. S., Roach, K., McKnight, A. J., MacLean, H. et al., (1994). The roles of legislation, education and reinforcement in changing road user behaviour. *Ontario Ministry of Transportation. Road Safety Research Office: Safety Policy Branch, Report No. 322*. Toronto, Canada.

Ludwig, T. D., & Geller, E. S. (1999). Behavior change among agents of a community safety program: Pizza delivers advocate community safety belt use. *Journal of Organizational Behavior Management, 19,* 3-24.

Martinko, M. J., White, J. D., & Hassell, B. (1989). An operant analysis of prompting in a sales environment. *Journal of Organizational Behavior Management, 10*, 93-107.

Naatanen, R., & Summala, H. (1976). *Road user behavior and traffic accidents*. Amsterdam: North Holland Publishing.

Nasar, J. L. (2003). Prompting drivers to stop for crossing pedestrians. *Transportation Research Part F, 6*, 175-182.

Reid, D. H., Luyben, P. D., Rawers, R. J., & Bailey, J. S. (1976). Newspaper recycling behavior. The effects of prompting and proximity of containers. *Environment and Behavior, 8*, 471-482.

Rothengatter, J. A. (1987). Current issues in road user research. In J. A. Rothengatter, K. Duncan, & J. Lepat (Eds.), *New technology and traffic safety*. Assen/Maastricht: Van Gorcum.

Singhal, D. & Wiesenthal, D. L. (2003). An assessment of the relationship between cellular telephone use while driving and motor vehicle accidents. Canadian Multidisciplinary Road Safety Conference XIII, June, Banff, AB.

Smith, J. M., & Bennett, R. (1992). Several antecedent strategies in reduction of an environmentally destructive behavior. *Psychological Reports, 70*, 241-242.

Solomon, L. J., Flynn, B. S., Worden, J. K., Mickey, R. M., Skelly, J. M., Geller, B. M., Peluso, N. W., & Webster, J. A. (1998). Assessment of self-reward strategies for maintenance of breast self-examination. *Journal of Behavioral Medicine, 21,* 83-102.

Stern, P. L., & Oskamp, S. (1987). Managing scarce environmental resources. In D. Stokols & I. Altman (Eds.), *Handbook of Environmental Psychology* (pp. 1043-1088). New York: Wiley.

Thyer, B. A., Geller, E. S., Williams, M., & Purcell, E. (1987). Community-based "Flashing" to increase safety belt use. *Journal of Experimental Education, 55*, 155-159.

Van Houten, R. (1988). The effects of advance stop lines and sign prompts on pedestrian safety in a crosswalk on a multilane highway. *Journal of Applied Behavior Analysis, 21*, 245-251.

Van Houten, R., & Malenfant, L. (1992). The influence of signs prompting motorists to yield before marked crosswalks on motor vehicle-pedestrian conflicts at crosswalks with flashing amber. *Accident Analysis and Prevention, 24*, 217-225.

Van Houten, R., Malenfant, L., & Rolider, A. (1985). Increasing driver yielding and pedestrian signalling with prompting, feedback, and enforcement. *Journal of Applied Behavior Analysis, 18*, 103-110.

Van Houten, R., & Nau, P. (1983). Feedback interventions and driving speed: A parametric and comparative analysis. *Journal of Applied Behavior Analysis, 12*, 253-281.

Van Houten, R., & Retting, R. A. (2001). Increasing motorist compliance and caution at stop signs. *Journal of Applied Behavior Analysis, 34*, 185-193.

Wagner, J. L., & Winett, R. A. (1988). Prompting one low-fat, high-fibre selection in a fast-food restaurant. *Journal of Applied Behavior Analysis, 21*, 179-185.

Wagstaff, M. C., & Wilson, B. E. (1988). The evaluation of litter behavior modification in a river environment. *Journal of Environmental Education, 20*, 39-44.

Wiesenthal, D. L., & Ford, D. (1990). Driver behaviour and prompts. *Highway Safety Conference, Ministry of Transportation, Toronto*, Ontario, May 1990.

Wiesenthal, D. L., & Ford, D. (1992). The effects of prompts on driver behaviour at stop signs. *Third annual conference on highway safety. Ministry of Transportation, Toronto, Ontario*, May 1992.

Winett, R., A., Stewart, G., & Majors, J. S. (1978). Prompting techniques to increase the return rate of mailed questionnaires. *Journal of Applied Behavior Analysis, 11*, 437.

Witmer, J. F., & Geller, E. S. (1976). Facilitating paper recycling: Effects of prompts, raffles, and contests. *Journal of Applied Behavior Analysis, 9*, 315-322.

In: Contemporary Issues in Road User Behavior and Traffic Safety ISBN:1-59454-268-6
Editors: D. Hennessy and D. Wiesenthal, pp. 199-214

Chapter 16

EARLY INDICATORS AND INTERVENTIONS FOR TRAUMATIC STRESS DISORDERS SECONDARY TO MOTOR VEHICLE ACCIDENTS

Connie Veazey
Houston Center for Quality of Care and Utilization Studies, Health Services Research and Development Service, Michael E. DeBakey Veterans Affairs Medical Center, Houston, Texas
Edward B. Blanchard
Center for Stress and Anxiety Disorders, University at Albany-State University of New York

INTRODUCTION

A large epidemiological study reported that there were16.4 million motor vehicle accidents (MVAs) in the United States in the year 2000 in which 5.3 million people were injured (Blincoe et al., 2002). The data from this study suggest that between 10% and 30% of people treated in an emergency department as a result of an MVA exhibit the symptoms of PTSD one year following their MVAs. Of the reported 4,259,000 treated in an emergency department this would have resulted in between 0.4 and 1.3 million people developing PTSD. Thus, PTSD following MVAs represents a problem of epidemic portion in the United States. In addition to the huge psychological impact of MVAs on survivors, there are both individual and societal economic losses as reflected by the $230.6 billion dollars in medical costs ($32.6 billion); property damage ($59 billion); loss of productivity ($81 billion); and other related costs ($58 billion). PTSD compounds economic losses as is reflected in the loss of productivity and other related costs.

Table 1. DSM-IV Criteria for PTSD and ASD (American Psychiatric Association, 1994)

	Posttraumatic Stress Disorder	Acute Stress Disorder
Criterion A (ASD & PTSD): Stressor	Exposed to traumatic event with both of following were present: 1) Event involved actual or threatened death or serious injury, or a threat to the physical integrity of self or others 2) Response involved intense fear, helplessness, or horror.	Exposed to traumatic event with both of following were present: 1) Event involved actual or threatened death or serious injury, or a threat to the physical integrity of self or others 2) Response involved intense fear, helplessness, or horror.
Criterion B (ASD): Dissociation		During or following the distressing event, 3 or more of following: 1) Subjective sense of numbing, detachment, or absence of emotional responsiveness 2) reduction in awareness of surroundings 3) derealization 4) depersonalization 5) dissociative amnesia (i.e. inability to recall important aspect of trauma)
Criterion B (PTSD) & Criterion C (ASD): Reexperiencing	One or more of following: 1) recurrent and intrusive recollections 2) recurrent distressing dreams 3) acting or feeling as if event were recurring 4) psychological distress on exposure to reminders 5) physiological reactivity on exposure to reminders	One or more of following: 1) recurrent images 2) recurrent thoughts 3) recurrent dreams 4) recurrent illusions 5) recurrent flashback episodes or sense of reliving the experience 6) distress on exposure to reminders

Table 1. DSM-IV Criteria for PTSD and ASD (American Psychiatric Association, 1994) (Continued)

	Posttraumatic Stress Disorder	Acute Stress Disorder
Criterion C (PTSD) & Criterion D (ASD): Avoidance and Numbing	Three or more of following: 1) avoidance of thoughts, feelings, or conversations associated with the trauma 2) avoidance or activities, places, or people associated with the trauma 3) inability to recall importance aspects of the trauma 4) diminished interest in significant activities 5) detachment or estrangement from others 6) restricted range of affect 7) sense of foreshortened future	*Marked* avoidance of stimuli that arouse recollections of the trauma: e.g. thoughts, feelings, conversations, activities, places, people
Criterion D (PTSD) & Criterion E (ASD): Arousal	Two or more of the following: 1) difficulty sleeping 2) irritability or outbursts of anger 3) difficulty concentrating 4) hypervigilance 5) exaggerated startle	*Marked* symptoms of anxiety or increased arousal: e.g. difficulty sleeping, irritability, poor concentration, hypervigilance, exaggerated startle response, motor restlessness
Criterion E (PTSD) & Criterion G (ASD): Duration	Duration of Symptoms in Criteria B, C, and D is more than 1 month.	Disturbance lasts for a minimum of 2 days and a maximum of 4 weeks and occurs within 4 weeks of the trauma.
Criterion F (PTSD) & Criterion F (ASD): Distress	The disturbance causes clinically significant distress or impairment in social, occupational, or other important areas of functioning.	The disturbance causes clinically significant distress or impairment in social, occupational, or other important areas of functioning or impairs the individual's ability to pursue some necessary task, such as obtaining necessary assistance or mobilizing personal resources by telling family members about the traumatic experience

The problem with focusing solely on PTSD following MVAs is that the functioning of trauma survivors in the immediate aftermath of a trauma is largely ignored due to the Diagnostic and Statistical Manual of Mental Disorders, Fourth Edition's (DSM-IV) Criterion E of PTSD which states the duration of symptoms must be more than 1 month (American Psychiatric Association, 1994). Assessing survivors within the first month following a trauma may help identify those who are at risk for the development of longer term maladjustment. Hence, Acute Stress Disorder (ASD) was added to the DSM-IV in order to identify trauma survivors warranting clinical diagnosis within 1 month of the trauma. The symptomatic criteria of ASD are almost identical to PTSD with two main exceptions. The first difference is the time criterion, with ASD diagnosed after 2 days, but within 1 month of the trauma, and PTSD not diagnosed until at least 1 month following the trauma. The second difference is the emphasis on dissociative symptoms in ASD. The person must experience at least three of the following dissociative symptoms: subjective sense of emotional numbing, a reduction in awareness of surroundings, derealization, depersonalization, and dissociative amnesia (American Psychiatric Association, 1994). See Table 1 for a comparison of the symptomatic criteria of these two disorders.

MVAs are a common trauma experienced by millions of people in the United States every year. In addition, millions of MVA survivors exhibit traumatic stress disorders (i.e. ASD and PTSD) as a result of their accidents. These disorders result in both economic and personal costs to survivors. Thus, it is useful to identify early indicators of such traumatic stress disorders following MVAs. In addition, the development of early interventions aimed at preventing chronic traumatic stress disorders is also warranted. First the prevalence and predictors of traumatic stress disorders are discussed with an attempt to identify survivors most at risk for the development of these disorders following MVAs. Then the early interventions aimed at preventing the development of traumatic stress disorders are discussed.

Prevalence of Acute Stress Disorder and Posttraumatic Stress Disorder following MVAs

The prevalence of Acute Stress Disorder (ASD) in MVA populations has been reported to range from 10.2% to 33% (Barton, Blanchard, & Hickling, 1996; Bryant, Harvey, Guthrie, & Moulds, 2000; Bryant & Harvey, 2003; Bryant & Panasetis, 2001; Harvey & Bryant, 1998; Harvey & Bryant, 1999a; Harvey & Bryant, 1999b; Holeva, Tarrier, & Wells, 2001; Mellman, David, Bustamante, Fins, & Esposito, 2001; Murray, Ehlers, & Mayou, 2002). Differing assessment techniques account for some of the disparities in the reported prevalence estimates. Until recently there were no valid measures of ASD. This resulted in studies utilizing non-standardized diagnostic approaches (Barton et al., 1996) or using measures intended for PTSD diagnosis (Holeva et al., 2001; Mellman et al., 2001; Murray et al., 2002). In studies utilizing standardized structured interviews specifically designed for ASD (Bryant et al., 2000a; Bryant & Harvey, 2003; Bryant & Panasetis, 2001; Harvey & Bryant, 1999a; Harvey & Bryant, 1999b), namely the Acute Stress Disorder Interview (ASDI) (Bryant, Harvey, Dang, & Sackville, 1998a), the prevalence estimates of ASD were highly similar 13% to 16.1%, with the exception of one study, which reported a 33% ASD rate (Bryant & Panasetis, 2001).

Table 2. Prevalence of ASD and PTSD in Studies Reviewed

Study	Measure	ASD	1 month	3-4 months	6 months	9 months	12 months	24 months
ASD &/or PTSD	Non-Standardized							
Barton et al. (1996) (3)	Self-report of 3 dissociative symptoms in personal MVA account CAPS[1]	23%		39%				
—	Self-Report	—	—	—	—	—	—	—
Holeva et al. (2001) (10)	Penn Inventory	21%			23%			
Murray et al. (2002) (12) Inpatient	PDS[2] SDQ[3]	12.5%	31.6%		19%			
Murray et al. (2002) (12) Outpatient		10.2%	28.3%		24.3%			
—	Structured Interview	—	—	—	—	—	—	—
Bryant et al. (2000) (4)	ASDI[4] CIDI-PTSD[5]	13.7%			21%			
Bryant & Harvey (2003) (5)	ASDI	13.5%			24.6%			
Harvey & Bryant (1999) (8)	ASDI CIDI-PTSD	13%			25.4%			30%
Harvey & Bryant (1999) (9)	ASDI	16.1%						
Mellman et al. (2001) (11)	Peritraumatic- Diagnostic Scale CAPS	16%		24%				
Bryant & Panasetis (2001) (6)	ASDI	33%						
PTSD Only	Self-Report	—	—	—	—	—	—	—
Ehlers et al. (1998) (13)	PSS[6]			23.1%			16.5%	
Frommberger et al. (1998) (14)	PSS			18.4%	8.6%			
------------------	Structured Interview	—	—	—	—	—	—	—
Blanchard et al. (1996) (16)	CAPS			39.2%	17.4%		18%	
Dougall et al. (2001) (17)	SCID-DSM-III-R PTSD Supplement		61%		45%		37%	
Freedman et al. (2002) (18)	CAPS		27%	17%				
Jeavons et al. (2000) (15)	PTSD-I[7]			8.3%	8%		8.6%	
Koren et al. (1999) (19)	SCID-III-R-NP						32.4%	
Shalev et al. (1998) (20)	CAPS		38%	23%				
Shalev et al. (1998) (21)	CAPS		29.9%	17.5%				
Ursano et al. (1999) (22)	SCID-DSM-III-R PTSD Supplement		34.4%	25.3%	18.2%	17.4%	14%	

[1] CAPS- Clinician Administered PTSD Scale; [2] PDS- Posttraumatic Diagnostic Scale ; [3] SDQ- State Dissociation Questionnaire; [4] ASDI- Acute Stress Disorder Interview;[5] CIDI-PTSD- Composite International Diagnostic Interview-Posttraumatic Stress Disorder Module;[6] PSS- Posttraumatic Stress Symptom Scale; [7] PTSD-I- The Post-traumatic Stress Disorder Interview.

The prevalence of PTSD following MVAs has been reported to range from 27% to 61% at one month post-MVA; 8.3% to 39.2% three to four months post-MVA; 8.6% to 45% six months post-MVA; 17.4% at nine months post-MVA; 8.6% to 32.4% one year post-MVA; and 30% at two years post-MVA (Blanchard, Hickling, Baron, & Taylor, 1996; Dougall, Ursano, Posluszny, Fullerton, & Baum, 2001; Ehlers, Mayou, & Bryant, 1998; Freedman et al., 2002; Frommberger et al., 1998; Jeavons, 2000; Koren, Arnon, & Klein, 1999; Shalev et al., 1998a; Shalev et al., 1998b; Ursano et al., 1999). Table 2 summarizes the percentages of MVA survivors meeting criteria for ASD and PTSD at each time point for each of the studies reviewed.

Relationship Between ASD and PTSD Development

When examining the diagnostic utility of ASD on PTSD, there are several measures of diagnostic utility including positive predictive power (PPP), negative predictive power (NPP), sensitivity, and specificity (Elwood, 1993). Sensitivity is the probability, given an individual meets PTSD, that they will have reported meeting ASD. This is related to PPP, which is the probability, given that the individual has reported ASD, that they will then meet criteria for PTSD. Specificity is the probability, given the person does not meet PTSD, that they did not meet ASD. NPP, then is the probability, given that the person does not meet ASD, that they will not meet PTSD.

While these diagnostic utility statistics could not be computed for all the studies reviewed, Table 3 summarizes the percentages of participants in the studies reviewed who had ASD and then went on to develop PTSD and the percentages of people with PTSD who reported ASD at the initial assessment. As can be seen the percentages of people with ASD who go on to develop PTSD is quite high ranging from 63% to 77.8% in the majority of the studies reviewed, with only one study reporting a percentage of less than 50 (Mellman et al., 2001). This is reflective of the general finding in the literature of high PPP of the ASD diagnosis on PTSD. That is the probability that someone will meet PTSD given that they met criteria for ASD is generally quite high. However, the percentages of participants meeting PTSD criteria, who met ASD criteria were quite lower, ranging from 22% to 59%. This is reflective of the general finding in the literature of the lower sensitivity of the ASD diagnosis on PTSD.

When examining which ASD symptoms clusters appear to be most beneficial as indicators of the later development of PTSD, several studies have found that NPP is strong for all ASD symptom clusters (Harvey & Bryant, 1998; Harvey & Bryant, 1999a). Given that someone does not report an ASD symptom cluster, they are less likely to later report symptoms of PTSD. However, PPP has been found to be strongest for ASD dissociation symptoms, "moderate" for the experiencing and avoidance symptoms, and weak for the arousal symptoms for both six month (Harvey & Bryant, 1998) and two year PTSD (Harvey & Bryant, 1999a). In addition, the PTSD symptoms, minus the time criterion, have been found to be as useful as the ASD symptoms in predicting chronic PTSD status (Harvey & Bryant, 1999a). Murray et al. (2002) found that persistent experience of dissociation at 4 weeks following an MVA was a better predictor of chronic PTSD at 6 months than peri-traumatic dissociation. In contrast to these findings, Mellman et al. (2001) found that ASD was not related to PTSD development at three months, and that early report of heightened arousal was the most significantly correlated ASD symptom with PTSD development at three months.

Table 3. Relationship of ASD and PTSD in Studies Reviewed

Study	Measure	% ASD	% PTSD	% ASD who develop PTSD	% PTSD who had ASD
	Non-Standardized				
Barton et al. (1996) (3)	Self-report of 3 dissociative symptoms in personal MVA account CAPS[1]	23	39		22
—	Self-Report	—	—	—	
Holeva et al. (2001) (10)	Penn Inventory	21	23	72	59
Murray et al. (2002) (12) Outpatient	PDS[2] SDQ[3]	10.2	28.3	77	?
—	Structured Interview	—	—	—	
Bryant et al. (2000) (4)	ASDI[4] CIDI-PTSD[5]	13.7	21	76	54
Bryant & Harvey (2003) (5)	ASDI CIDI-PTSD	13.5	24.6	70	48
Harvey & Bryant (1998) (7)	ASDI CIDI-PTSD	13	25.4	77.8	39
Harvey & Bryant (1999) (8)	ASDI CIDI-PTSD	13	30	63	29
Mellman et al. (2001) (11)	Peri-traumatic Dissociation Scale. CAPS	16	24	40	33

[1] CAPS- Clinician Administered PTSD Scale.
[2] PDS- Posttraumatic Diagnostic Scale.
[3] SDQ- State Dissociation Questionnaire.
[4] ASDI- Acute Stress Disorder Interview.
[5] CIDI-PTSD- Composite International Diagnostic Interview-Posttraumatic Stress Disorder Module.

While the diagnosis of ASD is a useful indicator of PTSD, the diagnostic utility findings indicate it is not the only factor contributing to the development of more chronic problems following MVAs, and it should not be viewed as a perfect predictor of the development of PTSD.

EARLY INDICATORS OF TRAUMATIC STRESS DISORDERS

In addition to the diagnosis of ASD, other factors have been explored as being indicative of greater risk for the development of PTSD or longer-term maladjustment following MVAs. Such factors include psychiatric indicators, prior trauma history, gender, injury and accident severity, responsibility for the MVA, and initial physiological arousal.

Psychiatric Indicators

Psychiatric indicators including a history of mood disorders, psychiatric treatment, and high levels of depression and anxiety symptoms have all been indicated as risk factors for ASD and PTSD following MVAs. Both a history of, and concurrent comorbid symptoms of depression are by far one of the strongest risk factors for both ASD and PTSD (Barton et al., 1996; Frommberger et al., 1998; Harvey & Bryant, 1999b). Of MVA survivors with PTSD, comorbid depression has been found in 44.5% at one month post accident, and in 43.2% at four months post accident (Shalev et al., 1998). In addition those with comorbid depression, exhibited higher overall symptom ratings and worse overall functioning than those without comorbid PTSD (Shalev et al., 1998). Anxiety symptoms in the initial phase following a MVA are also related to the development of PTSD (Frommberger et al, 1998.). A history of psychiatric treatment has been found to be related to both acute stress severity and ASD itself (Harvey & Bryant, 1999b). Thus, it appears warranted to examine these psychiatric indicators in the early phase following an MVA, especially depression, to help delineate survivors who may be at risk for PTSD.

Prior Trauma History

A prior trauma history has been identified as a risk factor for the development of PTSD following MVAs. It makes intuitive sense that the experience of a greater number of traumas places one at a greater risk for developing a traumatic stress disorder. Research with MVA survivors has documented that trauma histories do, in fact, place people at greater risk for the development of ASD and PTSD. The experience of prior MVAs has been found to place people at risk for ASD development following MVAs (Harvey & Bryant, 1999b). In addition, a history of PTSD from any prior trauma has been found to place people at risk for both ASD (Harvey & Bryant, 1999b) and PTSD (Blanchard, Hickling, Taylor, & Loos, 1995) following MVAs. There is evidence to indicate that a high percentage of MVA survivors have a history of prior trauma. Freedman et al. (2002) found that 93% of MVA survivors had experienced some type of prior traumatic event. A prior trauma history and experience of previous traumatic stress disorders appear to be useful indicators of the development of traumatic stress disorders following MVAs.

Injury Severity

There is some disparity in findings of the relationship between injury severity and the development of traumatic stress disorders following MVAs. In studies examining the relationship between injury severity and ASD development (Harvey & Bryant, 1998; Harvey & Bryant, 1999b), injury severity was not related to ASD development. However, in studies examining the relationship between injury severity and the development of PTSD (Blanchard et al, 1997; Dougall et al., 2001; Ehlers et al., 1998; Frommberger et al., 1998; Jeavons, 2000; Mellman et al., 2001; Murray et al., 2002), the results have been mixed. With several studies (Blanchard et al., 1997; Dougall et al., 2001; Frommberger et al., 1998; Jeavons, 2000) supporting the relationship between injury severity and PTSD development. However, in one of these studies (Dougall et al., 2001) the correlation between injury severity scores and PTSD was negative. Finally, several studies (Ehlers et al., 1998; Mellman et al., 2001; Murray et al., 2002) found injury severity scores not predictive of PTSD status. While there is likely some relationship between physical injury and development of traumatic stress disorders, this relationship is complex and as the research indicates not always direct. Due to the findings that injury severity is not related to ASD, but does play a role in the development of PTSD, the effects of physical functioning on psychological status seem to delayed and not evident in the immediate aftermath of an MVA.

Gender

The relationship of gender to the experience of traumatic stress disorders has been a topic of much discussion. A study examining the experience of prior trauma and gender differences in a sample of MVA survivors (Freedman et al., 2002), found no gender difference in the number of prior traumas experienced. However, there was a difference in the type of prior trauma, with females more likely to have experienced burglary, rape, and sexual assault and males more likely to have experienced witnessing injury, witnessing corpses, combat, and rock throwing. There was no gender difference in the prevalence or recovery from PTSD at 1 or 4 months post-MVA. In contrast to this study, a higher percentage of females than males were found to meet criteria for ASD (females 23%; males 8%) and PTSD at 6 months post-MVA (females 38%; males 15%) (Bryant & Harvey, 2003). Female gender has also been found to be a predictor of both one month (Dougall et al., 2001) and three month (Ehlers et al., 1998) PTSD status. Therefore, the likelihood of experiencing prior trauma does not appear to be biased by gender. The findings indicate that the experience of traumatic stress disorders is related female gender in some MVA populations.

Attribution of Responsibility

It has been hypothesized that trauma survivors who feel as if they had some control in the experience of their traumatic events, fair better psychologically than those who feel they had no control in the experience of a trauma. Delahanty et al. (1996) examined the role of responsibility for an MVA on the likelihood of developing PTSD. Two groups of crash survivors were examined including one group of survivors who stated they were responsible for the accident and one group who named another person as being responsible for the accident. The group who named someone else as being responsible for the accident was marginally more likely to meet PTSD than the group who indicated that they were responsible for the MVA. Thus, lack of personal control may contribute to the development of PTSD following MVAs.

Physiological Arousal

Biological models implement physiological arousal both, peritraumatic and in the immediate aftermath of the trauma, as playing a role in the development of traumatic stress disorders. According to such models, high physiological arousal is associated with a blunted cortisol response in certain trauma survivors, which places them at greater risk for developing a traumatic stress disorder. Cortisol is released during times of stress to help control bodily arousal, and if an insufficient amount is released then physiological arousal is high. The paradigms for studying physiological arousal have been the measurement of both cortisol levels and vital signs (heart rate (HR) and blood pressure (BP)) of survivors following the experience of a trauma.

Research has supported the assertion of lower levels of cortisol in the immediate aftermath of MVAs being related to the development of PTSD (Delahanty et al., 2000; McFarlane, Atchison, & Yehuda, 1997). However, the research examining early vital signs following a trauma has found mixed results. The majority of studies (Bryant et al., 2000a; Shalev et al., 1998) have found that survivors with higher arousal, as indicated by HR following MVAs, are more likely to develop PTSD. However, one study (Blanchard, Hickling, Galovski, & Veazey, 2002) has found the exact opposite, i.e. MVA survivors with higher arousal, as indicated by HR following the MVA, were less likely to develop PTSD. The relationship between such arousal and ASD is a bit more complex. Bryant et al. (2000a) found that those with ASD had lower arousal than those with sub-ASD, which included people who exhibited all of the symptom clusters of ASD except dissociation. This finding was explained by the assertion that dissociation dampens such early arousal.

The majority of the evidence at this time suggests that early physiological arousal is a risk factor for the development of traumatic stress disorders following MVAs. However, there is no good indication as to how to identify those with high arousal, for example how to set a cut score on HR to identify who is at risk.

Early Interventions

There are three general approaches to early intervention for stress disorders following MVAs, and these include psychological debriefing (PD), Cognitive Behavioral Therapy (CBT), and medication interventions.

Psychological Debriefing

The aim of PD is to allow trauma survivors an early opportunity to discuss their experience as well as educate survivors about reactions they may experience following trauma. There is an effort in PD to normalize traumatic stress reactions. There are four studies that have focused on examining the effectiveness of PD in preventing negative outcomes following MVAs (Bordow & Porritt, 1979; Conlon, Fahy, & Conroy, 1999; Hobbs, Mayou, Harrison, & Worlock, 1996; Mayou, Ehlers, & Hobbs, 2000).

The first study (Bordow & Porritt, 1979) was a controlled comparison of a social work crisis intervention including emotional, practical, and social support. The intervention group (n = 30) was limited to a maximum of 10 hours. It was compared to a delayed contact control group (n = 30) and an assessment only group (n = 10). The follow-up time point was three months following the MVA. The group receiving the intervention had the least percentage of "poor" outcomes on measures of adjustment. There were no specific measures of psychiatric disorders such as depression, ASD, or PTSD. The authors did not report how they determined "poor" outcomes on the measures utilized in the study.

In the largest and most scientifically rigorous trial (Hobbs et al., 1996), a standardized, one-hour session of PD (n = 54) including encouragement of emotional expression and cognitive processing related to the trauma, advice about common emotional reactions, advice about the beneficial effects of talking about the trauma, and advice about an early and graded return to normal road travel was compared to a control condition (n = 52). The measures included the Impact of Events Scale (IES) and the Brief Symptom Inventory (BSI). At four months, there was no reduction in scores for either group, with the PD group exhibiting worse scores on two subscales of the BSI than the control group. In a three-year follow-up of this study (Mayou et al., 2000), the PD group (n = 30) exhibited worse outcomes on the BSI, travel anxiety, pain, physical problems, overall level of functioning, and financial problems than the control group (n = 31). Of participants exhibiting high initial IES scores, those in the PD group showed no recovery while those in the control group did show a reduction in high IES scores. These results suggest that a standardized PD protocol may prove detrimental to the long-term functioning of MVA survivors.

A final randomized controlled trial compared PD (n = 18) to a control (n = 52) condition (Conlon et al., 1999). The PD condition was a standardized, thirty minute counseling session including encouragement of expression of emotions and processing of cognitions related to the MVA, education about posttraumatic stress symptoms, and advice on coping strategies. Measures included the IES and Clinician Administered PTSD Scale (CAPS). There was no difference between the groups at three months post MVA.

At this time, there is not adequate evidence to support the use of PD as an early intervention to prevent the development of traumatic stress disorders, and in the largest study (Hobbs et al., 1996; Mayou et al., 2000), there is evidence that PD contributes to worse longer-term psychological functioning in MVA survivors.

Cognitive Behavioral Therapy (CBT)

At this time, CBT represents the most promising early intervention for the prevention of the development of PTSD in MVA survivors. There have been two randomized controlled trials reporting on the utility of CBT to treat ASD and prevent PTSD in trauma survivors (Bryant, Harvey, Dang, Sackville, & Basten, 1998b; Bryant, Sackville, Dang, Moulds, & Guthrie, 1999). Bryant et al. (1998b) treated a sample of trauma survivors exhibiting ASD, as measured by the ASDI. Fifty eight percent of the sample included MVA survivors with the remainder of the sample including survivors of industrial accidents (42%). This randomized controlled trial compared CBT (n = 12) to a Supportive Counseling (SC) (n = 12) condition. Each condition consisted of five, ninety-minute individual sessions. The CBT condition included education, progressive muscle relaxation, imaginal exposure, cognitive restructuring, and graded in vivo exposure. The SC condition included education, general problem solving skills, and unconditional support. Outcome measures included the Impact of Event Scale (IES), Beck Depression Inventory (BDI), State Trait Anxiety Inventory (STAI), and Composite International Diagnostic Interview-PTSD module (CIDI-PTSD). Participants were assessed both post-treatment and at six months following the end of treatment. At post-treatment fewer participants in the CBT group (8%) than in the SC group (83%) met criteria for PTSD. This difference continued at the six-month follow-up with fewer CBT participants (17%) than SC participants (67%) exhibiting PTSD.

A second randomized controlled trial (Bryant et al., 1999) compared three interventions: Exposure plus Anxiety Management (E+AM) (n = 15), Exposure Only (E) (n = 14), and SC (n = 16). This sample included trauma survivors meeting ASD criteria on the ASDI and was composed of 47% MVA survivors and 53% nonsexual assault survivors. Each intervention group consisted of five, ninety-minute individual sessions. Participants were assessed both post-treatment and six months following the end of treatment. The E+AM intervention included education, breathing retraining, progressive muscle relaxation, self-talk to manage anxiety, prolonged imaginal exposure, cognitive restructuring, in vivo exposure, and relapse prevention. The E intervention included education, prolonged imaginal exposure, cognitive restructuring, in vivo exposure, and relapse prevention. The SC intervention included education, general problem solving skills, and unconditional support. At post-treatment, fewer participants in the E+AM (20%) and the E (14%) conditions met criteria for PTSD than in the SC (56%) group. This finding persisted at the six-month follow up with fewer participants in the E+AM (23%) and E (15%) conditions meeting criteria for PTSD than in the SC (67%) group.

Thus, CBT interventions are proving to be more effective in preventing the development of PTSD than supportive counseling techniques. In addition, these two studies started with trauma survivors exhibiting ASD, and the percentage of those developing PTSD at 6 months 15%-23% is much less than the expected 63% to 77.8% of trauma survivors who initially have ASD and then develop PTSD (Bryant et al., 2000a; Bryant & Harvey, 2003; Harvey & Bryant, 1998; Harvey & Bryant, 1999a; Holeva et al., 2001; Mellman et al., 2001; Murray et al., 2002).

Medication Intervention

Based on the hypothesis that early physiological arousal plays a role in the development of traumatic stress disorders, it is also speculated that medication treatment aimed at dampening early arousal may prevent the development of such disorders. A recent study (Pitman et al., 2002) examined the use of propranolol (a β-adrenergic receptor blocker) as a prophylactic agent to the development of PTSD in a sample of trauma survivors, including 71% MVA survivors. Propranolol (n = 18) was administered in a double-blind fashion with a starting 40 mg oral dose within 6 hours of the traumatic event and continued for a period of 10 days (40 mg four times a day) followed by a 9-day taper period. There were 23 participants in the placebo condition. There were no differences between the two groups in CAPS scores or percentage of participants meeting PTSD at one or three-months post-trauma. The participants in the placebo condition were more likely to respond physiologically to reminders of the trauma at the three months follow-up. While this study does not support the hypothesis that administration of a medication aimed at dampening post-traumatic arousal prevents the development of PTSD, it does support the notion that dampening such early arousal leads to a reduction in later arousal to cues of a trauma. At this time, more research is needed to determine if such early pharmaceutical intervention may prevent the development of PTSD.

CONCLUSIONS

Given the millions of MVAs occurring in the United States each year, there are millions of people at risk for the development of traumatic stress disorders following these accidents. While the ability to predict which MVA survivors will develop PTSD is not possible at this time, and given the complex nature of this disorder will likely never be possible, there is research to support indicators which place survivors at risk for developing traumatic stress disorders following an MVA. Currently, the early experience of traumatic stress symptoms is likely the best indicator of longer-term maladjustment following an MVA. However, there is not a perfect correlation between ASD and early PTSD and the development of chronic PTSD. Early symptoms lack sensitivity. While survivors who experience ASD and early PTSD are more likely to develop chronic PTSD than those who do not experience such symptoms, there remains a sizable portion of those who develop chronic PTSD who did not exhibit these symptoms in the early phase following the MVA.

Non-specific psychological factors also place MVA survivors at greater risk for the experience traumatic stress disorders. Such factors include a history of depression, anxiety, and other Axis I disorders as well as a history of psychiatric treatment. In addition, high scores on early measures of psychological distress such as depression and anxiety have also been shown to indicate worse outcome. In addition to these other early indicators include a history of prior trauma and prior experience of the traumatic stress disorders, female gender, injury severity, early physiological arousal.

Early attempts aimed at identifying those at risk for traumatic stress disorders should include an assessment of these indicators. Such an assessment should include measures of traumatic stress symptoms such as the Acute Stress Disorder Scale (ASS) (Bryant, Moulds, & Guthrie, 2000b) or the PTSD Checklist (PCL) (Weathers, Litz, Herman, Huska, & Keane, 1993). A measure of early dissociation such as the Peritraumatic Dissociative Experiences Scale (Marmar, Weiss, & Metzler, 1997) is also warranted. Measures of depression and anxiety symptoms should be included such as the Beck Depression Inventory: Second Edition (BDI-II) (Beck, Steer, & Brown, 1996) and the State Trait Anxiety Inventory (STAI) (Spielberger, Gorushch, Lushene, Vagg, & Jacobs, 1983). An assessment of survivors' prior trauma histories and past psychiatric functioning is also warranted. Assessments utilizing these multiple measures would help identify trauma survivors across the multiple risk factors, and would help identify those most suited to early interventions.

The most promising early intervention at this point is CBT. With its emphasis on education, imaginal and *in vivo* exposure, cognitive restructuring, and anxiety management, CBT targets all of the symptom clusters of traumatic stress disorders, and has proven useful in preventing the development of chronic PTSD when administered in the early phase following MVAs. At this time, early PD is not warranted and should be utilized with caution as research suggests that this intervention may actually lead to worse long-term outcomes following MVAs. Finally, the use of early medication interventions aimed at dampening physiological arousal in the early period following a trauma is still in the early experimental phase, and future research is needed before recommendations regarding the use of such medications can be made.

References

American Psychiatric Association. *Diagnostic and Statistical Manual of Mental Disorders* (fourth edition). (1994). Washington DC.

Barton, K. A., Blanchard, E. B., & Hickling, E. J. (1996). Antecedents and consequences of acute stress disorder among motor vehicle accident victims. *Behaviour Research and Therapy, 34,* 805-813.

Beck, A. T., Steer, R. A., & Brown, G. K. (1996). *Beck Depression Inventory: Second Edition Manual.* The Psychological Corporation.

Blanchard, E. B., Hickling, E. J., Barton, K. A., & Taylor, A. E. (1996). One-year prospective follow-up of motor vehicle accident victims. *Behaviour Research & Therapy, 34*, 775-786.

Blanchard, E. B., Hickling, E. J., Forneris, C. A., Taylor, A. E., Buckley, T. C., Loos, W. R., et al. (1997). Prediction of remission of acute posttraumatic stress disorder in motor vehicle accident victims. *Journal of Traumatic Stress, 10*, 215-234.

Blanchard, E. B., Hickling, E. J., Galovski, T., & Veazey, C. (2002). Emergency room vital signs and PTSD in a treatment seeking sample of motor vehicle accident survivors. *Journal of Traumatic Stress, 15,* 199-204.

Blanchard, E. B., Hickling, E. J., Taylor, A. E., & Loos, W. (1995). Psychiatric morbidity associated with motor vehicle accidents. *Journal of Nervous & Mental Disease, 183*, 495-504.

Blincoe, L. J., Seay, A. G., Zaloshnja, E., Miller, T. R., Romano, E. O., Luchter, S., et al. (2002). *The Economic Impact of Motor Vehicle Crashes 2000.* Report of the U. S. Department of Transportation National Highway Traffic Safety Administration, Washington DC.

Bordow, S., & Porritt, D. (1979). An experimental evaluation of crisis intervention. *Social Science and Medicine, 13A*, 251-256.

Bryant, R. A., & Harvey, A. G. (2003). Gender differences in the relationship between acute stress disorder and posttraumatic stress disorder following motor vehicle accidents. *Australian and New Zealand Journal of Psychiatry, 37*, 226-229.

Bryant, R. A., Harvey, A. G., Dang, S. T., & Sackville, T. (1998a). Assessing acute stress disorder: Psychometric properties of a structured clinical interview. *Psychological Assessment, 10,* 215-220.

Bryant, R. A., Harvey, A. G., Dang, S. T., Sackville, T., & Basten, C. (1998b). Treatment of acute stress disorder: a comparison of cognitive-behavioral therapy and supportive counseling. *Journal of Consulting and Clinical Psychology, 66*, 862-866.

Bryant, R. A., Harvey, A. G., Guthrie, R. M., & Moulds, M. L. (2000a). A prospective study of psychophysiological arousal, acute stress disorder, and posttraumatic stress disorder. *Journal of Abnormal Psychology, 109*, 341-344.

Bryant, R. A., Moulds, M. L., & Guthrie, R. M. (2000b). Acute stress disorder scale: A self-report measure of acute stress disorder. *Psychological Assessment, 12*, 61-68.

Bryant, R. A., & Panasetis, P. (2001). Panic symptoms during trauma and acute stress disorder. *Behaviour Research and Therapy, 39*, 961-966.

Bryant, R. A., Sackville, T., Dang, S. T., Moulds, M., & Guthrie, R. (1999). Treating acute stress disorder: An evaluation of cognitive behavior therapy and supportive counseling techniques. *American Journal of Psychiatry, 156,* 1780-1786.

Conlon, L., Fahy, T. J., & Conroy, R. (1999). PTSD in ambulant RTA victims: A randomized controlled trial of debriefing. *Journal of Psychosomatic Research, 46*, 37-44.

Delahanty, D. L., Herberman, H. B., Craig, K. J., Hayward, M. C., Fullerton, C. S., Ursano, R. J., et al. (1996). Acute and chronic distress and posstraumatic stress disorder as a function of responsibility for serious motor vehicle accidents. *Journal of Consulting & Clinical Psychology, 65,* 560-567.

Dougall, A. L., Ursano, R. J., Posluszny, D. M., Fullerton, C. S., & Baum, A. (2001). Predictors of posttraumatic stress among victims of motor vehicle accidents. *Psychosomatic Medicine, 63*, 402-411.

Ehlers, A., Mayou, R. A., & Bryant, B. (1998). Psychological predictors of chronic posttraumatic stress disorder after motor vehicle accidents. *Journal of Abnormal Psychology, 107*, 508-519.

Elwood, R. W. (1993). Psychological tests and clinical discriminations: Beginning to address the base rate problem. *Clinical Psychology Review, 13*, 409-419.

Freedman, S. A., Gluck, N., Tuval-Mashiach, R., Brandes, D., Peri, T., & Shalev, A. Y. (2002). Gender differences in responses to traumatic events: A prospective study. *Journal of Traumatic Stress, 15,* 407-413.

Frommberger, U. H., Stieglitz, R. D., Nyberg, E., Schlickewei, W., Kuner, E., & Berger, M. (1998). Prediction of posttraumatic stress disorder by immediate reactions to trauma: A prospective study in road traffic accident victims. *European Archives of Psychiatry and Clinical Neurosciences, 248*, 316-321.

Harvey, A. G., & Bryant, R. A. (1998). The relationship between acute stress disorder and posttraumatic stress disorder: A prospective evaluation of motor vehicle accident survivors. *Journal of Consulting and Clinical Psychology, 66*, 507-512.

Harvey, A. G., & Bryant, R. A. (1999a). The relationship between acute stress disorder and posttraumatic stress disorder: a 2-year prospective evaluation. *Journal of Consulting and Clinical Psychology, 67*, 985-988.

Harvey, A. G., & Bryant, R. A. (1999b). Predictors of acute stress following motor vehicle accidents. *Journal of Traumatic Stress, 12*, 519-525.

Hobbs, M., Mayou, R., Harrison, B., & Worlock, P. (1996). A randomised controlled trial of psychological debriefing for victims of road traffic accidents. *British Medical Journal, 313,* 1438-1439.

Holeva, V., Tarrier, N., & Wells, A. (2001). Prevalence and predictors of acute stress disorder and PTSD following road traffic accidents: Thought control strategies and social support. *Behavior Therapy, 32*, 65-83.

Jeavons, S.(2000). Predicting who suffers psychological trauma in the first year after a road accident. *Behaviour Research and Therapy, 39,* 499-508.

Koren, D., Arnon, I., & Klein, E. (1999) Acute stress response and posttraumatic stress isorder in traffic accident victims: A one-year prospective, follow-up study. *American Journal of Psychiatry, 156*, 367-373.

Marmar, C. R., Weiss, D. S., & Metzler, T. J. (1997). *The Peritraumatic Dissociative Experiences Questionnaire.* In Wilson, J. P., & Keane, T. M. (eds.) *Assessing Psychological Trauma and PTSD.* New York: The Guilford Press.

Mayou, R. A., Ehlers, A., & Hobbs, M. (2000). Psychological debriefing for road traffic accident victims: Three-year follow-up of a randomised controlled trial. *British Journal of Psychiatry, 176,* 589-593.

McFarlane, A. C., Atchison, M., & Yehuda, R. (1997). The acute stress response following motor vehicle accidents and its relation to PTSD. *Annals of the New York Academy of Sciences, 821*, 437-441.

Mellman, T. A., David, D., Bustamante, V., Fins, A. I., & Esposito, K. (2001). Predictors of post-traumatic stress disorder following severe injury. *Depression and Anxiety,* 14, 226-231.

Murray, J., Ehlers, A., & Mayou, R. A. (2002). Dissociation and post-traumatic stress disorder: Two prospective studies of road traffic accident survivors. *British Journal of Psychiatry, 180,* 363-368.

Pitman, R. K., Sanders, K. M., Zusman, R. M., Healy, A. R., Cheema, F., Lasko, N. B., et al. (2002). Pilot study of secondary prevention of posttraumatic stress disorder with propranolol. *Biological Psychiatry, 51*, 189-142.

Shalev, A. Y., Freedman, S., Peri, T., Brandes, D., Sahar, T., Orr, S. P., et al. (1998a). Prospective study of posttraumatic stress disorder and depression following trauma. *American Journal of Psychiatry, 155*, 630-637.

Shalev, A. Y., Sahar, T., Freedman, S., Peri, T., Glick, N., Brandes, D., et al. (1998b). A prospective study of heart rate response following trauma and the subsequent development of posttraumatic stress disorder. *Archives of General Psychiatry, 55*, 553-559.

Spielberger, C. D., Gorushch, R. L., Lushene, R. E., Vagg, P. R., & Jacobs, G. A. (1983). *Manual for the State-trait Anxiety Inventory*. Palo Alto, CA: Consulting Psychologists Press.

Ursano, R. J., Fullerton, C. S., Epstein, R. S., Crowley, B., Kao, T. C., Vance, K., et al. (1999). Acute and chronic posttraumatic stress disorder in motor vehicle accident victims. *American Journal of Psychiatry, 156*, 1808-1810.

Weathers, F. W., Litz, B. T., Herman, D. S., Huska, J. A., & Keane, T. M. (1993). *The PTSD Checklist: Reliability, validity, & diagnostic utility*. Paper presented at the Annual Meeting of the International Society for Traumatic Stress Studies. San Antonio, TX, October.

In: Contemporary Issues in Road User Behavior and Traffic Safety ISBN:1-59454-268-6
Editors: D. Hennessy and D. Wiesenthal, pp. 215-224

Chapter 17

SUPPLEMENTAL SPEED REDUCTION TREATMENTS FOR RURAL WORK ZONES

Eric D. Hildebrand, Frank R. Wilson and James J. Copeland
Transportation Group, University of New Brunswick

INTRODUCTION

In September 2002, the Province of New Brunswick announced their mandate to complete the remainder of the Trans-Canada Highway network as a four-lane divided highway. New Brunswick drivers have a reduced familiarity with operating their vehicles on a high-speed, four-lane, rural arterial highway. This is due to a predominantly rural population and the fact that the majority of the existing provincial highway system is comprised of rural two-lane, two-way arterial highways. Collision data, provided by the Department of Transportation, demonstrate a relatively low number of collisions and fatalities at rural highway temporary work areas in recent years. However, with the completion of a large number of kilometers of rural divided highway designed to high-speed standards (75 mph or 120 km/h), the need to better manage speeds through temporary work zones will be crucial.

One technique to analyze the traffic safety conditions at temporary work areas in New Brunswick would be to evaluate past accident collision experiences and attempt to model future collisions in order to circumvent potential problems. However, this method is data intensive, procedures are laborious, and as identified by Wang, Hughes, Council and Paniati (1996), critical voids exist in current collision databases and procedural methods to collect these data. Therefore, in recognition of traffic safety concerns in the study area, this research addressed speed management strategies for rural highway temporary work zones through field evaluation of select traffic safety enhancements.

Installation information and documentation of supplementary traffic control devices and their effect on vehicle speed management is available from studies in the United States and Europe where inconsistent results were documented. Several provincial agencies in Canada have included documentation on the use of specific safety enhancement devices at construction work zones. However, it is not clear if these recommended practices are based on local evaluation and experience, or based on results from other jurisdictions. As a result, further study was undertaken relative to conditions at rural highway temporary work zones within the study area. The objective of the project was to evaluate select safety enhancements that effectively reduce mean and 85th percentile vehicle speeds without compromising safety by inadvertently increasing speed variance. The measures chosen for study included portable changeable message signs, portable rubber rumble strips, transverse pavement markings (sometimes referred to as optical speed bars), and fluorescent orange construction sign materials.

FINDINGS FROM PREVIOUS STUDIES

An extensive literature search was conducted in order to synthesize previous studies conducted in both North America and elsewhere (Copeland, 2003). Issues relating to operating speed and posted speed limits, collision data analysis experience, and the application of specific supplementary traffic control devices were quantified.

A related study undertaken by Sargeant (1994), reviewed vehicle speed behavior and variability at temporary work zones in New Brunswick. Speed data were collected from 37 daytime locations at 15 Department of Transportation sites involving temporary work area conditions. Sargeant concluded that the 85th percentile speed did not significantly change when the normally posted speed limit was reduced at a temporary construction work zone. Warning signs (construction orange in color), the only device used in the Sargeant study, alone did not significantly reduce vehicle speeds.

Knowles, Persaud, Parker and Wilde (1997) attempted to assimilate results of several Canadian studies that reviewed the relationship between vehicle speeds and traffic safety. Their conclusions were that much of the research was anecdotal and lacked significant evidence. Research studies in Sweden by Nilsson (1990), and in Denmark by Christensen (1981) evaluated the reduction of maximum speed limit and its effect on traffic safety. Results of both studies were similar where Nilsson found that there was a decrease in mean speeds of vehicles and injury severity and Christensen found a decrease of mean speeds by 4-9 km/h and a 20% decrease of injury collisions. These studies did not account for the effect of increased enforcement and public awareness campaigns on the decrease of maximum speed limits.

Solomon (1964) published research work on conclusions of the relationship between speed variance and traffic safety. Essentially, it was concluded that the more a motorists' speed deviates from the mean, the greater their risk of collision. This premise that speed variance compromises safety was later confirmed by Hauer (1971). This study reviewed the number of passive and active maneuvers and how these relate to Solomon's U-shaped curve. Speed variance and risk of collision were again confirmed by more recent research performed by Harkey, Robertson and Davis (1990) and Fildes, Rumbold and Leening (1991).

After reviewing the literature on speed studies relating to traffic safety and evaluation studies of supplementary traffic control devices, two significant conclusions emerge. The first finding suggests that collision rates drop with reduced posted speed limits. The second conclusion (more concrete) is that reducing speed limits, for example, in a transition into a temporary work zone, often increases speed differentials and variability. This results in a higher collision rate thereby worsening traffic safety conditions for motorists.

A study to determine Portable Changeable Message sign effectiveness was performed by Wang, Dixon and Jared (2002). This study reviewed the effectiveness of a sign, supplemented with radar, at temporary work zones. It was concluded that, on average, the changeable message signs with radar reduced speeds by 11-13 km/h. Furthermore, variance of the observed data decreased in response to the sign use. The study was conducted over a three-week period and values of speed reduction and variance were sustained over this period. It was therefore concluded that a novelty effect did not exist for changeable message signs.

Conditions presented by a temporary work zone appear ideal for a safety-enhancing device such as a portable rumble strip pattern. However, a review of the effectiveness of rumble strips applied at temporary work zones by the Federal Highway Administration (FHWA) and Noel, Sabra and Dudek (1989) indicate that there are a limited number of studies and their findings are inconsistent. Harwood (1993) reinforces these conclusions by stating "…the evidence as to whether rumble strips are effective as a speed control device in work zones is inconclusive" (p. 21). Although, Meyer (2000) also found them to be ineffective, Fontaine and Carlson (2001) found them to be somewhat effective in rural applications. They found mean passenger car speeds were reduced by 1.6 and 3.2 km/h while heavy vehicles were observed to reduce speed by 4.8 and 6.4 km/h. The greatest speed reduction for all vehicles was observed immediately downstream of the rumble strips.

Optical treatments installed at highway temporary work zones are intended to capture a driver's attention to make them aware or more alert to potentially hazardous roadway conditions. A form of positive guidance, optical treatments have existed for many years, but have not been widely used in North America. In 1982, the City of Calgary conducted an experiment where transverse optical speed bars were employed on a highway exit ramp to reduce the potential for collision (Liebel & Bowron, 1984). The Before-and-After study observed speeds at a point 150 m from the terminus of the ramp. Results showed that there was a reduction of 2.1 km/h in the mean speed and a reduction of vehicles exceeding 80 km/h by 1.4 percent. Based on these findings, the researchers concluded that these pavement markings might reduce crash severity. Another Before-and-After study conducted by Agent (1980) recorded collision data and vehicle speeds at a hazardous horizontal curve on a two-lane, two-way road. Transverse bars were implemented as a warning device to enhance safety. Average vehicle speeds were reduced by 15.1 km/h one week after installation, and by 10.8 km/h after a six-month period. These speed reductions were found to be statistically significant. More research and field evaluation is required to determine optimal conditions of application for transverse optical speed bar enhancements. In addition, research is needed to determine an effective pattern of markings that are applicable to temporary work areas.

Hummer and Scheffler (1999) undertook a field evaluation of fluorescent orange sign sheeting at temporary work zones on high-speed highways (90–105 km/h) in North Carolina. The objectives of that study were to determine if the use of fluorescent orange sign sheeting would affect driver behavior, measured in the form of aggressive vehicle maneuvers, percent of vehicles in the left lane (lane where closure occurred), mean vehicle speed and speed variance. Mean vehicle speed reductions were not statistically significant and were observed to increase by 1.6 km/h, while variability of speeds decreased. It was concluded that the use of fluorescent orange sign sheeting at temporary work zones on high-speed facilities is recommended and the higher cost of material outweighs the safety benefits of reduced collision frequencies. A more recent study of the field performance of fluorescent orange colors on static temporary condition signs was conducted by Wang et al. (2002). The effects of speed reduction, speed variability and novelty effect were recorded in Georgia at three construction work zones. It was determined that speeds were reduced by 2-5 km/h, however, to the detriment of safety, speed variance increased during daylight conditions.

STUDY METHODOLOGY

At the onset of the project, an inventory of safety-enhancing products were assembled that had the potential to achieve the following criteria: speed reduction, decreased operating speed variability, economical feasibility (both purchase costs and maintenance costs) and ease of installation and removal. An iterative process began that narrowed the number of safety-enhancing devices to four. This process involved research of past experiences with a particular product, its approval by the research team, its approval by the New Brunswick Department of Transportation (in terms of monetary and labor implications) and finally, its availability. The four selected supplementary traffic control devices are listed in Table 1. A summary of site characteristics and sampling scheme is also provided. In all, 20 sets of observations were undertaken.

Table 1. Site Observation Summary

Supplementary Traffic Control Device Type	Number of Observations		Highway Classification	Reduced Posted Speed Limit
	Day	Night		
Portable Changeable Message Sign	8	2	RAD 120	70 km/h
Portable Rubber Rumble Strips	4	0	RAD120/RAU100	70 km/h
Transverse Pavement Markings	2	2	RAD 120	70 km/h
Fluorescent Orange Sign Sheeting	1	1	RAD120	Not Reduced

RAD120 – rural, arterial, divided, 120 km/h design speed.
RAU100 – rural, arterial, undivided, 100 km/h design speed.

In this research, the measurement of the effectiveness of supplementary traffic control devices was performed using a typical Before-and-After study procedure. The 'Before' condition consisted of radar measurements of operating speeds at established rural highway temporary work zones. These temporary work areas were designed based on the recommendations contained in the Work Area Traffic Control Manual (New Brunswick Department of Transportation, 1994). The 'After' condition was represented by recording operating speeds of vehicles after a selected supplementary traffic control device was installed at the same temporary work zone. The differences in the results from the 'Before' and 'After' phase demonstrate the relative effectiveness or ineffectiveness of a particular safety-enhancing device. Typical data sets comprised of approximately 100 speed observations during daytime testing. Under night conditions, sample sizes were reduced to 50 observations due to lower volumes of traffic. Generally, the location of recorded observations were taken upstream or prior to advance construction signing, immediately upstream of the supplementary traffic control device, and downstream of the traffic safety-enhancing location.

Study Findings

Data were collected consistent with a traditional Before-and-After study format. Several statistics were chosen as indicators of the effectiveness for the supplementary traffic control devices. These included mean vehicle operating speed, 85th percentile operating speed, 15 km/h pace, percent of vehicles in pace, standard deviation of mean operating speeds, coefficient of variation and a measure of kurtosis. Part of the analysis process was to test the statistical significance of the Before-and-After results at each test site. Tests for comparison of sample means and sample variances were both undertaken using a 5% significance level.

Portable Changeable Message Signs

A portable changeable message sign (PCMS) was tested for effectiveness as a supplementary traffic control device at ten rural highway temporary work zones. Specific characteristics were measured between 'Before' and 'After' phases to demonstrate whether the PCMS was an effective safety-enhancing device.

A summary of changes in operating speed characteristics after the application of the PCMS at temporary work zones are contained in Table 2. This table summarizes results of analyses using observational operating speeds of all vehicles traveling through rural highway temporary work zones. Mean and 85th percentile speeds were reduced on average by 4.6 km/h and 5.7 km/h, respectively. However, measures of variability, standard deviation and percent of vehicles in pace, increased by 0.29 km/h and decreased by 1.3%, respectively. These values represent an average for all ten temporary work zones where the PCMS was installed. Uniformity of results between the ten work zone configurations was not observed and must be considered when interpreting these results. Eight of ten sites demonstrated a statistically significant reduction of mean speed and three of ten sites demonstrated a significant reduction in variance. Six sites demonstrated a significant increase in variance. There were, however, no significant differences between the day or night condition results.

Table 2. PCMS Effect on Speed Characteristics

Work Zone	Change in Mean Speed (km/h)	Change in 85th Percentile Speed (km/h)	Change in Percent in Pace (%)	Change in Std. Deviation (km/h)
Rte 1 – site 1	-2.8*	-5.2	-6.0	+0.74*
Rte 1 – site 2	-3.3*	-4.7	+6.0	+0.26
Rte 1 – site 3	-3.6*	-6.3	-3.5	-0.61*
Rte 1 – site 4	-4.5*	-1.8	-7.0	+1.41*
Rte 1 – site 5	-7.4*	-10.0	+2.0	-0.25*
Rte 1 – site 6	-6.8*	-7.7	+12.0	+0.54*
Rte 1 – site 7	-1.2	-2.2	-12.0	+0.64*
Rte 2 – site 1	-9.2*	-5.8	-11.0	+1.97*
Rte 2 – site 2	+0.7	-3.8	+11.0	-2.10*
Rte 2 – site 3	-8.0*	-9.0	-4.0	+0.31*
Average	-4.6 km/h*	-5.7 km/h*	-1.3%	+0.29 km/h

* Statistically significant at 5% significance level.

Overall the application of the portable changeable message sign at rural highway temporary work zones appeared to improve safety conditions for motorists based on the analysis of specific speed characteristics. This conclusion is consistent with past research findings that demonstrate an increase in traffic safety through the reduction of mean speed (Christensen, 1981; Nilsson, 1990). The average mean speed reduction was statistically significant, however, the average operating speed variability increased, on average, at the ten sites, but was not statistically significant. Based on the work of Solomon (1964), the PCMS installation can therefore be considered likely to decrease a motorist's risk of collision. A study by Wang et al. (2002) was confirmed in terms of the PCMS installation effect on operating speed reduction, however, the reduction in speed variance was not confirmed.

PORTABLE RUBBER RUMBLE STRIPS

A summary of changes to operating speed characteristics following the application of portable rubber rumble strips at temporary work zones is presented in Table 3. Based on average data from the four test sites, rumble strips reduced the mean and 85th percentile speeds by 6.9 km/h and 9.5 km/h, respectively. The standard deviation of operating speeds was reduced by an average of 0.86 km/h. The average mean speed, 85th percentile speed and percent of vehicles in pace were found to improve statistically. The average standard deviation of the four test sites was not found to be statistically significant. Overall, the application of portable rubber rumble strips at rural highway temporary work zones appeared to improve safety conditions for motorists traveling through the work zones. This suggestion is based on the speed-safety relationships established by Solomon (1964), Christensen (1981) and Nilsson (1990). Research study findings relating to the evaluation of portable rumble strips confirmed previous work by Fontaine and Carlson (2001), yet contradicted the findings of Meyer (2000).

Table 3. Rumble Strip Effect on Speed Characteristics

Site	Change in Mean Speed (km/h)	Change in 85th Percentile Speed (km/h)	Change in Percent in Pace (%)	Change in Std. Deviation (km/h)
Rte 1 – site 1	-9.0*	-10.1	+3.0	-1.13*
Rte 1 – site 2	-5.6*	-5.0	-2.0	+0.81*
Rte 10 – site 1	-7.4*	-12.5	+15.2	-1.57*
Rte 10 – site 2	-5.7*	-10.5	+9.2	-1.56*
Average	-6.9 km/h*	-9.5 km/h*	+6.4%*	-0.86 km/h

* Statistically significant at 5% significance level.

TRANSVERSE (OPTICAL) SPEED BARS

The transverse speed bars were tested for effectiveness as a supplementary traffic control device at one rural highway temporary work zone. Testing of this device was conducted over a five-week period to determine immediate, and long-term impacts. Results of the analysis of the research data illustrated the novelty effect of the transverse speed bars and compared effectiveness during day and night conditions. A summary of the findings from the analyses after the application of the transverse speed bars at temporary work zones is presented in Table 4. This Table summarizes results of analyses using observational operating speeds of vehicles traveling through the rural highway temporary work zones on Route 1 in southern New Brunswick.

Results of the Before-and-After analysis for application of the transverse speed bars demonstrate a slight decrease in the average mean operating speed and a statistically significant reduction in standard deviation. The results show an improved effectiveness during night conditions when compared to results from data collected during daylight hours. Mean and 85th percentile speeds were reduced (statistically significant) on average by 3.4 km/h and 3.8 km/h, respectively. Measures of variability, percent of vehicles in the 15 km/h pace and standard deviation increased by 2.6% and decreased significantly by 0.94 km/h, respectively. These values represent an average of the four samples (two during day conditions, two during night conditions) collected in the Route 1 temporary work zone. One of four samples demonstrated a significant reduction of mean speed. This occurred during night conditions. Conversely, all four sites demonstrated a significant reduction in variance.

Table 4. Transverse (Optical) Speed Bar Effect on Speed Characteristics

Site	Change in Mean Speed (km/h)	Change in 85th Percentile Speed (km/h)	Change in Percent in Pace (%)	Change in Std. Deviation (km/h)
Rte 1 – site 1, day	-2.4	-3.2	-5.0	-0.24*
Rte 1 site 2, day	+0.6	-0.5	+3.0	-1.55*
Rte 1 – site 1, night	-7.7*	-7.4	+9.0	-1.44*
Rte 1 –site 2, night	-4.0	-3.9	+3.5	-0.53*
Average	-3.4 km/h*	-3.8 km/h*	+2.6%	-0.94 km/h*

* Statistically significant at 5% significance level.

Further comment can be made on the illusionary effect that is associated with transverse speed bars. The design and layout used at the temporary work zones did not appear to provide a strong impact based solely on mean speed reduction. However, it is evident from the results that there is likely an increased level of safety for motorists during night conditions. This may be attributed to increased contrast of the retro-reflective marking tape at night versus daytime.

Fluorescent-Orange Sign Sheeting

Fluorescent orange sign sheeting was tested for effectiveness as a supplementary traffic control device at one rural highway temporary work zone, with four test observation periods being completed. Specific characteristics were measured between 'Before' and 'After' phases to demonstrate whether the fluorescent signing was an effective safety-enhancing device. A summary of results of the analysis after the application of the fluorescent orange sign sheeting at temporary work zones is contained in Table 5. This table summarizes results of the analysis of observational operating speeds of vehicles traveling through the rural highway temporary work zone on Route 1. Study results of the Before-and-After analysis for application of the fluorescent orange signs demonstrate mixed results of safety condition indicators at rural highway temporary work zones. Average mean and 85th percentile speeds were reduced by 3.8 km/h and 0.9 km/h, respectively, and were not found to be statistically significant. Standard deviation increased by 3.22 km/h and the percent of vehicles in the 15 km/h pace decreased by 8.5%, both were found to be statistically significant. These values are an average of both test sites recorded (one during day conditions, one during night conditions) at the Route 1 temporary work zone. Only the mean speed under night conditions demonstrated a statistically significant improvement.

Overall, the application of the fluorescent orange sign sheeting at the rural highway temporary work zones did not appear to improve traffic safety conditions for motorists based on the analysis of specific speed characteristics. A relatively large increase in operating speed variance was observed resulting in a potential increased risk of collision based on conclusions of Solomon (1964). Conversely, the findings of Nilsson (1990) and Christensen (1981) relate to a potential increase in traffic safety due to a reduction of average mean speed. Specific studies evaluating the performance of fluorescent orange sign sheeting were confirmed on one account and contradicted on another. The Wang et al. (2002) research conclusions were confirmed as speeds were reduced and variability increased, however, the findings of Hummer and Scheffler (1999) were quite different than observed for this research.

Table 5. Fluorescent Orange Sign Effect on Speed Characteristics

Site	Change in Mean Speed (km/h)	Change in 85th Percentile Speed (km/h)	Change in Percent in Pace (%)	Change in Std. Deviation (km/h)
Rte 1 – site 1	-1.1	0.0	-9.0	+2.68
Rte 1 – site 2	-6.5*	-1.8	-8.0	+3.76
Average	-3.8 km/h	-0.9 km/h	-8.5%*	+3.22 km/h*

* Statistically significant at 5% significance level.

It must be noted that this evaluation was limited to one test location, under varied light conditions. Consideration, when interpreting the results, should be given to the positive guidance and human factors elements that are provided by the installation of fluorescent sign sheeting. A study by Schnell, Bentley, Hayes and Rick (2001) demonstrated an increased recognition distance improving the reaction times for motorists. Human factors and positive guidance were not included in the scope of this research.

Conclusions and Discussion

This research has shown that the PCMS, portable rubber rumble strips, and transverse (optical) speed bars are all able to statistically reduce mean operating speeds through rural temporary work zones. A reduction in operating speeds has the potential to decrease both accident frequency and severity. Mean speed reductions were found to be 4.6 km/h, 6.9 km/h and 3.4 km/h, respectively. Portable rubber rumble strips were shown to produce the most significant reduction, likely due to the tactile response the driver experiences. In fact, the 85th percentile speed (which is used to establish speed limits) was reduced by 9.5 km/h, on average, at those sites employing portable rubber rumble strips. Furthermore, both the portable rubber rumble strips and transverse speed bars were able to significantly lessen the dispersion of individual speeds as measured by either standard deviation or the percent of vehicles in the pace. The result should be improved safety levels.

The performance of fluorescent orange sign sheeting yielded mixed results. Daytime results showed no statistical differences, however, a significant reduction in mean speed of 6.6 km/h was observed during nighttime periods. Unfortunately, the dispersion of individual speeds was found to significantly increase, thereby possibly countering any gains in mean speed reduction. The transverse (optical) speed bars were found to yield better results at nighttime. This is likely due to the increased contrast between the retro-reflective pavement markings and a darkened environment. This performance characteristic could be useful if these treatments are used for high incident areas that are problematic at night (e.g., in advance of poorly delineated curves or intersections). Furthermore, they may provide a useful supplemental treatment for those work zones that are active at nighttime.

While this study has observed changes in driver behavior that are quantitative, what has not been evaluated are more subjective impacts such as changes in driver alertness/awareness. It is quite possible that with exposure to any of these supplementary treatments, drivers may become more attentive and watch for workers or equipment throughout the work zone. This behavioral change may not be necessarily reflected in vehicle speed. Consequently, the net impact any of these measures has on safety may be underestimated. Further study is needed in this regard. There is always concern that impacts of novel treatments might wane with either wide-spread or extended use. More research is needed to explore the long-term effects of application, effectiveness of installation at various locations within a work zone, or methods for shorter installation /removal times. Currently, many work area traffic control manuals do not discuss the use of supplementary traffic control devices to enhance safety at temporary work zones. It is recommended that policies be updated to reflect and accommodate the possible use of these devices given the positive findings of this and previous studies.

REFERENCES

Agent, K. R. (1980). Transverse pavement markings for speed control and accident reduction. *Transportation Research Record, 773*, 11-14. Washington, D.C.: Transportation Research Board.

Christensen, J. (1981). The effects of general speed limits on driving speeds and accidents in Denmark. *Proceedings of the International Symposium on the Effects of Speed Limits on Traffic Accidents and Fuel Consumption*. Organization of Cooperation and Development: Dublin, Ireland.

Copeland, J. J. (2003). Speed management strategies for rural temporary work zones. *Masters thesis, University of New Brunswick, Fredericton, New Brunswick, Canada.*

Fildes, B. N., Rumbold, G., & Leening, A. (1991). Speed behaviour and driver's attitude towards speeding. *Report No. 16. Clayton, Victoria, Australia: Monash University Accident Research Centre for Victoria State Road Authority.*

Fontaine, M. D., & Carlson, P.J. (2001). Evaluation of speed displays and rumble strips at rural-maintenance work zones. *Transportation Research Record, 1745*, 27-38. Washington, D.C.: Transportation Research Board.

Harkey, D. L., Robertson, H. D., & Davis, S.E. (1990). Assessment of current speed zoning criteria. *Transportation Research Record, 1281*, 40-51. Washington, D.C.: Transportation Research Board.

Harwood, D. W. (1993). Synthesis of highway practice: Use of rumble strips to enhance safety. *National Cooperative Highway Research Program Synthesis, 191*. Washington, D.C.: Transportation Research Board, National Research Council.

Hauer, E. (1971). Accidents, overtaking and speed control. *Accident Analysis and Prevention, 3*; 1-14.

Hummer, J. E., & Scheffler, C. R. (1999). Driver performance comparison of fluorescent orange to standard orange work zone traffic signs. *Transportation Research Record, 1657*, 55-62. Washington, D.C.: Transportation Research Board.

Knowles, V., Persaud, B, Parker, M., & Wilde,G. (1997). Safety, speed and speed management: A Canadian review (File No. ASF 3261-280). Ottawa, ON: Road Safety and Motor Vehicle Regulation, Transport Canada.

Liebel, D. J., & Bowron, D. J. (1984). Use of optical speed bars to reduce accidents – The Calgary experience. *Proceedings of the International Transport Congress*. Ottawa, Canada.

Meyer, E. (2000). Evaluation of orange removable rumble strips for highway work zones. *Transportation Research Record, 1715*, 36-42. Washington, D.C.: Transportation Research Board.

New Brunswick Department of Transportation. (1994). *Work area traffic control manual*. Fredericton, New Brunswick: N.B. Department of Transportation Maintenance and Traffic Branch.

Nilsson, G. (1990). *Reduction in the speed limit from 110 km/h to 90 km/h during summer 1989: Effects on personal injury accidents, injured and speeds* (VTI Rapport 358A). Linkoping, Sweden: Swedish Road and Traffic Research Institute.

Noel, E. C., Sabra, Z. A., & Dudek,C. L. (1989). *Work zone traffic management synthesis: Use of rumble strips in work zones* (Report No. FHWA-TS-89-037). McLean, Virginia: Federal Highway Administration, U.S. Department of Transportation.

Sargeant, S. (1994). Safety evaluation of reducing speeds in construction zones. Unpublished undergraduate senior report, Department of Civil Engineering, University of New Brunswick, Fredericton, New Brunswick.

Schnell, T., Bentley, K., Hayes, E., & Rick, M. (2001). Legibility distances of fluorescent traffic signs and their normal color counterparts. *Proceedings of the 80th Annual Meeting of the Transportation Research Board* (Publication No.01-2417). Washington, D.C.: National Academy of Sciences.

Solomon, D. (1964). *Accidents on main rural highways related to speed, driver and vehicle*. Washington, D.C.: U.S. Department of Commerce, Bureau of Public Roads.

Wang, J., Hughes, W. E., Council, F. M., & Paniati, J. F. (1996). *Investigation of highway work zone crashes: What we know and what we don't know. Transportation* Research Record, 1529, 54-62. Washington, D.C.: Transportation Research Board.

Wang, C., Dixon, K. K., & Jared, D. (2002). Evaluation of speed reduction strategies for highway work zones. *Proceedings of the 82nd Annual Meeting of the Transportation Research Board* (Publication No. 2003-2099). Washington, D.C.: National Academy of Sciences.

PART 6
ENGINEERING/HUMAN FACTORS

In: Contemporary Issues in Road User Behavior and Traffic Safety ISBN:1-59454-268-6
Editors: D. Hennessy and D. Wiesenthal, pp. 227-243

Chapter 18

IS IT SAFE TO USE A CELLULAR TELEPHONE WHILE DRIVING?

David L. Wiesenthal and Deanna Singhal
York University

INTRODUCTION

The sight of a motorist in animated conversation on a cellular telephone has become increasingly more common the world over. The cellular telephone has become a ubiquitous part of modern day life and, as such, it was inevitable that it would be used by drivers. Indeed, in North America, it is referred to as a mobile telephone more commonly than a wireless telephone. Currently, in North America, only Newfoundland and New York State have banned the use of hand-held cellular telephones while operating a motor vehicle, but since 1999, every American state has considered some form of legislative regulation of cellular telephones in vehicles, with 42 states considering legislation during the 2003 session (Sundeen, 2003). Sundeen reports that 40 countries either restrict or prohibit the use of wireless telephones while driving. Drivers in the Czech Republic, France, the Netherlands, and the United Kingdom using cellular telephones, who crash, may be liable to being fined. German drivers and those in the UK may suffer loss of insurance coverage should they collide while operating their cellular telephones. Recently, it has been estimated that 128 million cellular telephones are in use in the United States (Cohen & Graham, 2003) with 11.2 million registered in Canada in 2002 (Harbluk, Trbovich, Noy, Eizenman & Lochner, 2003). Sundeen (2003) reports a doubling of cellular telephone ownership by Americans since 1998. With Americans spending an estimated 500 million hours each week in their vehicles, and considering the estimate that the average American spends more than 300 hours each year in their automobiles (Sundeen), the potential for cellular telephone conversations is considerable.

Canada's Traffic Injury Research Foundation (TIRF) conducted a nationwide survey and found that 40% of Canadians believed that driver distraction constituted a serious problem, with two-thirds of respondents describing cellular telephone use by drivers to be a "'serious" or "extremely serious" problem (Beirness, Simpson & Pak, 2002). The TIRF survey estimated that 4.3 million drivers placed cellular telephone calls while driving over a one-week period and found that cellular telephone users tend to have the following characteristics: they are male, younger drivers, have a job requiring driving, live in urban areas, consume alcohol and then drive, and have received a traffic ticket (Beirness et al., 2002). As these drivers are already in the high-risk category of motorists, their use of cellular telephones notwithstanding (Evans, 1991), the fact that they are riskier drivers further complicates concluding that cellular telephone operation while driving is hazardous.

The public is becoming more aware of the dangers of inattention associated with cellular telephone use while driving and are seeking government regulations to control such hazards ("Drivers back ban," 2002). With such a large number of cellular telephones in use in the United States and an increasing number of people using them while commuting, it is essential to understand the risks they pose. Legislation may be often driven by social and political concerns rather than by the actual danger that cellular telephones may present. If the public, the media, and the police have identified cellular telephone use by motorists as posing a threat, then banning their operation by drivers seems a likely outcome.

Observational studies of cellular telephone use by drivers has indicated that 2.7% of Michigan drivers were using their hand-held cell phones while driving during the day (Eby & Vivoda, 2003). In comparison, 1.5% of drivers in Perth, Australia (Hornberry, Bubnich, Hartley & Lamble, 2001) and 3% of American drivers (Utter, 2001) were seen using their telephones while operating vehicles.

Figure 1 shows the growing popularity of cellular telephones nationally in Canada plotted against Ontario collision data (comparable national data was not readily available). It is reasonable to assume that at least a fraction of cellular telephone calls are placed by a vehicle operator while the automobile or truck is in motion. Yet the graph shows a *decrease* in accidents (adjusted for kilometrage) for the same time period, characterized by an explosive growth in the use of cellular telephones. Despite the increase in the Canadian cellular telephone possession rate over the last ten years (E-Community Link, 2003), the incidence of motor vehicle collisions has decreased (Ministry of Transportation, 2003). This presents a paradoxical situation, as it has been suggested that cellular telephone use while driving is detrimental, yet this data suggests otherwise. A number of factors could explain why this risk is not reflected in this data. First, not all cellular telephone users are necessarily drivers, nor are drivers necessarily making calls from their vehicles. Second, other factors which cause collisions, such as driving under the influence of alcohol, may be on the decline. Third, it is difficult to determine if an individual involved in the accident was using a cellular telephone while driving.

Research investigating possible risks inherent in operating cellular telephones while driving has employed diverse methodologies. The data produced from these methodologies often are contradictory, further complicating the desire for a simple statement regarding cellular telephone safety that would guide public policy. Epidemiological research, accident database analysis, laboratory simulations, and instrumented vehicles have all contributed to the growing literature on possible hazards.

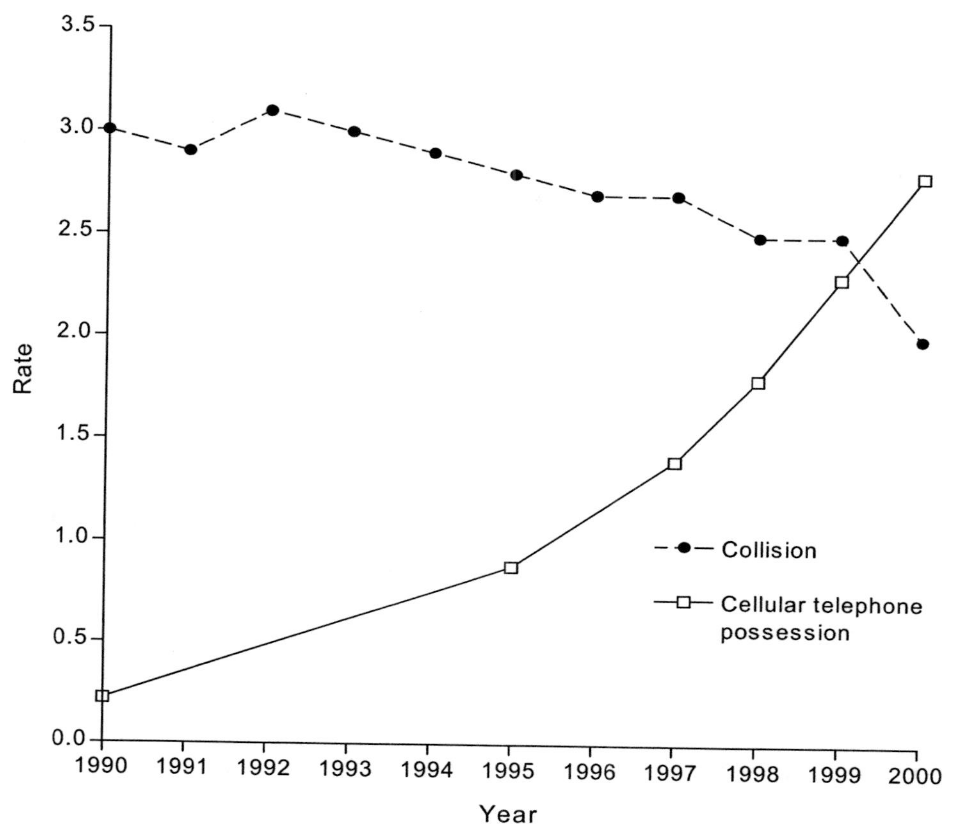

Figure 1. Changes in Accident and Cellular Telephone Possession Rate from 1990-2000 (Note: Cellular possession rate per 10 Canadians. Collision rate per one million kilometres travelled for Ontario (Ministry of Transportation, 2003).

EPIDEMIOLOGICAL RESEARCH

Redelmeier and Tibshirani (1997) investigated the relationship between cellular telephone use and accident involvement in Toronto drivers employing a case-crossover design where each driver was their own control. Case-crossover designs are used in epidemiology to assess transient risk where each subject ("case") serves as their own control. The investigators examined the cellular telephone records of 699 drivers who reported only property damage collisions at official accident reporting stations in Toronto. Billing records indicated that 170 drivers had used their mobile telephones during a 10 minute period prior to their collision. Compared to a control period (days prior to the crash), collision risk was 4.3 times higher when a telephone was used (Redelmeier & Tibshirani, 1997). More specifically, they found the relative risk to increase 4.8-fold for calls, on either a hand-held or hands free model, placed within 5 minutes of the collision in comparison to a 1.3 risk of collision when the calls were placed 15 minutes prior to the crash.

Maclure and Mittleman (1997) have criticized Redelmeier and Tibshirani's (1997) research sample size as lacking statistical power to adequately test the safety differential between the use of hands-free and hand-held cellular telephones. Maclure and Mittleman (1997) believe the relative risk increases to as high as 6.8-fold. However, caution needs to be taken when interpreting these findings. For example, the researchers did not control for the total amount of time driving and there is evidence from the Canadian TIRF survey, that cellular telephone users are driving greater distances and are already in high-risk categories (Beirness et al., 2002). Perhaps the individuals in the study were driving five times further or more often than others. The authors also assumed that drivers were aware of the exact time of their mishap. Furthermore, there is no information that indicates that the drivers with telephones were at fault for causing the collision, or that a distressing telephone conversation (e.g., learning of bad news) produced an emotional reaction that impaired the driver's safe operation of the vehicle. Also, we do not know if using a cellular telephone might have reduced driver stress (see Wiesenthal, Hennessy & Totten, 2000), which in turn, could have averted a collision. Being able to telephone an individual at a driver's destination and inform them that a delay has occurred on the road might reduce the motivation for speeding or other forms of reckless driving. Further clouding interpretation is the problem of estimating the proportion of drivers who safely use mobile telephones *without* experiencing collisions.

Violanti and Marshall (1996) utilized a case-control design to analyse data from drivers in New York State who filed accident reports because of property damage, personal injury, or both. This design compares affected individuals (the "cases") with controls, matched on particular characteristics. In case-control designs, the independent variables are the characteristics of interest, with the dependent variables being the resulting differences across these groups. In the Violanti and Marshall study, accident victims constituted the cases, with the controls being a random sample of New York State drivers who had been accident-free for at least 10 years and matched for the same geographic area of residence. The data indicated that talking more than 50 minutes per month while in a motor vehicle was associated with a 5.6-fold greater risk of a traffic accident.

In other research, Violanti (1998) also employed a case-control design with cases being Oklahoma drivers killed in crashes and controls being crash survivors. Both the presence and use of a cellular telephone was seen to increase the probability of a fatality. The presence of a cellular telephone doubled the risk of a fatality, while cellular telephone use increased that risk by a factor of nine. Interpretation of this data is complicated by the lack of specification of a hand-held or hands-free device and by the absence of any information regarding other distractions that might have affected the safe operation of the vehicle. Cher, Mrad, and Kelsh, (1999) have criticized Violanti's data analysis pointing out that (1.) driver characteristics should not have been entered into the regression analysis, (2.) the absence of a crude odds ratio is a cause for concern, and (3.) the lack of information on the control group, along with its disproportionately large size in relation to the small number of drivers killed while using their telephone, renders Violanti's interpretation of the results problematic.

While epidemiological designs are useful in establishing relationships, they do not necessarily provide information on cause and effect. While sudden cardiac deaths may increase following snowstorms (Chowdhury et al., 2003), it would be fallacious to conclude that snowstorms cause heart attacks. The majority of snow shovelers clearing their driveways do not collapse with heart attacks; rather, those with coronary artery disease, vulnerable to exertion, are often the victims of sudden cardiac death during snow removal efforts. Thus, while epidemiological designs may show an increased likelihood of mishap when cellular telephones are used by drivers, they are unable to provide any information on drivers who are able to safely operate their vehicles while on the telephone and they do not indicate what type of driving situation or roadway characteristic may also relate to accident causation. It may also be true that drivers using cellular telephones are engaging in other risky behaviors as well. In Eby and Vivoda's (2003) observational study of Michigan motorists using hand-held cellular telephones, it was noted that these motorists were less likely to be wearing seat belts than non-telephone using drivers. In the Canadian TIRF survey, it was seen that wireless telephone users are high-risk drivers: they are young, male, have jobs requiring driving, are urban, consume alcohol and drive, and have received a traffic ticket (Beirness et al., 2002).

Non-Epidemiological Analyses of Accident Databases

With police agencies recording vehicle crashes in large state-wide databases, accessible to outside researchers, it is possible to examine a variety of potential causes for collisions to determine the *relative* risk involved in cellular telephone use in comparison to other types of driver distraction. Reinfurt, Huang, Feaganes and Hunter (2001) examined North Carolina's database, created by the state highway patrol, which featured a new recording form that assessed the involvement of mobile telephones as a factor in accident causation. Reinfurt et al. found that over a three month period, 11 crashes out of 6,686 (0.16% or 1 in every 600 crashes), involved a cellular telephone.

Stutts, Reinfurt, Staplin and Rodgman (2001) described driver distractions based upon North Carolina accident records from 1995-1999. Using or dialling a cellular telephone accounted for a mere 1.5% of the recorded mishaps and was eighth in their list of distractions. Outside people/objects/events (29.4%), radio/cassette/CD adjustments (11.4%), vehicle occupants (10.9%), moving object in the vehicle (4.3%), other objects/devices (2.9%), vehicle climate controls (2.8%), and eating/drinking (1.7%) were all more commonly mentioned as distractions that caused the crash. In a later report, Stutts and her colleagues (Stutts et al., 2003), reported that cellular telephone use was cited as a distraction in only 5.2% of crashes recorded from 1999-2000 in a Pennsylvania database. A comparable analysis of Virginia's vehicle crash database indicated that cellular telephone use represented 3.9% of the 2,919 accidents caused by distraction (Glaze & Ellis, 2003). While the Virginia data indicates a slightly higher figure for cellular telephone involvement in accidents, it is also ranked eighth in the Virginia study following fatigue/sleep (17.0%), looking at roadside incidents (13.1%), looking at scenery (9.8%), passengers (8.7%), radio/cassette/CD operation (6.5%), daydreaming (4.3%), and eating/drinking (4.2%).

Stevens and Minton (2001) analyzed a database of 5,740 vehicle accidents where fatalities occurred over the years 1985-1995 in England and Wales. Based upon survivor testimony and evidence at the accident scene, distraction was believed to have been a factor in 2% of all accidents. Of the various distractions, interactions with passengers constituted the largest category (26%), with "other" (e.g., dealing with insects, putting on or removing gloves, sneezing, etc.) comprising 20%, adjusting the entertainment system, 19%, eating, drinking, or smoking representing 17% of accidents, and reading maps or dealing with vehicle controls each constituting another 8% of distraction related accidents. Mobile telephone use (3%) was the lowest category of distraction related mishaps.

A caveat to consider is that these database studies cover a period of time when cellular telephone ownership was not as high as is presently the case, so it might be expected that an analysis of more recent crashes would reveal a greater proportion of drivers colliding while operating their wireless devices. There are also problems related to social desirability effects in not reporting the use of cellular telephones to the police, the absence of information from witnesses, and the general problem of establishing the exact time of the crash to match it to the time a call was ongoing.

Simulation Research: Cognitive Interference

Driving simulators have been used as an alternative to 'real world' driving. Though they may have less ecological validity, they provide a safe environment, especially under taxing cognitive circumstances. Also, performance on a wide range of factors, relating to accident involvement, can be tested more extensively and accurately on a simulator than in a real vehicle, e.g., speed and off-road occurrences (Haigney, Taylor, & Westerman, 2000). Valverde (1973) suggested that duplication, in whole, of operational equipment is not required in order for a simulator to have training value. The validity of any simulation device is a relevant issue, as even the best of these devices differ from actual driving. Unlike operating an actual vehicle, drivers in simulation research need not fear death, injury or property damage, which might enhance the likelihood of test drivers taking greater risks in a simulator than in their own automobiles.

If cellular telephones serve to distract drivers, it is reasonable to investigate the effects of a variety of distractions upon safe vehicle operation. Practicality, as well as ethical considerations, dictates the use of simulation methods for data collection. Research on the cognitive interference of cellular telephone use has used a variety of devices to simulate driving. Regardless of the equipment employed, similar detriments in driving performance have been found.

Simple Cursor Pursuit Tracking

To investigate the effects of conversation on driving, Strayer and Johnston (2001) used a simple joystick pursuit tracking task as the primary task, assuming that this simulation imposes similar attentional demands as steering a vehicle. The dynamic of steering in a vehicle, however, is much different from joystick manipulation and a pursuit tracking task eliminates other task requirements of driving, such as braking and accelerating with foot pedals. Subjects manipulated the joystick to align a cursor with a moving target. When subjects saw a red target, they were instructed to press a "brake" button on the top of the joystick. While performing this task, subjects were either engaged in a conversation with the experimenter, on a handheld or hands-free telephone, about the Clinton impeachment or the Salt Lake City Olympic Committee bribery scandal (experimental group) or listening to the radio or to a book on tape (control groups).

Although there were no significant impairments for the control groups, missed red targets doubled for the experimental group and they exhibited a longer reaction time (RT) to the appearance of the red target. These findings suggest that conducting conversations while performing a simple pursuit tracking task can cause impairments. Strayer and Johnston (2001) generalized this finding to driving. The experimental (dual) condition decrements were similar for both models of telephone, suggesting that safety problems exist even when operating a hands-free telephone.

In a second experiment, Strayer and Johnston (2001) used pursuit tracking with two levels of difficulty, one with more unpredictable events and the other with a more circuitous course. The pursuit task was performed alone or with a second task, either shadowing or word generation. In shadowing, the experimenter read a word aloud with the subject repeating that word. Word generation, which entailed a word being read aloud by the experimenter with the subject vocalizing a new word beginning with the last letter of the first word, was believed to be more attention demanding as it imposed a higher cognitive load than shadowing.

When the tracking task was performed alone, the more unpredictable/difficult condition produced the significantly higher tracking error, indicating that the two levels differed in difficulty. In the dual condition, an interaction was found, demonstrating that the easier shadowing task did not produce an increase in tracking error, but word generation did, with the greatest increase occurring in the most difficult tracking condition. This suggests that more cognitively taxing telephone conversations would be most detrimental under difficult driving conditions. In more recent research, Strayer, Drews and Johnston (2003) have noted that drivers operating a simulator experienced "inattention blindness", i.e., they were less likely to notice peripheral aspects (e.g., billboards) of the roadway environment.

Driving Simulator Use

Alm and Nilsson (1994) investigated the effects of cognitive load that cellular telephone use imposes upon driving by having subjects perform the Working Memory Span Test (Baddeley, Logie, Nimmo-Smith, & Brerefon, 1985) while driving a sophisticated simulator, consisting of a Volvo 740 vehicle cab with a moving base system and wide-angle visual presentation of the driving environment. The memory task involved the presentation of sentences, in the format of "X does Y", to which subjects responded "Yes" if the sentence was sensible and "No" if it was not. Following the presentation of five sentences, subjects were asked to recall the last word of each of those sentences. The number of correct sentence judgements and correctly recalled words served as performance measures. Subjects drove along either a straight (easy driving) or curvy (difficult driving) road without the additional cognitive task, i.e., control condition, or simultaneously performing the memory task. During the course of driving, a red square appeared in the left-hand corner of the visual scene and subjects had to produce a brake response.

Lateral position, speed of driving, and subject's perceived workload, calculated from the NASA-TLX, was also measured. The NASA-TLX questionnaire was designed to assess the subject's perception of workload required to complete a task (Hart & Staveland, 1988). Various components of workload, such as mental demand, physical demand, time pressure, performance, effort, and frustration, are assessed. In the current experiment, all components of workload revealed a significant increase when driving was performed in conjunction with the memory task versus the control condition. Also, subjects were significantly more frustrated when they were performing the memory task during the hard driving condition.

The researchers found a significant increase in braking RT and a significant decrease in speed when the two tasks were combined. Contrary to expectation, these findings were only obtained in the easy driving condition. Lateral position (i.e., more deviations from the driving lane) was the only variable that significantly increased for the hard driving condition. Alm and Nilsson (1994) explained the obtained results by hypothesizing that their subjects were able to prioritize their task. If driving was the primary task, concentration would have been greatest under hard conditions and performance decrements may not have been noticeable. There was a nonsignificant tendency for the memory task to be performed more poorly under the hard driving condition, suggesting that attention was expended in concentrating on driving. Alm and Nilsson concluded that the effects of non-driving tasks are dependent upon the priority accorded by drivers.

Alm and Nilsson (1995) performed a replication using a Saab 9000 simulator. In both the control and experimental conditions, subjects encountered continuous oncoming traffic and were forced into vehicle-following situations 16 times. During 4 of the 16 times, critical safety events occurred. The events involved the lead vehicle braking and the subject responding with a braking action. In the experimental condition, subjects performed the Working Memory Span Test simultaneously. Subjects were instructed to drive as they normally would, maintaining a constant speed of 90 km/hr. Brake RT and ratings for all workload components of the NASA-TLX, except physical demand, were significantly higher for the experimental group. Also, despite the increase in brake RT, subjects did not increase their headway. The researchers suggested that driving in a vehicle-following situation, while using a cellular telephone, may increase the chance of an accident.

To further investigate how cellular telephone conversations interfere with driving, Strayer, Drews, Crouch and Johnston (in press) had subjects drive a sophisticated simulator while carrying on a conversation. The simulator included a dashboard, steering wheel, accelerator, and brake pedals, like that of a Ford Crown Victoria sedan. The driving condition was either easy, with only one lead vehicle on a multi-lane highway, or difficult, with 32 distractor vehicles passing in the left-hand lane. The lead vehicle would randomly brake, to which the subject had to produce a brake response. In the dual condition, subjects were engaged in a conversation with the experimenter, involving a topic in which they had previously expressed interest. Brake RT, brake offset time (the total amount of time the brake was depressed), and the time required for subjects to reach their minimum speed all significantly increased in the dual condition. Poor performance was exacerbated by high traffic density along with the subjects' tendency to increase their following distance in an attempt to compensate for sluggish driving behavior.

Consiglio, Driscoll, Witte and Berg (2003) employed a within-subjects design, comparing conversations with another passenger, talking over a cellular telephone (either hand-held or hands-free), listening to a radio, or experiencing a control condition without a cellular telephone or other auditory stimulation. Consiglio et al. devised a laboratory simulation of the braking response, which required their participants to lift their foot from an accelerator pedal and depress a brake pedal as quickly as possible when a red light was activated. The mean RT indicated that conversations, regardless of whether they were in-person or over a cellular telephone (regardless of whether the telephones were hand-held or hands-free) resulted in slower reactions, but radio listening did not significantly affect RT.

INSTRUMENTED VEHICLE USE

Instrumented vehicles have also been used to conduct research on the cognitive interference of cellular telephones with driving. Brookhuis, DeVries, and De Waard (1991) had subjects perform a paced serial addition task (PSAT) on the telephone while driving an instrumented vehicle behind a lead vehicle. The arithmetic task required subjects to listen to integers, spoken one at a time, and then add them. This was used to impose a fixed cognitive load on the subject while driving. The subjects drove through light motorway traffic and on a busy ring road.

Subjects' mental workload, measured by heart rate, heart rate variability, and effort, experienced a significant increase when the arithmetic task and driving were performed together versus separately. RT to speed variations of the lead vehicle significantly increased when the two tasks were combined, although, to compensate, most subjects increased their following distance. Brookhuis et al. (1991) concluded that, when operating a cellular telephone, the driver should keep ample distance from other vehicles and drive at a moderate speed in the slowest lane.

The combination of type of road and arithmetic task produced a significant interaction, with respect to rear-view mirror checks. In the most difficult driving condition (heavy ring road traffic), subjects reached a minimum level of attention that was not lowered with the addition of the arithmetic task, (i.e., number of rear-view mirror checks remained constant). In the least difficult driving condition (light motorway traffic), there was ample margin for a decrease to occur, i.e., rear-view mirror checks decreased as the driving difficulty increased. This suggests that additional cognitive load may be manageable under easier driving conditions, but not under more difficult ones.

Lamble, Kauranen, Laakso and Summala (1999) had subjects drive an instrumented vehicle behind a lead vehicle while performing a PSAT. Detection ability and brake RT were of the same magnitude as those for dialling numbers while driving, i.e., detection ability was impaired by approximately 0.5 s and brake RT by approximately 1 s. Lamble et al. pointed out that these latencies are three times larger than the deterioration that De Waard and Brookhuis (1991) found for drivers who were under the influence of alcohol, at just the legal limit. This suggests that cellular telephone use could be even more detrimental to driving than being under the influence of alcohol.

Cooper, Zheng, Richard, Vavrik, Heinrichs and Siegmund (2003) tested a wide age range of drivers on a closed track employing an instrumented vehicle. Drivers were presented with recorded messages requiring a response to simulate the information processing demands of a cellular telephone conversation. The information processing task resulted in a degradation of driving performance when the driver was required to weave the vehicle around obstacles and engage in left turns. Performance was less affected when simpler driving situations (stopping on an amber signal) were encountered, but older drivers were more impaired when listening to the recorded stimuli in terms of stopping at traffic signals. Under wet pavement conditions, drivers listening to the recorded presentation responded with riskier driving.

To compare the relative cognitive demands inherent in the use of different types of cellular telephones (hand-held or hands-free telephones, with either an external loudspeaker or an earphone), Matthews, Legg and Charlton (2003) had a diverse group of experienced motorists drive on a rural highway in New Zealand while operating each type of telephone, and rate their experience using the NASA-TLX. Speech intelligibility was highest for the earphone equipped hands-free telephone and lowest using the external speaker. Workload was perceived to be highest for the hands-free external speaker telephone and lowest for the hands-free telephone equipped with an earphone.

Patten, Kircher, Östlund, and Nilsson (2004) had professional drivers operate an instrumented Volvo automobile on a highway for a round trip from Linköping to Norrköping, Sweden, a distance of 74 km. During the trip, a series of 8 telephone calls were placed to the drivers who used both hands-free (employing an external speaker) and hand-held cellular telephones. A conversational task, with three levels of difficulty, was presented, along with a peripheral detection task that required the driver to activate a microswitch on the steering wheel when a light emitting diode's reflection was seen on the windshield. As the complexity of the conversation increased, the RT for detecting the stimuli at the driver's periphery also increased, but no differences between the hand-held or hands-free telephones were obtained. The authors cautioned that the complexity of the conversation represents a threat to the safe operation of the vehicle.

Lamble et al. (1999) had male and female drivers in an instrumented vehicle follow a test automobile on a highway outside of Helsinki, Finland. The lead vehicle decelerated while the drivers in the test car engaged in a counterbalanced telephone dialling task, a cognitive task (involving memory and addition), or a control condition, which just involved focusing on the automobile they were following. Drivers performing the cognitive task were seen to reduce their braking RT by 0.5 s and almost 1 s in time-to-collision. Similar impairments were noted when the drivers performed the dialling task while simultaneously attending to the road. Though the authors concluded that hands-free or voice interface telephones could confer a safety advantage, it would not eliminate the cognitive interference effects.

Hancock, Lesch and Simmons (2003) had drivers operate an instrumented Ford Crown Victoria automobile on a closed track in Massachusetts while engaged in a simulated cellular telephone task intended as a distraction. Drivers had to respond to a red light on a traffic signal flashed shortly after the presentation of a distracting call on a simulated cellular telephone. This dual task situation resulted in a decrement of 15% in stopping following the onset of the red light. Older drivers and women were more adversely impaired by the distracting task. Drivers were seen to stop more quickly during the distraction, indicating a harder braking response to the traffic signal, which served to compensate for their reduced RT in braking.

In subsequent research involving the same test track (Lesch & Hancock, 2004), drivers were asked to indicate their confidence in dealing with distraction when driving while operating a cellular telephone. Their actual driving performance was compared to their perceived ability. Male drivers' perceptions of their driving skill was reflective of their actual ability, but confidence was unrelated to performance for the female motorists. In the case of the older women, as confidence increased, a *decrease* in their actual driving performance occurred. Lesch and Hancock (2004) point out the risk inherent in the belief that drivers may overestimate their ability to both drive and use a cellular telephone.

Harbluk and her colleagues (Harbluk, Noy & Eizenman, 2002; Harbluk et al., 2003; also see Harbluk & Trbovich, this volume) have employed eye-tracking cameras and video recorders to study where drivers were looking while they drove their own vehicles in the left lane of a 4 km stretch of a busy 4-lane city street in Ottawa, Ontario. The section of roadway had 14 signalized intersections and had a 50 km/hr speed limit. The drivers either had to solve demanding arithmetic problems presented to them via a hands-free cellular telephone or did not experience any additional cognitive demands placed upon them while driving. When solving the demanding problems, drivers were less likely to glance at traffic signals and look to the right.

To follow up their examination of driver distraction in relation to collisions contained in the North Carolina database, Stutts and her associates (Stutts et al., 2003) placed a set of three video cameras and a recording device in the vehicles of 70 drivers in Chapel Hill, North Carolina and Philadelphia, Pennsylvania. A total of three hours of driving data was content analyzed for each driver. Distractions were seen to be a common experience in everyday driving. About one-third of all drivers were recorded talking on their cellular telephones (30.0%), 27.1%, dialled their telephones and 15.7% were observed answering their telephones. Other distractions that exceeded cellular telephone operation consisted of drivers manipulating audio controls (91.4% of drivers), speaking with passengers (77.1%), eating or drinking (71.4%), grooming themselves (45.7%), and engaging in either reading or writing (40%). In terms of the total amount of driving time, cellular telephones accounted for only 1.3% of *all* 70 of the drivers' time compared to 15.3% spent talking to passengers. For the 24 drivers using wireless telephones (34% of the sample), telephones were in use for 3.8% of their driving time. In terms of distraction from passengers, children accounted for about four times the distraction that adults represented, with the figure doubling to an eight-fold distraction when drivers had infants in their vehicles. Dialling, answering, and talking on cellular telephones were seen to significantly reduce the proper gripping of the steering wheel and significantly affect where the drivers' eyes were focussed.

It should be noted that the volunteers were driving on familiar roadways, which would be expected to represent a lower cognitive demand than wayfinding on unfamiliar roads or searching for a specific address. Although Stutts et al. (2003) did not include measures of cognitive workload, it may be assumed that familiar locations are easier for drivers to process than unfamiliar ones. This omission makes this project difficult to compare to research assessing cognitive demands, as it fails to specify which distractions have the greatest effect on workload. Although cameras recorded a variety of measures of driving performance (e.g., whether hands were holding the steering wheel, direction of focus, and lane deviations), these aspects are only intuitively linked to crash causation.

In general, research concerning the cognitive effects of cellular telephone operation while driving, regardless of the type of driving equipment used, has found decrements in various driving behavior measures. Strayer and Johnston (2001) state "that cellular-phone use disrupts performance by diverting attention to an engaging cognitive context other than the one immediately associated with driving…and when attention is diverted from the driving context, the appropriate reactions to unpredictable events will be impaired" (Strayer & Johnston, 2001, p.466). Despite bans on handheld units, driving performance can still suffer as a result of cellular telephone conversation. Therefore, cognitive interference (Brown, Tickner & Simmonds, 1969) remains the most crucial source of interference of a cellular telephone with driving.

Methodological Considerations in Cellular Telephone Interference Research: Secondary/Dual Task Methodology

Research investigating the effects of cellular telephone use on driving performance has utilized a form of dual task methodology, where two tasks are performed simultaneously. Depending on the experimental design and the instructions given to subjects, one must be careful how results are interpreted. In order to perform two tasks simultaneously, one must be able to share processing resources efficiently, which are limited in their availability and capacity (Wickens, 1984), between the two tasks. According to Wickens, each task requires a certain amount of resources from a "pool" of available resources. When performing two tasks simultaneously, the amount of resources required to perform both tasks optimally may exceed that available and, as a result, performance on a task, or both tasks, may suffer. Increasing the level of difficulty or complexity of a single task increases the amount of resources required to optimally perform that task.

Measuring the *actual* amount of processing resources required to complete a task is not as simple as measuring an individual's perceived workload, as can be done with the NASA-TLX. Secondary or dual task methodology has been used to assess operator workload and processing resource capacity. In this situation, two tasks are performed simultaneously, with one of the tasks being designated as the primary task. In a scenario where one is interested in the cognitive loading effects of a secondary task on the primary task, i.e., the cognitive loading effects of cellular telephone use on driving performance, a loading task paradigm is appropriate.

Loading Task Paradigm

In the loading task paradigm, subjects are to maintain performance on the secondary task, even if primary task decrements occur (O'Donnell & Eggemeier, 1986). Secondary task performance is measured to ensure subjects are maintaining performance criteria and any primary task decrements are used as an index of secondary task workload. Increases in secondary task loading would be reflected in greater primary task decrements. Measured performance on the individual tasks is required as a baseline for assessing the effects of the dual task scenario. When subjects are not instructed to attend to one task over another, they attempt to perform both tasks optimally, i.e., concurrent performance, and individual differences can arise because of strategy formation. A point relevant to previous research is that, when the experimenter is unclear as to where the subject's attention is being placed, the interpretation of results becomes extremely difficult (O'Donnell & Eggemeier, 1986).

Conclusion

After examining the divergent conclusions that the different methodologies have revealed, one is reminded of the poem by John Godfrey Saxe (2004), *The Blind Men and the*

Elephant, based upon an old Indian parable. The reader will recall that a group of blind men inspect an elephant for the first time. Each touches a different body part and assumes that part is the defining essence of the beast. Much the same statement could be made about reaching a firm conclusion about the risks inherent in drivers' use of cellular telephones. Differing conclusions may be drawn from the diverse methodologies: studies employing epidemiological research designs yield data indicating an increased risk for collisions, while studies using large databases indicate little risk in using wireless telephones, especially in relation to other risks that drivers routinely expose themselves to when operating radios or adjusting ventilation controls. Simulation and instrumented vehicle research consistently finds a diminution of attention related to various types of cognitive loads imposed upon test drivers. Advocates of legislation banning the operation of these devices will find support for their views in the existing literature, as will opponents of prohibition.

In evaluating the possible risks inherent for drivers placing telephone calls while operating their vehicles, several variables need to be discussed. We have summarized the issue of workload and the evaluation of the complexity of dual task demands. Not all telephone calls require the same amount of information processing, which might be expected to serve as a moderating variable affecting driver performance. Placing a call stating that a driver will arrive later for dinner requires less workload than a salesperson placing a complex order with different delivery dates and locations.

Speaking a second language may place a greater demand than conversing in a mother tongue. Given the research on circadian rhythms and driver fatigue (see Arnedt, Wilde, Munt & MacLean, 2001; Brown, 1994; Corfitsen, 1994; Fell & Black, 1997; Fletcher & Dawson, 2001; Horne & Reyner, 2001; Johns, 2000; Lenné, Triggs & Redman, 1997; Lisper, Erikson & Fagenström, 1979), it is quite likely that driving while using a cellular telephone will have different levels of risk as a function of the motorist's circadian rhythm and fatigue interacting with the cognitive workload of the telephone call. Factors within the driver such as age, driving experience, current stress level (Hennessy & Wiesenthal, 1997; Hennessy & Wiesenthal, 1999; Hennessy, Wiesenthal & Kohn, 2000; Wiesenthal, Hennessy & Totten, 2000), familiarity with local roadways, and the specific telephone (i.e., handheld versus hands-free) as well as the vehicle, should relate to the level of risk confronting the vehicle operator.

The type of roadway encountered might also be expected to relate to the driver's cognitive workload. Superhighways with few curves and an absence of signalized intersections might be more conducive to conversations than curving city streets with traffic lights and high levels of both pedestrian and vehicle traffic. The level of traffic congestion, the number of lanes, weather, and lighting conditions all represent contextual variables that may moderate or mediate risk levels. As might be suspected, the potential is enormous for these factors to interact in complex ways.

The environment within the vehicle undoubtedly plays a role. While distraction from passengers may exist (and, in the case of children and infants, may represent a considerable distraction), passengers may be sensitive to roadway dangers and suspend their conversation with the driver, unlike those speaking to a driver on a telephone who are unable to adjust their conversation to what is happening on the road. Finally, the risk levels inherent in a driver's use of cellular telephone must be assessed *in relation* to other risks that are routinely encountered in drinking coffee, adjusting the car heating controls, and putting a compact disc into a dashboard player.

Chapman and Schofield (1998) have found that cellular telephones are frequently used to report a variety of emergency situations. This safety enhancing aspect of mobile telephone use needs to be balanced by the possibility of risk. In a national Australian telephone survey of cellular telephone users, it was observed that 1 in 8 users have used their telephones to report a traffic accident, 1 in 4 have reported dangerous situations, 1 in 16 have reported medical emergencies (unrelated to collisions), 1 in 20 have reported a crime, and 1 in 45 have reported being lost or experiencing a boating danger. The most common reported "emergency" situation (with 63% of cellular telephone owners reporting this use) was for drivers telephoning from the highway that they would be delayed in reporting for appointments. Eighty percent of cellular users in this group reported frequently calling to advise that they would be late. Chapman and Schofield (1998) reported that 66% (2.11 million cellular owners) reported that they were able to drive slower and relax following such a call, which might be related to a reduction in highway mishaps.

Currently, the existing body of research has examined the issue of driver distraction when using a cellular telephone as the *only* distraction confronting the vehicle operator. It is also important to consider *other* activities that may be occurring while the wireless telephone call is made. Drivers may be note-taking or glancing at printed material while conducting their conversations. If a crash happened, would the distraction that caused the event be note-taking, reading, the use of the telephone or a combination of these factors? To answer this question, future research will have to discover where, when, and how the telephone is actually used by drivers. Observational research is needed to clarify what behaviors the driver is emitting. Do drivers gesture when speaking on the telephone? How frequently might they remove their hands from the steering wheel to do so? From the standpoint of public policy, it may be difficult to implement a complete ban on cellular telephone use by drivers because of the convenience such devices afford, along with the sense of safety that these devices provide for drivers, who perceive themselves as vulnerable to collisions, breakdowns, flat tires, and other roadway mishaps. Research endeavours need to be directed to studying how to maximize the benefits achieved from the use of these communication devices, while minimizing the risk that distraction will cause the cellular operator to crash.

Acknowledgements

The authors wish to acknowledge the assistance of Judy Manners for her meticulous word processing and the bibliographic, technical and scholarly assistance of Christine Wickens.

References

Alm, H., & Nilsson, L. (1994). Changes in driver behaviour as a function of handsfree mobile phones: A simulator study. *Accident Analysis and Prevention, 26,* 441-451.

_____ . (1995). The effects of a mobile telephone task on driver behaviour in a car following situation. *Accident Analysis and Prevention, 27,* 707-715.

Arnedt, J. T., Wilde, G. J. S., Munt, P. W., & MacLean, A. W. (2001). How do prolonged wakefulness and alcohol compare in the decrements they produce on a simulated driving task? *Accident Analysis and Prevention, 33,* 337-344.

Baddeley, A. D., Logie, R., Nimmo-Smith, I., & Brerefon, N. (1985). Components of fluent reading. *Journal of Memory and Language, 24,* 119-131.

Beirness, D. J., Simpson, H. M., & Pak, A. (2002). *The Road Safety Monitor: Driver Distraction* (No. 02A). Ottawa, ON: Traffic Injury Research Foundation.

Brookhuis, K. A., De Vries, G., & De Waard, D. (1991). The effects of mobile telephoning on driving performance. *Accident Analysis and Prevention, 23,* 309-316.

Brown, I. D. (1994). Driver fatigue. *Human Factors, 36,* 298-314.

Brown, I. D., Tickner, A. H., & Simmonds, D. V. C. (1969). Interference between concurrent tasks of driving and telephoning. *Journal of Applied Psychology, 53,* 419-424.

Chapman, S., & Schofield, W. N. (1998). Lifesavers and Samaritans: Emergency use of cellular (mobile) phones in Australia. *Accident Analysis and Prevention, 30,* 815-819.

Cher, D.J., Mrad, R. J., & Kelsh, M. (1999). Cellular telephone use and fatal traffic collisions: A commentary. *Accident Analysis and Prevention, 31,* 599.

Chowdhury, P. S., Franklin, B.A., Boura, J. A., Dragovic, L. J., Kanluen, S., Spitz, W., et al. (2003). Sudden cardiac death after manual or automated snow removal. *American Journal of Cardiology, 92,* 833-835.

Cohen, J. T., & Graham, J. D. (2003). A revised economic analysis of restrictions on the use of cell phones while driving. *Risk Analysis, 23,* 5-17.

Consiglio, W., Driscoll, P., Witte, M., & Berg, W. P. (2003). Effect of cellular telephone conversations and other potential interference on reaction time in a braking response. *Accident Analysis and Prevention, 35,* 495-500.

Cooper, P. J., Zheng, Y., Richard, C., Vavrik, J., Heinrichs, B., & Siegmund, G. (2003). The impact of hands-free message perception/response on driving task performance. *Accident Analysis and Prevention, 35,* 23-35.

Corfitsen, M. T. (1994). Tiredness and visual reaction time among young male nighttime drivers: A roadside survey. *Accident Analysis and Prevention 26,* 617-624.

De Waard, D., & Brookhuis, K. A. (1991). Assessing driver status: A demonstration experiment on the road. *Accident Analysis and Prevention, 23,* 297-307.

Drivers back ban on using cellphones while driving. (2002, March 26). *Toronto Star,* pp. A1, A8.

Eby, D. W., & Vivoda, J. M. (2003). Driver hand-held mobile phone use and safety belt use. *Accident Analysis and Prevention, 35,* 893-895.

E-Community Link. Telecommunications: Communication infrastructure. Retrieved March 9, 2003, *http://www.ecomlink.org/Default.htm.*

Evans, L. (1991). *Traffic safety and the driver.* New York: Van Nostrand Reinhold.

Fell, D. L., & Black, B. (1997). Driver fatigue in the city. *Accident Analysis and Prevention, 29,* 463-469.

Fletcher, A., & Dawson, D. (2001). Field-based validations of a work-related fatigue model based on hours of work. *Transportation Research Part F, 4,* 75-88.

Glaze, A. L., & Ellis, J. M. (2003). *Pilot study of distracted drivers.* Richmond, VA: Transportation and Safety Training Center, Center for Public Policy, Virginia Commonwealth University.

Haigney, D. E., Taylor, R. G., & Westerman, S. J. (2000). Concurrent mobile (cellular) phone use and driving performance: Task demand characteristics and compensatory processes. *Transportation Research Part F, 3,* 113-121.

Hancock, P. A., Lesch, M., & Simmons, L. (2003). The distraction effects of phone use during a crucial driving maneuver. *Accident Analysis and Prevention, 35,* 501-514.

Harbluk, J. L., Noy Y. I., & Eizenman, M. (2002). *The impact of cognitive distraction on driver visual behaviour and vehicle control.* Ottawa, ON: Transport Canada Ergonomics Division, Road Safety Directorate and Motor Vehicle Regulation Directorate.

Harbluk, J. L., Trbovich, P., Noy, I., Eizenman, M. & Lochner, M. (2003). Driver visual behaviour at intersections: *The impact of cognitive distraction.* Canadian Multidisciplinary Road Safety Conference XIII, Banff, Alberta, June 2003.

Hart, S. G., & Staveland, L. E. (1988). Development of NASA-TLX (Task Load Index): Results of empirical and theoretical research. In P. A. Hancock, & N. Meshkati (Eds.), *Human mental workload.* Amsterdam: Elsevier Science Publishers.

Hennessy, D. A., & Wiesenthal, D. L. (1997). The relationship between traffic congestion, driver stress and direct versus indirect coping behaviors. *Ergonomics, 40,* 348-361.

_____ . (1999). Traffic congestion, driver stress, and driver aggression. *Aggressive Behavior, 25,* 409-423.

Hennessy, D. A., Wiesenthal, D. L., & Kohn, P. M. (2000). The influence of traffic congestion, daily hassles, and trait susceptibility on state driver stress: An interactive perspective. *Journal of Applied Biobehavioral Research, 5,* 162-179.

Hornberry, T., Bubnich, C., Hartley, L., & Lamble, D. (2001). Drivers' use of hand-held mobile phones in western Australia. *Transportation Research Part F, 4,* 213-218.

Horne, J., & Reyner, L. (2001). Sleep-related vehicle accidents. Some guides for road safety policies. *Transportation Research Part F, 4.* 63-74.

Johns, M. W. (2000). A sleep physiologist's view of the drowsy driver. *Transportation Research Part F, 3,* 241-249.

Lamble, D., Kauranen, T., Laakso, M., & Summala, H. (1999). Cognitive load and detection thresholds in car following situations: Safety implications for using mobile (cellular) telephones while driving. *Accident Analysis and Prevention, 31,* 617-623.

Lenné, M. G., Triggs, T. J., & Redman, J. R. (1997). Time of day variations in driving performance. *Accident Analysis and Prevention, 29,* 431-437.

Lesch, M. F., & Hancock, P. A. (2004). Driving performance during concurrent cell-phone use: Are drivers aware of their performance decrements? *Accident Analysis and Prevention, 36,* 471-480.

Lisper, H., Erikson, B., & Fagerström (1979). Diurnal variation in subsidiary reaction time in a long-term driving task. *Accident Analysis and Prevention, 11,* 1-5.

Maclure, M. & Mittleman, M. A. (1997). Cautions about car telephones and collisions. *New England Journal of Medicine, 336,* 501-502.

Matthews, R., Legg, S., & Charlton, S. (2003). The effect of cell phone type on drivers subjective workload during concurrent driving and conversing. *Accident Analysis and Prevention, 35,* 451-457.

Ministry of Transportation. (2003). Ontario Road Safety Annual Report 2001. Toronto, ON: Publishing Ontario. Retrieved March 9, 2003, *http://www.mto.gov.on.ca/english/safety/orsar/orsar01/.*

O'Donnell, R., & Eggemeier, F. T. (1986). Workload assessment methodology. In K. R. Boff, L. Kaufman, & J. P. Thomas (Eds.). *Handbook of Perception Vol. II, Cognitive Processes and Performance.* New York: John Wiley and Sons.

Patten, C. J. D., Kircher, A., Östlund, J., & Nilsson, L. (2004). Using mobile telephones: Cognitive workload and attention resource allocation. *Accident Analysis and Prevention, 36,* 341-350.

Redelmeier, D. A., & Tibshirani, R. J. (1997). Association between cellular-telephone calls and motor vehicle collisions. *New England Journal of Medicine, 336,* 453-458.

Reinfurt, D. W., Huang, H. F., Feaganes, J. R., & Hunter, W. W. (2001). Cell phone use while driving in North Carolina. Highway Safety Center, University of North Carolina, Chapel Hill, North Carolina.

Saxe, J. G. *The blind men and the elephant.* Retrieved July 10, 2004, www.wordfocus.com/word-act-blindmen.html.

Stevens, A., & Minton, R. (2001). In-vehicle distraction and fatal accidents in England and Wales. *Accident Analysis and Prevention, 33,* 539-545.

Strayer, D. L., Drews, F. A., & Johnston, W. A. (2003). Cell phone induced failures of visual awareness during simulated driving. *Journal of Experimental Psychology, Applied, 9,* 22-23.

Strayer, D. L., & Johnston, W. A. (2001). Driven to distraction: Dual-task studies of simulated driving and conversing on a cellular telephone. *Psychological Science, 12,* 462-466.

Strayer, D. L., Drews, F. A., Crouch, D. T., & Johnston, W. A. (in press). Why do cell phone conversations interfere with driving? To appear in W. R. Wallace & D. Herrmann (Eds.), *Cognitive technology: Transforming thought and society.* Jefferson, NC: McFarland & Co.

Stutts, J. C., Feaganes, J., Rodgman, E., Harmlett, C., Meadows, T., Reinfurt, D., et al. (2003). Distractions in everyday driving. AAA Foundation for Traffic Safety: Washington, DC.

Stutts, J. C., Reinfurt, D. W., Staplin, L., & Rodgman, E.A. (2001). *The role of driver distraction in traffic crashes.* Washington, DC: AAA Foundation for Traffic Safety.

Sundeen, M. (2003). *Cell phones and highway safety.* Denver, CO: National Conference of State Legislatures. Retreived July 5, 2004, *www.ncsl.org/programs/esnr/cellphoneupdate1203.htm.*

Utter, D. (2001). *Passenger vehicle driver cell phone use: Results from the Fall 2000 National Occupant Protection Use Survey.* National Highway Traffic Safety Administration.

Valverde, H. H. (1973). A review of flight simulator transfer of training studies. *Human Factors, 15,* 510-522.

Violanti, J. M. (1998). Cellular phones and fatal traffic accidents. *Accident Analysis and Prevention, 30,* 519-524.

Violanti, J. M., & Marshall, J. R. (1996). Cellular phones and traffic accidents: An epidemiological approach. *Accident Analysis and Prevention, 28,* 265-270.

Wickens, C. D. (1984). Processing resources in attention. In R. Parasuraman (Ed.), *Varieties of attention.* Orlando: Academic Press.

Wiesenthal, D. L., Hennessy, D. A., & Totten, B. (2000). The influence of music on driver stress. *Journal of Applied Social Psychology, 30,* 1709-1719.

In: Contemporary Issues in Road User Behavior and Traffic Safety ISBN:1-59454-268-6
Editors: D. Hennessy and D. Wiesenthal, pp. 245-250

Chapter 19

COGNITIVE DISTRACTION: ITS EFFECT ON DRIVERS AT INTERSECTIONS

Joanne L. Harbluk
Transport Canada
Patricia Trbovich
Carleton University

INTRODUCTION

With the advent of new technologies in vehicles, drivers can access many forms of information (e.g., email, address books, web pages) from a variety of sources (e.g., cell phones, PDAs, driver support systems) while they drive. The information provided by some of these sources, such as traffic, weather, navigation and route guidance information, may aid drivers by supporting the task of driving. Other types of information provided by these sources facilitate activities not related to driving such as office tasks (e.g., phone and email), location based commercial information (e.g., hotels, restaurants) and entertainment.

In-vehicle information systems (IVIS) are becoming increasingly popular in vehicles and their functionality is expanding. They bring with them an increasing concern that the use of these devices while driving can impair driver performance and increase crash risk. Cell phones are currently the most popular type of information device used in vehicles and their use is widespread. There were 15-million cell phone users in Canada in 2003 (Canadian Wireless Telecommunications Association, 2004) and in a recent survey 23% of respondents indicated that they used a cell phone while driving during the previous week (Beirness, Simpson, & Desmond, 2002). The use of in-vehicle information systems is expected to expand as new technologies and applications are developed.

In an effort to find safer alternatives to systems that require visual attention to input information manually or to read visual output, system designers have turned to speech-based interfaces. It is commonly assumed that speech-based technologies for in-vehicle devices do not distract drivers because the drivers are not required to take their eyes off the road or their hands off the steering wheel. Some studies have demonstrated that there may be benefits associated with speech-based systems compared with systems requiring manual input (Gellatly & Dingus, 1998; Ranney, Harbluk, & Noy, 2005). Nevertheless, voice-based interactions are not effortless and they have the potential to place cognitive demands on drivers (Goodman, Tijerina, Bents, & Wierwille, 1999; Harbluk, Noy, & Eizenman, 2002; Lee, Caven, Haake, & Brown, 2001; Strayer, Drews, & Johnson, 2003).

The focus of the present study was cognitive distraction, the sort that might arise in the course of interacting with an in-vehicle information system that could be considered "hands-free", requiring only speech interaction and no manual interaction. Drivers drove a vehicle on-road while listening and responding to cognitive tasks presented to them via a hands free cell phone. The task did not require them to look away from the road or to manipulate the phone manually. A cell phone was chosen because it represents the most common form of this type of technology currently used by drivers, although parallels can be drawn to other interactive speech-based interfaces.

Driving through intersections is one of the most complex conditions drivers encounter. Many different complex perceptions, decisions and actions are required to successfully negotiate intersections. Drivers must be aware of traffic signal changes, pedestrians and on-coming traffic. Estimates vary, but crashes at intersections have been shown to account for as many as 50% of all crashes involving injury. A 1998/99 cross-Canada survey concluded that intersection crashes were a factor in 13.1% of fatal collisions resulting in 415 fatalities, 46,802 injuries and a cost of $1.2 billion annually (Canadian Council of Motor Transport Administrators, 2002). Clearly intersections are a relatively dangerous feature of the road traffic environment.

The measures of interest were the drivers' eye glances as they drove on a busy city road through intersections, when they were performing complex cognitive tasks and when they drove with no additional task. While the analyses reported in this paper focus specifically on the intersections, a more general analysis of the impact of performing demanding cognitive tasks on drivers' visual behavior and vehicle control can be found in Harbluk et al. (2002).

METHOD

Participants

Twenty-one participants (9 women and 12 men), aged 21-34 years took part in the study. All were experienced drivers (minimum of 5 years experience) who drove more than 10,000 km annually. They were recruited via an advertisement in the local Ottawa newspaper and were paid for their participation.

Procedure

The test route was a 4-km stretch of a busy 4-lane city street in Ottawa, Canada with 14 signalized intersections and a posted speed limit of 50 km/h. Drivers were required to drive in the left lane and not to change lanes as they drove. All driving took place under clear and dry summer conditions between the hours of 9:30 am and 11:45 am on weekday mornings.

The drivers drove the route under conditions where they listened to and answered demanding cognitive questions (e.g., 47 + 38) and while not performing any additional tasks. The drivers were not required to look away from the road or to interact manually with the cell phone to hear or answer the questions. The cognitive tasks were communicated via hands-free cell phone query and response. The drivers wore head-mounted eye-tracking equipment (VISION 2000, El Mar) that enabled the collection of data on their visual behavior. The cell phone (Nokia 5160) remained in the cradle mounted to the right of the console for the duration of the study. The cell phone microphone was secured to the pillar at the top left of the windshield and the speaker was mounted under the dash.

RESULTS

Eye tracking data were obtained for the drivers as they drove through the 14 signalized intersections on the route. Data where drivers stopped at intersections for red lights were not included in the analyses. Data for one subject were not available for analysis due to a recording error.

The primary comparisons of interest were differences in the drivers' visual behavior as they drove through the intersections with or without performing the demanding cognitive task. Specific analyses focussed on the mean number of glances made to the traffic lights, the mean percent times that drivers completely shed the task of looking at the traffic lights, and the mean number of glances made to the right and to the left as they approached (7 *sec.* prior to entering the intersection) and passed through the intersections.

First, clear differences in visual behavior as indicated by the glances to the traffic lights were observed under the two task conditions. More specifically, as shown in Figure 1, when driving and performing the demanding cognitive task, drivers made significantly fewer glances ($M = 1.70$, $SE = \pm 0.18$) to the traffic lights compared to driving without the task ($M = 2.68$, $SE = \pm 0.29$), $F(1,19) = 21.34$, $p < .001$.

Second, in some cases, drivers did not inspect the signal lights at all. The percent time drivers completely shed the task of looking at the traffic lights, under the cognitive task condition ($M = 21.9\%$, $SE = \pm 4.53$), was significantly larger than in the no-task condition ($M = 7.8\%$, $SE = \pm 2.56$), $F(1,19) = 10.95$, $p < .004$.

Finally, there were differences in the ways that drivers inspected the environment around the intersection as they approached and passed through it as shown in Figure 2. When performing the cognitive task, drivers reduced the frequency with which they looked at the area to the right ($M = 1.34$, $SE = \pm 0.16$) compared with the no-task condition ($M = 1.70$, $SE = \pm 0.18$), $F(1,19) = 4.65$, $p < .05$. The drivers, however, were relatively consistent with respect to the frequency with which they looked to the left, for both the cognitive task ($M = 1.48$, $SE = \pm 0.20$) and no-task conditions ($M = 1.34$, $SE = \pm 0.14$). Since the drivers were required to drive in the left lane, most of their monitoring of traffic around them would be focussed on the right lane beside them rather than the area to the left.

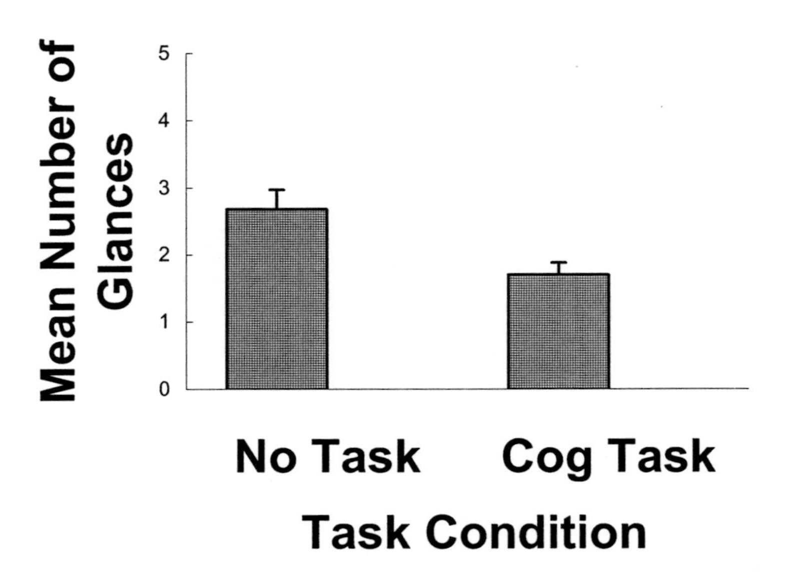

Figure 1. Mean number of glances (± SE) to traffic lights.

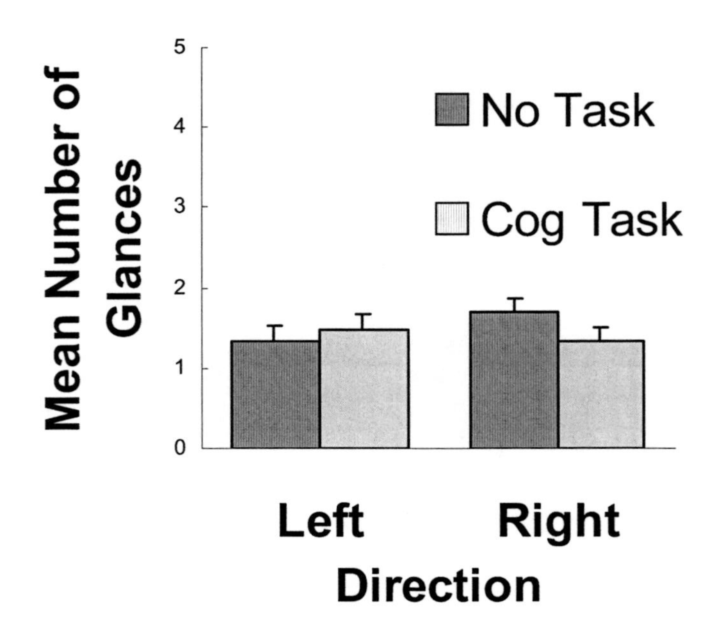

Figure 2. Mean number of glances (± SE) to the left and right while passing through Intersections.

DISCUSSION

The appeal of using speech-based interfaces for in-vehicle information systems is that they enable drivers to maintain a forward look out while keeping both hands on the steering wheel. These types of interfaces are seen as a preferable alternative to interfaces that require manual input and direct the driver's visual attention away from the road.

The purpose of this study was to examine the impact on drivers' visual behavior of using an interface that required only speech interaction, in this case a hands free cell phone. The drivers preformed demanding cognitive tasks in this manner as they drove a real vehicle on a busy city road. The specific analyses of interest focussed on drivers' eye glance behavior as they drove through the intersections along the route.

Intersections represent one of the most complex conditions that drivers encounter. Consequently, driving instruction courses recommend that when approaching an intersection, drivers should maintain proper surveillance of intersection area for hazards, vehicles, pedestrians, and signals if present. The results of the present day study indicated that, even though drivers were required to take their eyes off the road, there were appreciable changes in drivers' eye glance patterns when they were engaged in demanding cognitive tasks while driving. An improved understanding of the ways in which drivers interact with hands-free, speech based interfaces should result in safer designs for in-vehicle information systems.

There are multiple reasons for crashes at intersections. Driver disregard of traffic and control signals is a known contributing factor. This study adds to our understanding of these crashes by documenting that drivers who are engaged in demanding cognitive tasks (even one that does not require them to look away from the road) no longer monitor the intersection environment as they normally do when they approach and pass through intersections.

While this study does not provide a direct connection between driver distraction and intersection crashes, the changes that were observed in driver behavior are consistent with poorer lookout and less awareness of the driving environment. An improved understanding of the ways in which drivers interact with hands-free, speech-based interfaces should result in safer designs. Understanding how to support the interactions with these interfaces is critical in minimizing driver distraction. The systematic application of human factors in product development would help to ensure these devices do not directly or indirectly increase the risk of collision or injury to vehicle occupants or other road users. The process would further enhance the usability and appeal of products because it would lead to the development of in-vehicle devices that match user needs in a way that is compatible with and suitable for driving.

REFERENCES

Canadian Council of Motor Transport Administrators Standing Committee on Road Safety Research and Policies. (2002, October), *Speed and intersection safety management*. Available online: Retrieved March, 2003, from *http://www.ccmta.ca/english/pdf/SISM%20Report%20October%202002.pdf*

Canadian Wireless Telecommunications Association (CWTA, 2005). Available online: Retrieved January, 2005, from *http://www.cwta.ca/CWTASite/english/industry.html#MobileTelephony*

Beirness, D. J., Simpson, H. M., & Desmond, K. (2002). *Road safety monitor 2002: Risky driving.* Traffic Injury Research Foundation (TIRF), Ottawa. Available online: Retrieved March, 2002, from *http://www.trafficinjuryresearch.com/whatNew/newsItemPDFs/RSM_Risky_Driv ing.pdf*

Gellatly, A. W., & Dingus, T. A. (1998). Speech recognition and automotive applications: Using speech to perform in-vehicle tasks. *Proceedings of the Human Factors and Ergonomics Society 42nd Annual Meeting*, Santa Monica CA, 1247-1251.

Goodman, M. J., Tijerina, L., Bents, F. D., & Wierwille, W. W. (1999). Using cellular phones in vehicle: Safe or unsafe? *Transportation Human Factors, 1,* 3-42.

Harbluk, J. L., Noy, Y. I., & Eizenman, M. (2002). *The impact of cognitive distraction on driver visual behaviour and vehicle control* (Transport Canada Tech. Rep. No. 13889 E). Retrieved March, 2002, from *http://www.tc.gc.ca/roadsafety/tp/tp13889/en/menu.htm*

Lee, J. D., Caven, B., Haake, S., & Brown, T. L. (2001). Speech-based interaction with in-vehicle computers: The effects of speech-based e-mail on drivers' attention to the roadway. *Human Factors, 43*, 631-640.

Ranney, T. A., Harbluk, J. L., & Noy, Y. I. (2005). The effects of voice technology on test track driving performance: Implications for driver distraction. In press, *Human Factors*.

Strayer, D. L., Drews, F. A., & Johnson, W. A. (2003). Cell phone induced failures of visual attention during simulated driving. *Journal of Experimental Psychology: Applied, 9*, 23-52.

In: Contemporary Issues in Road User Behavior and Traffic Safety ISBN:1-59454-268-6
Editors: D. Hennessy and D. Wiesenthal, pp. 251-262

Chapter 20

THE USE OF EVENT DATA RECORDERS IN THE ANALYSIS OF REAL-WORLD CRASHES: TALES FROM THE SILENT WITNESS

Kevin McClafferty and Paul Tiessen
Multi-Disciplinary Accident Research Team, University of Western Ontario
Alan German
Road Safety and Motor Vehicle Regulation Directorate, Transport Canada

INTRODUCTION

It has long been recognized that human error precipitates the vast majority of motor vehicle collisions. While major efforts have been expended to enhance the crashworthiness of motor vehicles and improve occupant protection systems, less attention has been paid to the challenge of reducing the incidence of collisions, or at least of reducing their severity. Central to the development of strategies to provide effective collision avoidance measures is a broad-reaching understanding of the significant factors involved in collision causation. Not the least of such factors are those relating to the individuals involved, the pre-crash human factors.

High-level data from real-world collisions regarding vehicle speeds, perception and reaction, braking and steering inputs, are fundamental to researchers in all disciplines conducting studies into pre-crash human factors. However, objective data on these important aspects of collisions are often not readily available. As a result of a lack of hard data on driver actions prior to impact, most studies of pre-crash factors in the real world have relied heavily on soft accounts from involved parties and independent witnesses.

Many late-model General Motors' light-duty vehicles are equipped with event data recorders (EDR) that act as impartial witnesses to collision events. These systems variously capture data relating to the crash pulse, the time history of the velocity change that occurs in the collision, and the pre-collision time history of a number of key vehicle parameters. Currently, only late-model General Motors' vehicles are equipped with EDR's that can record pre-crash data elements retrievable with a publicly-available tool.

The EDR forms part of the sensing and diagnostic module (SDM) that controls deployment of the airbag system. As previously noted, EDR's can capture certain information relating to the pre-crash and/or the crash phases of motor vehicle collisions. The precise nature of the data record is dependent on the type of SDM with which a specific year, make and model of vehicle is equipped. EDR's can be interrogated by means of a Crash Data Retrieval (CDR) tool, allowing the stored data to be retrieved, analyzed and reported (Chidester, Hinch, Mercer, & Schultz, 2001).

Pre-crash information consists of the vehicle's speed (mph), engine speed (rpm), throttle position (%), and the status of the brake light switch (on or off) in one-second increments, for a period of five seconds prior to the event that triggered the recording. In addition, the EDR indicates the status of the driver's seat belt buckle switch (buckled or unbuckled) at the time of the event. It is these particular data elements that are of specific interest in the present work. Data relating to vehicle speed, engine speed, percentage throttle and brake switch status are stored in a buffer that is capable of storing five values of each data element. Values are recorded at one-second intervals, with the most recent values superceding the oldest values. When the SDM senses a certain vehicle deceleration (approximately 2g), the airbag deployment algorithm is enabled to monitor the crash pulse in detail and determine if the airbags need to be fired. Subsequent to algorithm enable (AE), the last five values of the pre-crash data elements in the buffer are stored in non-volatile memory for subsequent retrieval. The pre-crash data are reported at one-second intervals, starting at five seconds prior to AE (t = -5s) and ending at one second prior to AE (t = -1s).

Data records may be captured for both airbag deployment events and certain non-deployment events. In general, a near-deployment event consists of vehicle deceleration sufficient to trigger the sensing algorithm, but not severe enough to require deployment of the airbag. Deployment level events that occur after the deployment event may also be recorded. The EDR can store up to two events. Deployment events cannot be overwritten or cleared from the SDM, while non-deployment events can be so overwritten.

It is important to note that there is no real-time clock associated with the EDR. The total number of ignition cycles (i.e. ignition key turned on and off) since vehicle assembly is stored as part of each recorded event. Normally, the number of ignition cycles at the time data is retrieved from the EDR is also reported. Comparison of the ignition cycles for a recorded event to the ignition cycles at download is used by collision investigators to associate, or disassociate, a data record with a specific collision.

Generally, once a vehicle's airbags have been fired as a result of a crash, the vehicle will not be driveable and will be towed from the scene. In such a case, the number of ignition cycles associated with the airbag deployment event will usually be very close to the number of ignition cycles recorded when the data is downloaded. In contrast, following a non-deployment event, the vehicle can often still be operated. Thus, if the number of ignition cycles associated with a non-deployment record is considerably less than the number of ignition cycles reported at the time of data retrieval, the non-deployment event may not be associated with a recent crash, and could have actually occurred at some time in the past. The situation is further complicated by a variety of options in the storage system for non-deployment events, the details of which are beyond the scope of the present paper. Suffice it to say that considerable care must be exercised in attributing stored records to specific collision events.

DATA INTERPRETATION

As a result of the manner in which these data elements are captured, considerable care must be given to their interpretation. For example, the fact that the EDR reports the driver's seat belt to have been buckled is not necessarily indicative that the seat belt was actually used by the driver. The system cannot discriminate between a properly restrained driver and an individual who is unbelted, but has deliberately fastened the seat belt buckle to defeat the warning system, and is sitting on the webbing.

Furthermore, power loss during the crash event may result in an unbuckled status being recorded when the driver was actually using the seat belt system. In such cases, the "unbuckled" status is usually a default state for the EDR's register which would have been overwritten by "buckled", as monitored by an on-board sensor, towards the end of the data recording process. If the vehicle's electrical system is compromised part way through the process of writing data to the EDR's registers, a default state for a given register may not be overwritten by the true value.

Similarly, the signals from the sensor and control modules are not necessarily synchronized in absolute time. Thus, the reported position of the throttle at any given second may not correspond exactly to the status of the application of the brake. It should be further noted that no indication is provided of how hard the brake pedal was applied. The driver may have been touching the brake pedal lightly, sufficient to illuminate the brake lights, but not hard enough to develop much deceleration.

Recent research has shown that the EDR-recorded vehicle speeds are very accurate under steady state conditions (Lawrence, Wilkinson, & Heinrich, 2003). However, it should be noted that the accuracy of the recorded vehicle speed can be affected if the tire size or final drive axle ratio of the vehicle has been changed from the factory build specifications. When assessing the accuracy of the speed data, one may also have to consider other factors such as wheel slip, anti-lock brake systems (ABS), traction control systems and vehicle yaw.

METHODOLOGY

A series of real-world motor vehicle collisions, subject to conventional in-depth investigation and reconstruction techniques, have been included in this study. The individual cases were drawn from Transport Canada's on-going collision investigation program. They include investigations focused on airbag deployment crashes, moderately-severe side impacts, and a trauma-based sample of children in collisions. The present series also includes a subset of cases from a human factors pilot study in which detailed and comprehensive interviews were conducted with involved drivers. The common element to the cases so gathered is the availability of pre-crash data elements downloaded from event data recorders present in at least one of the collision-involved vehicles. The present study includes 76 such collisions that involved late-model General Motors' vehicles.

In this demonstration project, no attempt was made to sample crashes in any systematic manner. The intention was merely to capture pre-crash data elements from EDR's to compare to information determined by conventional research techniques. By such means, the utility and limitations of these data with regard to identifying and quantifying issues related to collision causation can be explored.

RESULTS

Collision Type

The 76 collisions examined in this series were comprised of 55 front airbag deployments and 21 non-deployments. Frontal impacts (n = 53) were the predominant crash type for the EDR-equipped vehicle in this series. Useful pre-crash data was recorded for side impacts (n = 22) over a wide spectrum of crash severity. A single rollover event was also recorded. Most of the cases in the series involved multiple vehicle collisions (n = 59). These collisions typically occurred at intersections (n = 51) and often involved vehicles making a left turn (n = 24) or vehicles approaching at right angles with one of the involved drivers failing to yield the right of way (n = 22). There were also several rear-end collisions (n = 5) where a slowly moving or stopped vehicle was struck. Inattentiveness, turning conflicts and view obstructions were common factors in the multiple vehicle collisions. Single vehicle collisions (n = 17) were less common in the series. Loss of directional control occurred prior to impact in 71% of the single vehicle collisions (n = 12).

Maximum Recorded Vehicle Speed

The maximum speed recorded by the EDR was compared to the posted speed limit. The maximum recorded speed occurred at the t = -5 s to AE sampling interval in 66% of the cases (n = 50). A maximum speed above the posted limit was recorded in 59% of the cases (n = 45). Speeding was more frequently observed in single vehicle collisions where 88% of the vehicles were exceeding the posted speed limit (n = 15). While most of the multiple vehicle collisions were urban events, the single vehicle collisions often occurred on rural roads with higher posted speed limits.

The EDR recorded a maximum speed 20 km/h or more above the posted limit in 29% of the total cases (n = 22) and 40 km/h or more above the posted limit in just 9% of the total cases (n = 7). In 71% (n =12) of the single vehicle collisions the maximum recorded speed was 20 km/h or more above the posted speed limit and in 29% the maximum recorded speed was more than 40 km/h above the posted speed limit (n =5).

A maximum speed 20 km/h or more below the posted limit was recorded in 18% of the cases (n = 14). These cases typically involved an EDR-equipped vehicle that was in the process of turning when the collision occurred.

Drivers' Speed Estimates

Where possible, driver-reported speeds were determined either by interview or from the police report. In 58 of the cases in our series, both the driver-reported speed and the maximum EDR-recorded speed were available. In 5% of these 58 cases (n = 3) the driver reported a speed 10 km/h or more above the posted speed limit. By comparison, the EDR recorded a maximum speed 10 km/h or more above the posted limit in 38% of the cases (n = 22) and a speed 20 km/h or more above the posted limit in 24% of the cases (n = 14).

Drivers' speed estimates were below the maximum EDR-recorded speed in 86% of the cases (n = 50). The drivers underestimated their speeds by 14 km/h, on average. Their speed estimates ranged widely from 20 km/h above the maximum EDR-recorded speed to 60 km/h below this maximum speed. The drivers' speed estimates were within 10 km/h of the maximum recorded speed in 40% of the cases. The drivers' speed estimates tended to be the most accurate when the recorded speed was near the posted speed limit. For driver-reported speeds of 40 km/h or less (n = 19), the recorded speed was higher by 10 km/h or more 58% of the time (n = 11).

Speeding and Crash Severity

The change in velocity (delta-V) experienced by the EDR-equipped vehicle ranged from 5 km/h to 65 km/h with an average value of 24 km/h. As might be expected, the crash severity, as measured by the delta-V, was observed to increase with the speed differential above the posted limit. Vehicles that sustained a delta-V of 24 km/h or less were travelling at an average speed of 56 km/h and the average maximum-recorded speed was 1 km/h below the speed limit. Vehicles that sustained a delta-V above 24 km/h were travelling at an average speed of 75 km/h and the average maximum-recorded speed was 14 km/h above the posted speed limit. Vehicles that were travelling near the speed limit tended to sustain less severe crashes than those vehicles where the speed differential was higher. While the average delta-V for the 24 cases where the recorded speed was within 10 km/h of the posted speed limit was 17 km/h, the average delta-V for the 22 cases where the recorded speed was 20 km/h or more above the limit was 34 km/h.

Pre-Impact Braking

Emergency braking prior to impact was a frequent occurrence. In 68% of the cases (n = 52), the brake switch status was recorded as on in the t = -1 s to AE sampling interval. In 24% of the cases (n = 18), the brake switch status was recorded as on in both the t = -1 s and t = -2 s to AE sampling intervals. In just 8% of the cases (n = 6), the brake switch status was recorded as on in the t = -1 s, -2 s and -3 s to AE sampling intervals. The brake switch status was recorded as off in all intervals in 20% of the cases (n = 15).

Restraint Use and Loading Evidence

The driver's belt switch circuit status was recorded by the EDR as being buckled in 87% of the cases (n = 66), while it was recorded as unbuckled in 13% of the cases (n = 10). In one case, the EDR recorded an unbuckled status yet loading evidence was found on the seat belt. The reason for this is unknown, however the loading evidence indicated that the seat belt was probably worn and buckled at the time of the crash. Due to collision-induced forces, loading marks are frequently present on seat belt systems when they are used. Loading evidence on the restraint system verified the buckled belt switch status in 26 of the 66 cases where the EDR recorded a buckled status.

There were 9 unbelted drivers in this series. Three of the 9 drivers reported that they were unbelted in an interview. In 2 cases driver interview was not possible. The remaining 4 unbelted drivers incorrectly reported that they were belted. However, the EDR recorded that the belts were unbuckled and there was no loading on the restraints.

Discussion

Accuracy of Witness Statements

In this study we compared driver statements regarding pre-crash speed and seat belt use to hard data recorded by the EDR. The drivers' speed estimates were generally found to be quite poor when compared to the recorded data. In 60% of the cases, the maximum recorded speed was greater than the drivers' speed estimates by over 20%. The drivers in this series underestimated their speed in comparison to the EDR in the vast majority of the cases. Drivers who were not in the process of turning, usually reported that their speed was close to the posted speed limit. Drivers seldom reported speeds much above the posted speed limit. In fact, only 5% of drivers reported vehicle speeds more than 10 km/h above the posted speed limit. However, speeding was very common and nearly half of the maximum recorded speeds were more than 10 km/h above the posted speed limit. The agreement between the drivers' speed estimates and the recorded speeds tended to be best when vehicles were actually travelling near the posted speed limit.

It has been known for many years that people are not particularly good at reporting numerical details such as vehicle speed (Gardner, 1933). In a test administered to Air Force personnel, who knew in advance that they would be questioned about the speed of a moving automobile, estimates ranged from 10 to 50 mph when the car they watched was actually going only 12 mph (Marshall, 1969).

Of course, speed limits and speedometers provide drivers with some advantage when estimating their speed prior to a collision. However, in many cases, this available information will not be useful to the driver in subsequent estimation of vehicle speed. In this series, accurate estimation of speed appeared to be particularly difficult for drivers when they were travelling at low speed.

The fallibility of eyewitness memory is becoming increasingly clear in the scientific literature where it has been reported that memories can be false or inaccurate even though the witnesses believe them to be true (Loftus & Hoffman, 1989). Studies have shown that memory of numerical details such as vehicle speed can be readily influenced by outside factors (Loftus & Palmer, 1974). Research indicates that memory is prone to error at several stages of information processing and, as such, these factors must be taken into account when assessing the potential accuracy of witnesses' statements.

While problems with perception and memory may well influence the accuracy of driver-reported speeds, witness bias is also a very important consideration. One of the major criteria for evaluating the credibility of an eyewitness is impartiality. It is reasonable to assume that many drivers involved in motor vehicle collisions will be somewhat less than impartial, particularly if they were at fault or breaking the law. Drivers that under-reported their speeds by the widest margins were typically travelling well over the posted speed limit prior to the collision. It is very likely that many of these drivers were intentionally misrepresenting their speed.

Similar results were found with seat belt use. Only 3 of the 7 unrestrained drivers who survived the collision reported that they were unbelted, while the other 4 reported that they were belted. This tendency for unrestrained drivers to report that they were belted after a collision was also observed in a study of collisions in the United States (Gabler, Hampton, & Hinch, 2004). This study compared the seat belt buckle status recorded by the EDR to seat belt usage reported by crash investigators. Seat belt usage was determined by physical loading evidence, when present, and from occupant interviews. In approximately 25% of the cases (67 of 273 cases), the investigators reported a belted driver while the EDR recorded that the belt status was unbuckled. While fallibility of memory may have been a factor, it is very likely that many of these unrestrained drivers were well aware that they were not belted.

The Role of Speeding

Real world field studies can yield tremendous insight into the factors that precipitate motor vehicle collisions. The strength of such in-depth studies is the quality of the data obtained, especially when these data are based on solid physical evidence. Softer data based on the recollections of involved parties can also provide important insight into pre-crash events. However, the current study shows that reliance on driver-reported information regarding speed or seat belt could result in substantial errors.

Police officers investigating motor vehicle collisions must often rely on the statements of the involved drivers in order to determine the events that caused the crash. The recollections of eyewitnesses are often critical for establishing criminal and civil liability following a motor vehicle collision. Road safety research also relies heavily on eyewitness evidence and much of our knowledge about driver pre-crash actions is based upon data obtained from drivers. However, inaccurate witness statements can lead to erroneous conclusions.

The *2001 Ontario Road Safety Annual Report* (Ontario Ministry of Transportation, 2001) showed just 5% of drivers involved in a crash were coded by police as exceeding the speed limit or traveling at a speed too fast for the conditions. If taken at face value, some may conclude from this statistic that excessive speed does not play a major role in motor vehicle collisions. However, in the vast majority of cases, the involved drivers were the sources of the data on speeding and, as we have observed in this study, involved drivers are often very unreliable sources of information on vehicle speed.

Over the years there have been many investigations into the role of speed in motor vehicle collisions. Most studies of speed and crash causation have relied heavily on witness reports or have been correlational in nature (USDOT, 1998). While both types of studies have yielded much useful data, there remain large gaps in knowledge, and the role of speed in crashes is still highly debated.

A recent Australian study of the role of speeding in real-world injury-producing collisions utilized data obtained through advanced crash reconstruction techniques (Lowden,

McLean, Moore, & Ponte, 1997). Using a case-control study design and in-depth collision reconstruction, the speeds of cars involved in casualty crashes were compared with the speeds of cars not involved in crashes. The study relied heavily on analysis of scene evidence. The authors found that cars involved in the casualty crashes were generally travelling faster than cars that were not involved. In 14% of casualty crashes, the vehicles were travelling faster than 80 km/h in a 60 km/h speed zone compared to less than 1% of those not involved in a crash.

The current study had some similar findings in that a large proportion of collisions in our series were exceeding the speed limit – often with a maximum speed 20 km/h or more above the posted limit. As might be expected, the incidence of speeding was greater in the severe crashes. For example, a maximum speed 20 km/h or more above the posted limit was recorded in 58% of the crashes where the delta-V experienced by the EDR-equipped vehicle was greater than 30 km/h. Similarly, a maximum speed 40 km/h or more above the posted limit was recorded in 29% of the crashes where the delta-V experienced by the EDR-equipped vehicle was greater than 30 km/h.

Note that only the speed of the EDR-equipped vehicle is considered in this study. It is likely that in some cases the non-EDR vehicle was exceeding the posted speed limit. Thus, while a maximum speed 20 km/h or more above the posted limit was recorded for 29% of the EDR-equipped vehicles, it is very likely that a vehicle was exceeding the speed limit in more than 29% of the crashes.

The current study has provided a glimpse of how EDR's might be used effectively to conduct detailed research on pre-crash speed. The preliminary results suggest that a large-scale study of real world collisions utilizing EDR pre-crash data could provide tremendous insight into the role of speed and the influence of collision avoidance actions in motor vehicle crashes. Preliminary findings suggest that the role of speed in crash causation and crash severity could be much more extensive than current knowledge suggests.

Pre-Impact Braking

The Australian study by Lowden et al. (1997) also examined the role of braking in real-world injury-producing collisions. They found that 71% of the 151 case vehicles skidded under emergency braking before the crash. The vehicles under study were rarely equipped with anti-lock braking systems (ABS). For comparative purposes, pre-crash braking was identified by the EDR in the t = -1 s to AE interval in 68% of the cases in the current study.

Pre-crash events, such as the occurrence and effect of vehicle braking, can often be determined with significant accuracy through in-depth investigation and analysis of real world motor vehicle collisions. While crash reconstruction is often an effective technique for determining pre-crash driver actions, it relies heavily on fleeting evidence obtained from the crash scene. Wet or snow-covered roads, high traffic volumes, anti-lock brakes, competing scene evidence and the cover of darkness make the collision reconstructionist's job very difficult.

Some automobiles equipped with anti-lock braking systems may leave traces of visible scuff marks (Wright, 1995). However, the marks are often very faint and may require special vantage points and lighting in order to be visible. The presence of these marks depends on the combination of road surface, speed, temperature, tires and ABS. The rate at which the scuff marks fade has been found to be extremely high. In the past, an understanding of driver action and vehicle movement was frequently based upon skid mark evidence. Even under carefully controlled conditions, conventional collision reconstruction methods that rely on skid mark length measurements may not be effective when a vehicle is ABS-equipped (Metz & Ruhr, 1990).

Pre-impact tire marks from the case vehicle were rarely observed by the investigating police officer in this series. This was not surprising since most of the vehicles were ABS-equipped. In those cases where there was significant pre-impact braking, there was usually insufficient physical evidence to allow accurate reconstruction and determination of speeds prior to crash avoidance. While impact speed could usually be calculated with reasonable accuracy, accurate calculation of initial speed was not possible when there was substantial pre-impact braking, but an absence of pre-impact tire marks. The importance of the EDR data in assessing liability will likely increase as ABS becomes more common in the Canadian fleet.

Determination of vehicle speed through reconstruction of the physical evidence is one of the major roles of the police reconstructionist. When speed is found to be excessive, criminal charges may result. Successful prosecution of criminal cases requires proof beyond reasonable doubt of recklessly high speed. Lack of physical evidence, competing interpretations of physical evidence and uncertainty in calculations are some of the barriers to successful prosecution. The EDR provides extensive supporting documentation that, when combined with competent reconstruction, should increase the probability of successful prosecution. However, it is readily foreseeable that in many cases the EDR will provide the only accurate physical evidence of vehicle speed. In the current series of cases, the EDR was found to be very useful at identifying instances of substantial speeding which might have been missed if one had relied on witness statements. While in-depth reconstruction of the physical evidence often provided substantial insight into pre-crash driver actions, there was often insufficient scene evidence to make accurate determinations of speed prior to collision avoidance actions. The EDR was often the only way an investigator could accurately determine speed prior to collision avoidance actions.

Limitations of EDR Data

While the quality of the EDR data has been shown to be reasonably good, the technology is still in its infancy and a strong measure of caution is required when interpreting the data. In one case the driver's seat belt buckle status was recorded as unbuckled, yet loading evidence on the restraint indicated that it was almost certainly being worn by the driver at impact. There was no obvious evidence of power loss in this case which could cause the buckle status to default to unbuckled. Similarly, in 3 cases, the recorded brake switch status was found to be reversed from the actual status. While this vehicle-based anomaly is now identified in the crash data retrieval software, there is a potential for other such anomalies, should they go undiscovered, to lead to erroneous conclusions.

There are numerous interpretation considerations with current EDR data and it is critical that the pre-crash data be carefully evaluated in conjunction with other evidence. For example, vehicle speed is normally supplied by the Powertrain Control Module (PCM) which monitors a sensor located at the output shaft of the transmission. Vehicle speed is related to the rotational speed of the output shaft through the differential gear ratio and the rolling radius of the tires. The PCM monitors the speed of the wheels and not the actual ground speed of the vehicle. In some cases the speeds recorded by the EDR may not be representative of the actual vehicle speed, particularly if the vehicle is in a yaw or there is a heavy throttle input on a slippery surface resulting in wheel slip.

Loss of directional control and yaw due to hard braking and steering were probable contributory factors in cases where the recorded speed losses appeared unusually high. A data resampling anomaly whereby data nominally in the t = -1 s last interval is actually sampled after the impact may also be a factor in some cases. In at least one case the maximum recorded speed was significantly higher than the true vehicle speed probably as a result of heavy throttle and significant wheel slip on wet grass prior to impact. Further research is needed to determine the precise influence of steering, braking and loss of control on the accuracy of the EDR speed data.

The EDR provided valuable pre-crash data over a variety of crash modes. Data collection was not restricted to frontal impacts and a high percentage of cases involved side impacts over a wide severity range. Data was also obtained for a single rollover collision although, in this case, as for several other single vehicle crashes, the initiation of loss of control was not captured by the EDR. Single vehicle collisions, such as rollovers, often involve loss of control where the time lapse between initiation of loss of control and impact may be significantly greater than 5 seconds. In these cases the recorded data in the EDR's limited buffer may not capture the initiation of loss of control. Review of the cases in this series, shows that an additional 5 to 10 seconds of pre-crash data would have been helpful for determining pre-impact speed and driver actions.

Event data recorders are excellent research tools that can lead to a much greater understanding of both pre-crash and crash events. However, the current systems have a number of limitations such that the data retrieved must be carefully interpreted in conjunction with conventional in-depth collision reconstruction techniques. This is especially important in the prosecution of criminal cases where a tendency to review the EDR data first and then reconstruct the crash could potentially lead to the wrong conclusions.

Promise for the Future

It is clear that event data recorders offer a path to significantly better understanding of many aspects of motor vehicle collisions. The current technology is limited in both its scope and application but, with the rapid pace of the development of electronic systems, this is likely to change in the not-too-distant future. One can envisage subsequent generations of systems that will provide many more data channels and considerably more comprehensive information on individual data elements. With such a wealth of data, the performance of complex vehicle-based safety systems, as well as the actions of vehicle operators prior to a crash, will likely be extremely well defined. However, the availability of such a powerful tool for understanding how collisions occur raises a number of privacy considerations with regard to the use of these devices. In particular, there is the question of data ownership and who can obtain access to the information. Is the operator of the vehicle the owner of any collision-related data? Can the owner of the data refuse to divulge the information to interested parties?

Certainly, safety researchers have no mandate to compel provision of such data. Rather, studies of crashes involving EDR's are usually carried out in a cooperative and permissive manner with vehicle operators. Typically, the vehicle driver is a study subject and consent is sought for an interview to collect specific details of interest, often in the form of a questionnaire. In the case of EDR-equipped vehicles consent is sought to include EDR data as part of the dataset. In other areas of collision investigation and reconstruction, such permission is not always forthcoming, yet EDR data can still be obtained and used by certain authorities. Recent jurisprudence has shown that police, acting under the authority of a search warrant or Coroner's warrant, can download data from an EDR and introduce this information into court proceedings as evidence. A parallel situation would seem to apply to insurance claims since policyholders have an obligation to give their insurer access to their collision-involved vehicle, and this would include access to any data contained in an EDR. Similarly, in the case of an individual involved in civil litigation, the other litigating party would no doubt seek the right of access to any related EDR data.

It appears likely that crash data retrieval will become an integral part of the crash investigation process. However, perhaps the greatest strength for this technology is as a research tool and this is certainly an area where the greatest safety benefits can be achieved. In particular, it is foreseeable that these data will greatly increase our knowledge of pre-crash factors in motor vehicle collisions which, in turn, will lead to the development of more effective safety countermeasures.

Thus, the major question that arises is the individual's right to privacy versus society's right to know in order to develop a safer transportation environment for all its citizens. Driving is already a highly regulated activity. Vehicles must comply with a host of safety regulations prescribing the availability and level of performance of collision avoidance and crashworthiness systems. Drivers require a license, subject to testing for competence, and requiring annual renewal. They must also insure their vehicle for at least third-party liability in the event of a crash. Drivers are required to obey traffic laws, and failure to do so may bring penalties. Similarly, failure to abide by the contractual terms of an insurance policy, such as driving while impaired, may result in denial of coverage in the event of a collision. Surely, there is room in such a comprehensive system for additional requirements relating to EDR's to ensure that real-world collision data can be obtained, analyzed, and used to implement new safety countermeasures.

One could consider a system where federal regulations required the installation of an EDR in every vehicle, and provincial laws required that police officers and insurance adjusters routinely had access to the associated data in order to expedite investigations and settle legal proceedings and claims. If such data were to become part of an expanded motor vehicle collision reporting system, the mass data could be made available, in an anonymous format, for review and analysis by a wide variety of specialists in order to address a broad range of safety concerns.

Already, EDR's are being used to understand the behaviour and performance of complex occupant protection systems such as seat belt systems with pretensioners and load limiters that deploy as a function of crash severity, and advanced airbags with dual deployment thresholds and dual stage inflators which are dependent on factors such as occupant presence, proximity to the airbag, seat belt use and crash severity. As such systems become more complex, with further technological refinements, and the addition of other restraint systems, such as side airbags and head curtains, the data available from on-board crash recorders will be critical to understanding the precise nature of the function of these safety systems. Similarly, there is a revolution underway in collision avoidance systems with a host of new devices coming to market such as anti-lock brakes, brake assist and speed reduction systems, adaptive cruise control, traction and stability control systems. As such vehicle-based collision avoidance measures become both more common and more complex, there will be increasing need for high quality pre-crash data to aid in system evaluation and development. While the technology of event data recorders is currently in its infancy, these devices hold much promise for increasing road and motor vehicle safety.

REFERENCES

Chidester, C., Hinch, J., Mercer, T. C., & Schultz, K. S. (2001). Recording automotive crash event data. *Available online at http://www-nrd.nhtsa.dot.gov/edr-site/dataformat.ht*ml.

Gardner, D. S. (1933). The perception and memory of witnesses. *Cornell Law Quarterly.*

Gabler, H .C., Hampton, C. E., & Hinch, J. (2004). *Crash severity: A comparison of event data recorder measurements with accident reconstruction estimates.* SAE-2004-01-1194.

Lawrence, J. M., Wilkinson, C. C., & Heinrichs, B. E. (2003). The accuracy of pre-crash speed captured by event data recorders. *SAE 2003-01-0889.*

Loftus, E. F., & Hoffman, H. G. (1989). Misinformation and memory. The creation of new memories. *Journal of Experimental Psychology: General, 118, 100-104.*

Loftus, E. F., & Palmer, J. C. (1974). Reconstruction of automobile destruction: An example of the interaction between language and memory. *Journal of Verbal Learning and Verbal Behaviour, 13, 585-589.*

Lowden, C. N., McLean, A. J., Moore, V. M., & Ponte, G. (1997). Travelling speed and the risk of crash involvement: Volume 1 – Findings. *NHMRC Road Accident Research Unit, The University of Adelaide. Available online at http://raru.adelaide.edu.au/speed.*

Marshall, J. (1969). Law and psychology in conflict. *New York: Anchor Books.*

Metz, D. L., & Ruhr, R. L. (1990). Skid mark signatures of ABS-equipped passenger cars. *SAE 900106.* Available online at *http://www-nrd.nhtsa.dot.gov/edr-site/index.html.*

Ontario Ministry of Transportation. (2001). Ontario Road Safety Annual Report 2001. Available online at *http://www.mto.gov.on.ca/english/safety/orsar/.*

United States Department of Transportation.. (1998)˙ Synthesis of safety research related to speed and speed limits. *Publication No. FHWA-RD-98-154. Available online at http://www.tfhrc.gov///safety/speed/spdtoc.htm*

Wright, B. (1995). Evidence of ABS scuff marks on the roadway. Available online at *http://www.tarorigin.com/art/absscuff/.*

In: Contemporary Issues in Road User Behavior and Traffic Safety ISBN:1-59454-268-6
Editors: D. Hennessy and D. Wiesenthal, pp. 263-274

Chapter 21

THE EFFECTIVENESS OF AIRBAGS FOR THE ELDERLY

Eric D. Hildebrand and Erica B. Griffin
Transportation Group, University of New Brunswick

INTRODUCTION

North American demographic forecasts indicate that the proportion of the population aged 65 years or greater will approximately double within the next twenty years (Rosenbloom, 2003; Statistics Canada, 2001). In addition to the swell in absolute numbers, a greater proportion of the aged will be drivers. Although older drivers are involved in relatively few collisions due to limited exposure, once involved in a crash they are more likely to sustain severe injuries or death (Cunningham, Coakley, O'Neill, Howard & Walsh, 2001). Several studies have confirmed that as people age, they are more likely to sustain serious or fatal injuries from the same severity crash (Bedard, Guyatt, Stones & Hirdes, 2002; Evans, 2001; Li, Braver & Chen, 2001; Mercier, Shelley, Rimkus & Mercier, 1997). Elderly drivers and occupants are especially at risk of thoracic region injuries due to increased bone fragility (Foret-Bruno, Trosseille, Page, Huere & LeCoz, 2001; University of Michigan, 2001; Wang et al., 1999). Currently, no motor vehicle safety standards in North America are designed to specifically address the needs of elderly persons.

The standards set by government to protect our population should be continually monitored and modified if required to reflect changing demographics. As people age, they become less able to overcome injuries due to increased frailty. The main goals of this research were to develop a better understanding of elderly injury outcomes and to establish whether the elderly warrant specific inclusion in the Motor Vehicle Safety Standard development process.

BACKGROUND

Since the introduction of airbags in the early 1990s, there has been a disproportionate number of air bag induced fatalities among small-statured women involved in low speed collisions. A study by the United States National Highway Traffic Safety Administration (NHTSA) found that deployed airbags inadvertently killed 38 adults between 1990 and 1997, 14 of them being small-statured females (National Highway Transportation Administration, 1998). To address this issue, Transport Canada is currently considering adopting modifications to their occupant restraint safety standards by including an anthropomorphic testing device (ATD) to represent a small-statured occupant.

While the majority of air bag deployments serve their intended purpose, some cause injuries beyond what is expected from superficial contact with the rapidly expanding air bag (Government of Canada, 2001). Current North American airbag systems are designed to deploy in crashes with frontal decelerations equivalent to that experienced in a 19 km/h crash into a fixed barrier (Segui-Gomez, 2002). In real world crashes, this translates into a "must fire" threshold around 24 km/h and a "guaranteed no fire" threshold around 14 km/h. At collision severities near the lower deployment threshold, the airbag has actually been found to be a greater injury risk than the level of injury reduction being afforded (Dalmotas, German, Hendrick & Hurley, 1995; Dalmotas, Hurley, German & Digges, 1996; German, Dalmotas & Hurley, 1998; Libertiny, 1995).

Currently, the 50th percentile male Hybrid III is the most widely used frontal crash test dummy around the world. This dummy represents a man of average size at 5'10" tall and 170 pounds. The 5th percentile female Hybrid III crash test dummy is approximately 4'11" in height and weighs 108 pounds.

The analysis undertaken by Transport Canada to determine the need for the 5th percentile ATD assumed that 50 percent of the females and all of the age 50+ population are best represented by the lower injury tolerance imposed by the 5th percentile adult female dummy (L.P. Lussier, personal communication, 2002). An underlying objective of this study was to determine whether or not there is need for this ATD to more specifically model the unique physical characteristics of elderly persons.

RESEARCH OBJECTIVES

This study utilized different collision datasets collected by the Road Safety and Motor Vehicle Regulation Directorate of Transport Canada to develop a better understanding of injury outcomes of elderly occupants. The focus of the study was a comparative analysis of injuries sustained by the elderly versus both their younger counterparts and small-statured females in similar collisions. The initial objective of this research was to determine how the injuries of elderly occupants in motor vehicle collisions differ from those of the general population and to develop a better understanding of the impact that airbags have had in reducing or exacerbating injuries in the elderly.

Proposed changes to Canada's occupant restraint systems in frontal impacts (Motor Vehicle Safety Standard 208) include new injury criteria and test procedures using an assortment of ATDs, including the 5th percentile adult female ATD. A further study premise is that a comparison of injury type/severity of elderly persons versus small-statured females will determine whether the introduction of the proposed 5th percentile adult female ATD in crash tests might provide beneficial outcomes specifically for elderly persons. Consequently, the results of the study provide a means to identify opportunities where Canadian Motor Vehicle Safety Standards might be enhanced to better accommodate older occupants.

STUDY APPROACH

This study utilized the Passenger Car Study and Air Cushion Restraint Study datasets developed by the Standards and Regulations Division (ASFBE), Road Safety and Motor Vehicle Regulation Directorate of Transport Canada. The Passenger Car Study (PCS) is a statistically representative sample of collisions involving passenger cars that occurred in Canada from 1984 to 1992. The PCS database was used in this study to describe injuries to front-seat restrained occupants resulting from frontal collisions of various severities. It provided the benchmark against which injury patterns to occupants in vehicles equipped with air cushion restraint systems could be compared.

The Air Cushion Restraint Study (ACRS) was initiated in October of 1993 to examine the injury experience of occupants protected by supplementary air bag systems (Dalmotas, 1998). This database was used to identify injury types and severities associated with collisions where an airbag had deployed. When examining the results of the comparative analyses synthesized in following sections, it is important to consider that all differences in injury patterns between groups may not be solely due to the introduction of airbags. Several other factors may affect the injury pattern differences between the PCS and ACRS datasets, such as the change in vehicle design, the change in fleet mix and changes in other aspects of the safety restraints. Nevertheless, the focus of this research explores the comparative differences between the elderly sub-group and the remaining population.

Based on inclusion criteria established for this study (Griffin, 2003) the number of useable records was reduced to 1,213 and 1,078 collisions from the Passenger Car Study (PCS) and Air Cushion Restraint Study (ACRS), respectively. Criteria for inclusion included restrictions such as frontal collision only, occupants had to be restrained, availability of all pertinent medical/collision information. A summary of the database dimensions is given in Tables 1 and 2. For the purposes of this research, all persons aged 14 to 64 years were referred to as "young" and their data were used as a benchmark to highlight where elderly injuries are over-represented. Elderly persons were defined as those aged 65 years and older. For this study, small-statured females were defined as any female 14 to 64 years of age with a height between 145 cm and 155 cm (4'9" to 5'1") inclusive and a mass of 42 kg to 57 kg (91.8 lb to 124.2 lb) inclusive.

Table 1. Passenger Car Study Data Summary

Occupant	Drivers	Right Front Passengers	Total
Young (14-64)	792	278	1070
Elderly (65+)	94	49	143
Total	886	327	1213
Small-statured females (14-64)	30	7	37

Table 2. Air Cushion Restraint Study Data Summary

Occupant	Drivers	Right Front Passengers	Total
Young (14-64)	827	182	1009
Elderly (65+)	54	15	69
Total	881	197	1078
Small-statured females (14-64)	21	10	31

Collision Severities

For analyses undertaken in this study, the impact-induced change in velocity (delta-V) was used as the indicator or metric for collision severity. The data were grouped into the following delta-V categories, consistent with collision severity breakdowns presented in previous Transport Canada analyses, including Dalmotas et al. (1996):

Minor severity collisions: Delta-V < 24 km/h (15 mph)
Moderate severity collisions: Delta-V = 24-39.9 km/h (15 – 25 mph)
High severity collisions: Delta-V = 40+ km/h (25+ mph)

The study findings are based on a subset of the Transport Canada datasets because only cases for which the delta-V's for the crashed passenger car were known could be included. Delta-V's are most often missing (about half of cases in this analysis) because the algorithm used by crash investigators in their computation cannot be used when data about the crash or vehicle are insufficient.

Injury Severity Assessment

The Abbreviated Injury Scale (AIS) is a six-point ordinal scale that classifies injuries by body region (National Highway Traffic Safety Administration, 1980). Injuries are ranked on a scale of 1 to 6, with 1 being minor, 5 critical, and 6 being a non-survivable injury. The AIS code represents the threat to life associated with an injury and is not meant to represent a comprehensive measure of severity.

To provide reliable assessments of overall injury severity, especially where medical knowledge or expertise is not available, the Maximum AIS (MAIS) is often used. The MAIS is the highest severity code AIS injury sustained by the occupant in the collision (Association for the Advancement of Automotive Medicine, 1990). This injury can be inflicted on any part of the body. An occupant can sustain more than one injury at the same maximum level; for example, if an occupant sustains several AIS 1 injuries but no injuries classified as higher than this, then the MAIS is still 1. The analyses gave precedence to head injuries if an occupant had a maximum head, chest injury, and neck injury at the same AIS level (National Highway Traffic Safety Administration, 2000). The Maximum AIS is used in this research to compare injuries in the PCS dataset to injuries in the ACRS dataset.

In the Air Cushion Restraint Study, injuries are coded using the "NASS 1993 Injury Coding Manual" and the "NASS 1993 Crashworthiness Data System Data Collection, Coding and Editing Manual" (National Highway Traffic Safety Administration, 1993), based on AIS-90. Injury severity is coded based on the AIS-90 scale, and the Maximum AIS value is determined. The Passenger Car Study (PCS) dataset uses the NASS Injury Coding Manual based on AIS-80 for injury coding. Based on the changes between the AIS scale revisions, it can be inferred that injuries in the Air Cushion Restraint Study sample may have been assigned lower MAIS injury scores than injuries of the same nature in the Passenger Car Study. This suggests that any increase in injury severity associated with the introduction of airbags may actually be slightly greater than presented in the results. Likewise, any decrease in injury severity for similar severity collisions from the Passenger Car Study to the Air Cushion Restraint Study may be slightly less than presented due to the variations in the AIS scales.

For each delta-V category, an *average* Maximum Abbreviated Injury Scale (MAIS) score was calculated. The Abbreviated Injury Scale is an ordinal scale, and a specific injury severity of 4.34, for example, does not actually exist. The average MAIS value was used only to compare injury severity among population groups. Changes in the calculated averages merely indicate shifts in the distribution of observed injury severities. In addition to considering the severity of injuries sustained in motor vehicle collisions, the location of the most severely injured body region was examined.

STUDY FINDINGS

Analyses of Elderly Airbag Injuries

The main research objective of developing a better understanding of airbag injury outcomes was achieved through detailed analyses of maximum and average injury scores sustained among the elderly. The scores were contrasted against younger aged occupants involved in motor vehicle collisions of various severities, emphasizing the changes resulting from the introduction of airbags. Furthermore, injury patterns were contrasted to identify regions where the aged have been shown to be particularly vulnerable. Note that the analyses in this study do not delineate the effects of second-generation airbags introduced in 1998, which reduce inflator peak pressure and/or rise rate and reducing airbag volume (Summers, Hollowell & Prasad, 2001). However, only 6 of the 1,078 records in the Air Cushion Restraint Study database extracted for analyses involve deployment of the redesigned airbags.

The Maximum Abbreviated Injury Score (MAIS) is the highest severity code AIS injury sustained by an occupant in a collision (Association for the Advancement of Automotive Medicine, 1990). In low severity collisions, only 26 percent of elderly persons sustained no injury when the airbag deployed, compared to 38 percent for their younger counterparts (see Figure 1). The data in Figure 1 also show that 20 percent of the elderly sustained MAIS 2+ injuries as compared to only six percent for the younger group. This finding suggests that in minor severity collisions, elderly persons are more susceptible to airbag injury than their younger counterparts.

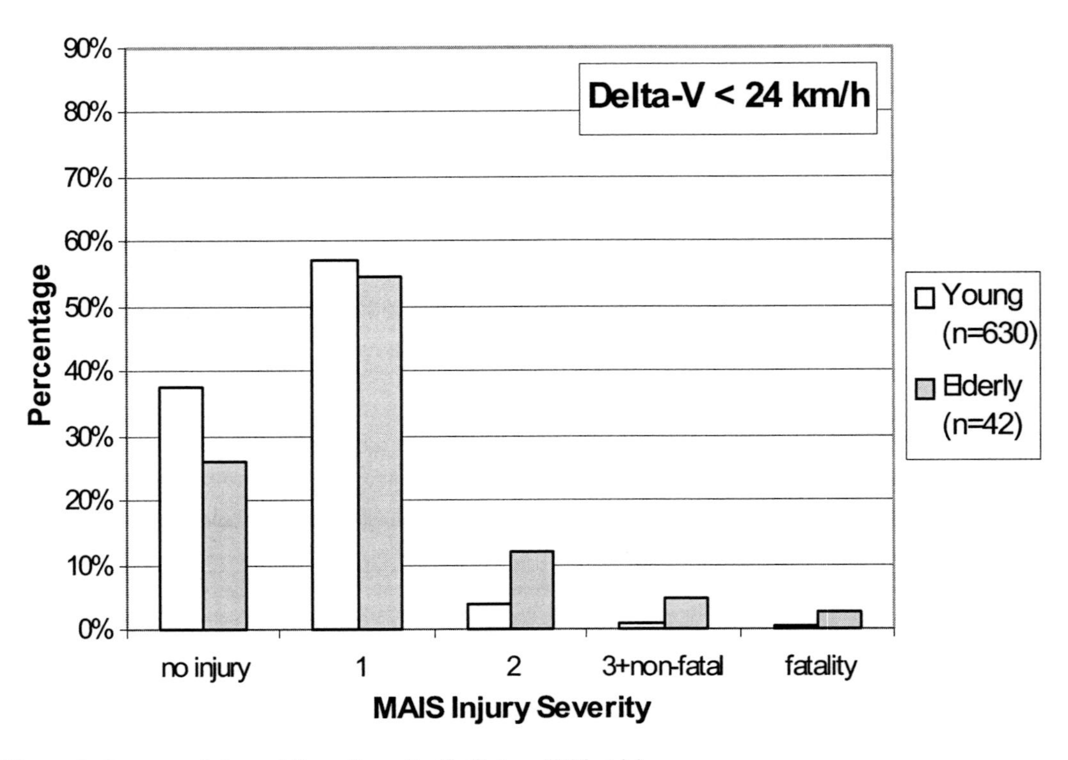

Figure 1. Occupant Injury, Minor Severity Collisions With Airbags.

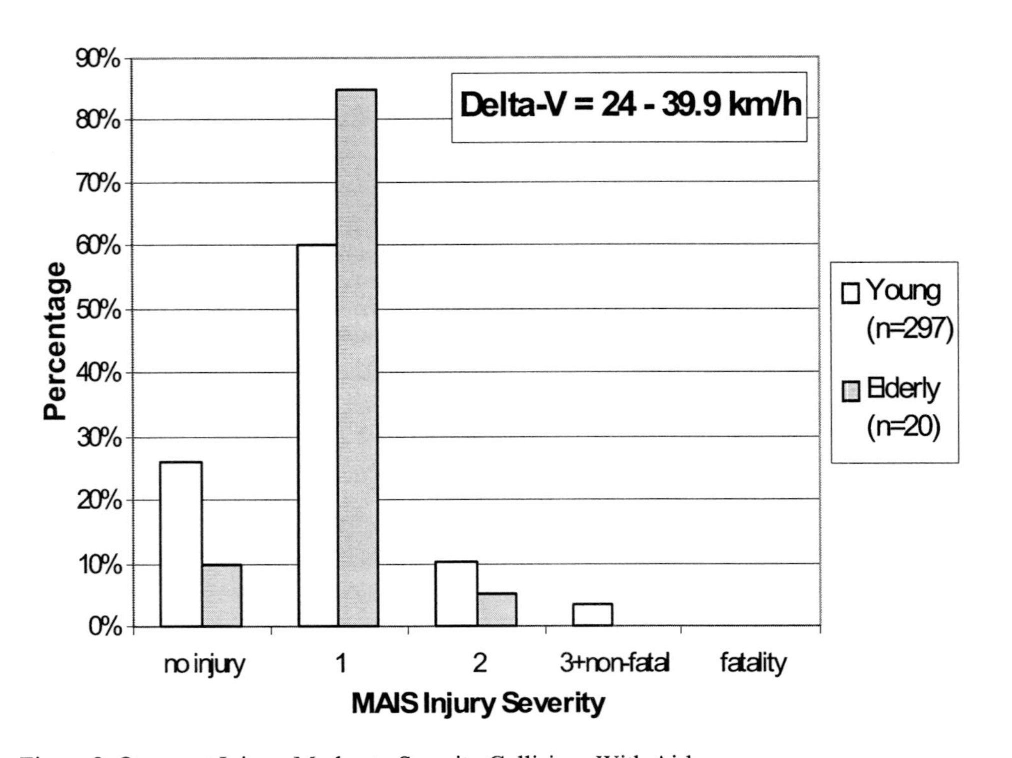

Figure 2. Occupant Injury, Moderate Severity Collisions With Airbags.

In moderate severity collisions with airbags, the data showed that overall the elderly sustained more severe injuries than their younger counterparts (see Figure 2). Only ten percent of elderly persons sustained no injury in these collisions, compared with 26 percent of the young group. The data also showed that the young group sustained fewer MAIS 1 injuries than the elderly occupants (60% versus 85%).

An alternative method used to examine injury severity was developed on an average MAIS score among age groups and collision severities. As mentioned, the Abbreviated Injury Scale is an ordinal scale, and a calculated average injury severity score of 4.34, for example, merely reflects the distribution of ordinal scores for a study group. It is used for comparative purposes. The data presented in Figure 3 provide a summary of the average MAIS by age group and collision severity. These data indicate that elderly persons sustain higher MAIS injuries in low severity collisions (delta-V < 24 km/h) than their younger counterparts. Average MAIS values for this collision severity were 1.03 and 0.70 for the elderly and younger group respectively, up from 0.78 and 0.56 without airbags. This substantial increase in average MAIS for the elderly indicates that elderly persons are sustaining more serious injuries in minor severity collisions with the airbag than without it. As discussed earlier, airbags sometimes tend to exacerbate rather than prevent injuries in minor severity collisions.

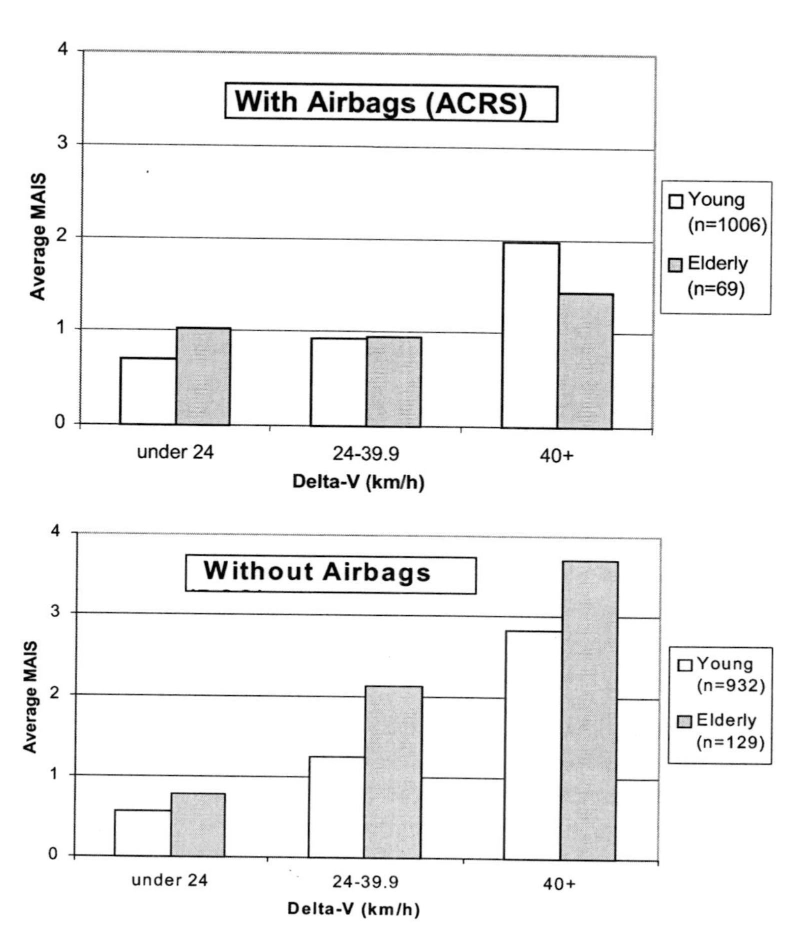

Figure 3. Average Occupant Injury by Collision Severity With and Without Airbags.

For moderate severity collisions (delta-V = 24-39.9 km/h), the elderly data showed a substantial decrease in average MAIS from 2.12 before airbags to 0.95 after airbags. This indicates that airbags are indeed effective in preventing injuries to the elderly in moderate severity collisions. The young group also showed to be protected by the airbag, with average MAIS for moderate severity collisions reducing from 1.25 without airbags to 0.93 with airbags. For severe collisions (delta-V = 40+ km/h), the data for the younger occupants showed an average MAIS of 1.96, and the elderly an average MAIS of 1.43. The introduction of the airbag in this case helped to provide a reduction in average MAIS of 0.88 and 2.27 for the young and elderly groups respectively.

Table 3. Distribution of the Most Severely Injured Body Regions by Collision Severity

Collision Severity	Injured Body Region	Without Airbags (percent)	With Airbags (percent)
Delta-V < 24 km/h			
Young (14-64)	Face	8	36
	Spine	34	5
	U. Extremity	19	32
Elderly (65+)	Face	4	40
	Thorax	38	7
	Spine	33	3
	U. Extremity	4	27
Delta-V = 24-39.9 km/h			
Young (14-64)	Face	13	41
	Thorax	19	16
	Spine	22	3
	U. Extremity	13	19
Elderly (65+)	Face	25	33
	Thorax	38	28
Delta-V = 40+ km/h			
Young (14-64)	Head	20	25
	Face	14	32
	L. Extremity	20	8
Elderly (65+)	Face	7	71
	Neck	0	14
	Thorax	50	14
	Spine	14	0

For this study, body regions were classified based on the Abbreviated Injury Scale Body Regions. Table 3 provides a summary of injured body regions by population group. The two body regions most frequently found to be the site of the most severe injury are identified in bold for each category. It was found that before the introduction of airbags, the elderly, unlike the base population, sustained the greatest percentage of serious injuries to the thorax in all collision severities (see Table 3). This high percentage of thoracic injuries in the elderly supports the literature on their susceptibility to chest region injuries due to bone frailty. With the introduction of airbags, both the young and elderly groups sustained the highest distribution of injuries to the face and upper extremities in all collision severities. Morris,

Barnes, Fildes, Bentivegna and Seyer (2001) found similar results and suggested that contact with the airbag would account for the increased number of upper extremity injuries.

Comparison of Elderly and Small-Statured Female Airbag Injuries

The second part of this research involved a comparison of elderly versus small-statured female airbag-induced injuries sustained in similar severity frontal collisions to extrapolate any potential benefits resulting from the proposed modifications to Section 208 of Canada's Motor Vehicle Safety Standards (specifically the inclusion of a 5th percentile female ATD). These data were also analyzed to determine whether commonalities exist between the injury patterns of small-statured females and elderly persons involved in similar severity collisions, the premise being that proposed changes to CMVSS 208 to include small-statured females may also benefit elderly persons. The following sections outline the key findings.

With the introduction of airbags, approximately 23 percent and 37 percent of the elderly and small-statured female group, respectively, experienced a shift from "no injury" to MAIS 1 in low severity collisions. The base population experienced a shift of only 14 percent. Further, about 20 percent of both the elderly and small-statured female groups sustained MAIS 2+ injuries in these collisions. These values are substantially higher than the six percent MAIS 2+ injury occurrence for the base population. This over-representation by both elderly and small-statured female occupants suggests that in low severity collisions, the elderly may benefit from the introduction of the small female ATD in crash tests.

In moderate severity collisions with airbag deployment, changes to motor vehicle standards to further increase protection for small females may again be found to benefit elderly occupants. In these collisions small females and elderly persons sustained similar patterns of injuries that were different than those of the base population. Among the elderly group, 85 percent sustained MAIS 1 injuries, with ten percent sustaining no injury. Similarly, 75 percent of small females sustained MAIS 1 injuries with eight percent being uninjured. Sixty percent of the younger age group sustained MAIS 1 injuries and 26 percent sustained no injury in these collisions.

The research data indicate that in minor severity collisions, all three groups experienced an increase in average MAIS with the introduction of airbags. Although these averages are based on ordinal AIS scores, it is clear that the elderly and small-statured females experience a greater overall shift in injury distributions to more severe levels in low-speed collisions involving airbags. The shift in injuries for both the elderly and small-statured females suggests that the proposed changes to CMVSS 208 to include a 5th percentile ATD may prove to be beneficial to the elderly as well.

The average MAIS sustained by small-statured females in minor severity collisions was almost twice as high (1.29 versus 0.71) with the airbag than without. This is consistent with that observed for elderly occupants in low severity collisions where an increase in average MAIS from 0.78 to 1.03 was observed with the introduction of airbags. The increases in average MAIS scores for the elderly and small-statured females with the introduction of airbags were substantially higher than that observed for younger occupants (0.56 to 0.70). Again, the relatively high frequency of injuries in low severity collisions with airbags for both the elderly and small-statured females suggests that the proposed changes to CMVSS 208 to include a 5th percentile ATD may prove to be beneficial to the elderly as well.

The study results suggest that when an airbag is deployed, small-statured females experience a high proportion of facial injuries (73%) in moderate severity collisions. Elderly occupants also sustained injuries to the face (33%) but suffered a high percentage of thoracic injuries (28%) where the small-statured females sustained none. These results clearly indicate the need for increased chest area protection for the elderly. In high severity collisions, the distribution of injuries to both small-statured females and the elderly group appears to have shifted from the head (33%) and the thorax (50%) to the face (71% and 80%, respectively) with the introduction of airbags. The small sample size for the small-statured female group in high severity collisions was so small (n = 7) that the results may be considered inconclusive.

Conclusions and Discussion

With Canada's aging population, it is extremely important that issues related to and affecting elderly persons be candidly considered. In the automotive industry, the federal government creates laws and regulations to set standards for the safety and environmental performance of new vehicles manufactured in or imported into Canada, and for used vehicles imported from the United States (Transport Canada, 1995). To be effective, these regulations must be enforced fairly, firmly and consistently across the nation. They should also be continually monitored and modified if required to reflect changing demographics and technological advances.

Proposed changes to Standard 208 should increase protection to small-statured females in motor vehicle collisions. It was found that these standard changes may also prove to reduce the number of airbag-induced injuries in the elderly, specifically in minor and moderate severity collisions.

Elderly drivers and passengers are more susceptible to airbag injuries than their younger counterparts in both minor and moderate severity collisions. This suggests that motor vehicle occupant protection standards could be enhanced to address the unique needs of elderly persons. Specifically, study results indicate a need for better chest area protection for the elderly in moderate severity collisions. These findings support the premise that an ATD designed to represent the elderly has potential for significant social benefit through injury prevention or mitigation.

Both the elderly and small-statured females sustained disproportionately more severe injuries than the base population in low severity collisions involving airbags. The predominant location for the most severe injury to the elderly in these low speed collisions was the facial area. These findings suggest that in minor collisions with airbag deployment, changes to motor vehicle standards to increase protection for small females may also be found to benefit elderly occupants. In these cases, Transport Canada's assumption that persons aged 50 years and older will benefit from the proposed changes to CMVSS 208 is likely correct, and a specific statement in the regulation change regarding elderly protection is justified.

The study findings for elderly persons suggest that in low severity collisions, the face and upper body are often injured by the deploying airbag. Dual-stage airbags were introduced in some 1998 model and newer vehicles and may prove to reduce the number of upper body injuries. The current research was based on data collected prior to the introduction of this technology. An area of further study would be to examine whether or not the change in airbag technology provides increased protection for elderly persons in motor vehicle collisions.

REFERENCES

Association for the Advancement of Automotive Medicine (AAAM). (1990). *Abbreviated injury scale – 1990 revision*. Des Plains, IL.

Bedard, M., Guyatt, G., Stones, M., & Hirdes, J. (2002). The independent contribution of driver, crash, and vehicle characteristics of driver fatalities. *Accident Analysis and Prevention, 34*.

Cunningham, C., Coakley, D., O'Neill, D., Howard, D., & Walsh, J. (2001) The effects of age on accident severity and outcome in Irish road traffic accident patients. *Irish Medical Journal, 94,* 169-171..

Dalmotas, D., German, A., Hendrick, B., & Hurley, R. (1995). Airbag deployments: The Canadian experience. *The Journal of Trauma: Injury, Infection and Critical Care, 38,* 476-481.

Dalmotas, D., Hurley, J., German, A., & Digges, K. (1996). Airbag deployment crashes in Canada. *Proceedings of the 15th International Technical Conference on the Enhanced Safety of Vehicles.* Melbourne, Australia.

Dalmotas, D. (1998). Assessments of airbag performance based on the 5th percentile hybrid III crash test dummy. *Proceedings of the 16th International Technical Conference on the Enhanced Safety of Vehicles.* Windsor, Ontario, Canada.

Evans, L. (2001). Age and fatality risk from similar severity impacts. *Journal of Traffic Medicine, 29*, 10-19.

Foret-Bruno, J. Y., Trosseille, X., Page, Y., Huere, J. F., & Le Coz, J. Y. (2001). Comparison of thoracic injury risk in frontal car crashes for occupant restrained without belt load limiters and those restrained with 6 kN and 4 kN belt load limiters. *Stapp Car Crash Journal, 45,* 205-224.

Garthe, E., Mango, N, & States, J. (1998). AIS unification: The case for a unified injury system for global use. *Proceedings of the 16th International Technical Conference on the Enhanced Safety of Vehicles (ESV)*: Windsor, Ontario, Canada. Retrieved October 19, 2002 from *http://www.nhtsa.dot.gov/esv /16/98S6O50.pdf*

German, A., Dalmotas, D., & Hurley, R. (1998). Airbag collision performance in a restrained occupant population. *Proceedings of the 16th International Technical Conference on the Enhanced Safety of Vehicles*: Windsor, Ontario, Canada.

Government of Canada. (2001). *Canada Gazette Part I, 135*(26). Ottawa, ON: Canadian Government Publishing, Public Works and Government Services Canada.

Griffin, E. (2003). *Considering the elderly in the development of Canadian motor vehicle safety standards*. Masters thesis, University of New Brunswick, Fredericton, New Brunswick, Canada.

Libertiny, G. (1995). *Airbag effectiveness – Trading major injuries for minor ones* (SAE Paper 95-871). Detroit, Michigan : Society of Automotive Engineers.

Li, G., Braver, E., & Chen, L. H. (2001). *Exploring the high driver death rates per vehicle-mile of travel in older drivers: Fragility versus excessive crash involvement.* Arlington, Virginia: Insurance Institute for Highway Safety.

Mercier, C., Shelley, M., Rimkus, J., & Mercier, J. (1997). Age and gender as *predictors of injury severity in head-on highway vehicular collisions. Transportation* Research Record, 1581, 37-46. Washington, DC: Transportation Research Board.

Morris, A., Barnes, J., Fildes, B., Bentivegna, F., & Seyer, K. (2001). *Effectiveness of ADR 69: A case-control study of crashed vehicles equipped with airbags.* Canberra: Department of Transport and Regional Services, Australian Transport Safety Bureau.

National Highway Traffic Safety Administration. (NHTSA). (1980). *Collision deformation classification* (SAE J224 MAR80). Warrendale, PA : Society of Automotive Engineers, Inc.

National Highway Traffic Safety Administration (NHTSA). (1993). *National accident sampling system (NASS) 1993 crashworthiness data system injury coding manual.* Washington, D.C.: U.S. Department of Transportation, National Center for Statistics and Analysis.

National Highway Traffic Safety Administration (NHTSA). (1998). *Advanced airbag technology assessment.* Washington, D.C.: Dowdy M., D. Ebbeler, E.H. Kim, N. Moore, R. Phen, and T. VanZandt.

National Highway Traffic Safety Administration (NHTSA). (2000). Chapter 6. *Final economic assessment FMVSS 208 advanced airbags.* Retrieved August 8, 2002 from *http://www.nhtsa.dot.gov /airbag/AAPFR/econ/*

Rosenbloom, S. (2003). *Older drivers: Should we test them off the road?* (Access No. 23). Berkeley, CA: University of California, Transportation Research.

Segui-Gomez, M. (2002). Setting optimal airbag deployment threshold. Harvard Injury Control Research Center Small Projects. Retrieved May 17, 2002 from *http://www. dveemedia/com/Hicrc /Researchprog/smallprj.html#prevention*

Statistics Canada. (2001). *Population by sex and age.* (CANSIM II Table 051-0001 and CANSIM Matrix 6900) Retrieved February 10, 2002 from *http://www.statcan.ca.*

Summers, L., Hollowell, W., & Prasad, A. (2001). Analysis of occupant protection provided to 50th percentile male dummies sitting mid-track and 5th percentile female dummies sitting full-forward in crash tests of paired vehicles with redesigned air bag systems. *Proceedings of the 17th International Technical Conference on the Enhanced Safety of Vehicles.* Amsterdam, The Netherlands.

Transport Canada. (1995). *The Canada motor vehicle safety act guidelines on enforcement and compliance policy.* TP 12957 (E). Ottawa, ON.

University of Michigan. (2001). *CIREN Program Report.* Ann Arbour, Michigan: CIREN Center.

Wang, S., Siegel, J., Dischinger, P., Loo, G., Tenenbaum, N., Burgess, A., Schneider, L., & Bents, F. (1999). The interactive effects of age and sex on injury patterns and outcomes in elderly motor vehicle crash occupants. *Proceedings of the 3rd Annual CIREN Conference.* San Diego, CA.

In: Contemporary Issues in Road User Behavior and Traffic Safety ISBN:1-59454-268-6
Editors: D. Hennessy and D. Wiesenthal, pp. 275-283

Chapter 22

The Role of Control Data in Crash Investigations: Haddon Revisited

Mary L. Chipman
Department of Public Health Sciences, University of Toronto, Canada

Introduction

Research in traffic crashes, their causes, and consequences has attracted scientists from many different backgrounds. A recent review of progress made in reducing crash and injury risk in the United States in the 20th century identifies contributions from engineering, public health, law and the social sciences (Waller, 2002). Much of the research that underlies this progress cannot be based on formal experiments, with randomly assigned treatments etc., but must rely on good observational studies. Each of the disciplines mentioned have well-developed observational methods, whether based on surveys, the careful analysis of administrative records or other forms of observation.

The first Director of the National Highway Traffic Safety Administration (NHTSA) in the United States was a public health physician, William Haddon Jr. His name has long been associated with two schemes for the prevention of injury: Haddon's Matrix (Haddon, 1972) and the Ten Countermeasure Strategies (Haddon, 1973). Before these papers appeared, however, Haddon and his colleagues had conducted a number of epidemiological studies investigating potential causes of injury. One of the best-known is the study of fatally injured pedestrians in Manhattan (Haddon, Valien, McCarroll, & Umberger, 1961). For each of 50 pedestrians who had been killed in traffic crashes, researchers returned to the site of the crash at the same time of day and day of the week within a few weeks of the crash date. They stopped four pedestrians of the same sex as the person killed, administered a brief survey and obtained a sample of breath for analysis. This study demonstrated that it was at least as risky to be a drunken pedestrian in Manhattan as to be a drunken driver. This was also one of the earliest examples of a case-control study used to examine risk factors for traffic crashes or associated injuries.

In this paper, I wish to describe another example where the design Haddon used has been followed. I also plan to discuss the application of case-control studies elsewhere in traffic injury research. Side impact crashes are an important contributor to the burden of road trauma. In the province of Ontario, Canada, crashes occurring at an angle, while turning or as a sideswipe accounted for 39% of all crashes and 38% of all crashes incurring either death or injury (ORSAR, 2001). Much investigative work has been done recently to develop and test standards for vehicle design and construction to reduce the risk of injury to vehicle occupants; NHTSA began conducting crash tests of side impacts in 1997 (NHTSA, 2003). For drivers, only being older and having an age-related medical condition have been consistently associated with side impact crashes (Pruesser, Williams, Ferguson, Ulmer, & Weinstein, 1998; Zhang, Fraser, Lindsay, Clarke, & Mao, 1998). Environmental factors, in particular the characteristics of intersections, also influence crash occurrence (Chipman, 2004). By selecting control vehicles at the site of the crash as Haddon had done, we could control for environmental variation and concentrate of the vehicle features associated with high or low risk of crash occurrence.

METHOD

Transport Canada sponsors crash investigation teams in eight cities and seven universities in Canada. The team in Toronto is based at Ryerson University. They reviewed the crashes investigated by that team from 1998 to 2003 and identified 26 that met the following criteria:

1. Two vehicle crashes, one striking the other on the left or right side at an angle within 45 degrees of perpendicular;
2. The struck vehicle less than 10 years old at the time of the crash, sustaining 'moderate' damage and with an occupant on the struck side and in the intrusion zone;
3. Both vehicles must be passenger vehicles used for personal transportation and with Ontario license plates.

The teams must work closely with local police services, that are responsible for completing crash reports on crashes that result in death, injury or property damage over a certain amount. The police play an essential role in the identification of crashes suitable for more detailed study. When notified of a crash that appears to meet Transport Canada criteria, team members visit the crash site, seek out and examine the vehicles and interview drivers, other occupants, attending police officers etc. The detailed investigation requires assessment of pre-existing circumstances, actions of the drivers, characteristics of the vehicles and the interactions that resulted in the crash and its consequences. The data collected are extensive and include environmental data from the site, measurements of damage to the vehicles, make, model, year and Vehicle Identification Number (VIN) of each vehicle, age, sex and seat position of vehicle occupants and injuries, if any.

Different methods were required to obtain information on control vehicles. As Haddon had done, investigators returned to the site of the crash at the same time of day, day of the week and time of year. Haddon returned within a few weeks of the pedestrian injury. For crashes occurring from late in 2001 to 2003, returning within four weeks after the crash was possible, but for crashes occurring before this, observers returned within four weeks of the anniversary of the crash. At each site, they noted the license numbers of up to four vehicle travelling on the same roads and in the same direction as the crashing vehicles, or up to eight vehicles per crash. They were instructed to take the first four passing vehicles that met most of the criteria given earlier; i.e., non-commercial vehicles with Ontario license plates. Selecting vehicles within 10 years of manufacture applied only to the struck vehicle in the crash and was judged too hard to assess at the roadside, and was not used. Other characteristics of the site were noted in non-crash conditions, notably traffic volume and average speeds of travel.

License numbers were used to obtain the VIN for each control vehicle from the Ministry of Transportation in Ontario (MTO). Encoded in the VIN is information on the make, model, year, curbweight and external dimensions of the vehicle. For many, information on engine size, number of cylinders and specific safety features was also provided. For the remainder, we counted the feature as present or absent depending on whether it was standard equipment or not available for vehicles of that model and year. Optional features were treated as missing values.

Analyses began with descriptive statistics, to compare the struck and striking vehicles in each crash, and to describe crashing vehicles and their controls. We also looked at crash characteristics and the settings in which these crashes occurred. To assess the associations between vehicle characteristics and crash risk, conditional logistic regression (Selvin, 2001) was required because crashing and control vehicles were matched on environmental factors. This method produces estimates of odds ratios to assess the difference in crash risk. For categorical factors, the odds ratio estimates the odds that a vehicle with a specific feature (e.g., has ABS) is one of the crashing vehicles relative to the odds that a vehicle without that feature is one of the crashing vehicles. For continuous variables, the odds ratio estimates the rate of change of the odds for a unit change (e.g., per litre of engine size) in the variable. When there is little or no association, the odds ratio will be close to 1.0. An important question was whether a vehicle characteristic had the same association with crash risk for bullet (striking) as for target (struck) vehicles. If it was (assessed by both examination of effect size and a test of significance), a single estimate of the odds ratio could be used. Otherwise, separate estimates for bullet and target vehicles were made. In reporting tests of significance, p-values for two-sided tests of significance have been given.

RESULTS

There were 26 crashes (52 vehicles) and 161 control vehicles identified. No control was possible for one vehicle in each of two crashes; these were crashes where one vehicle was exiting a private driveway, and controls were obtained only for the second vehicle, on a public road, in each crash. A complete set of four controls was obtained for 30 of the remaining 50 vehicles, three controls for 6 vehicles, two for 9 vehicles and one control for the remaining 5 vehicles.

What were the environments in which these crashes occurred? The crash investigation team at Ryerson University was working in a predominantly urban environment, with some suburban but no rural crashes, and few crashes occurring on a provincial highway. In this respect, their situation was similar to that of Haddon, returning to the streets of Manhattan to identify controls where a pedestrian had been killed. Although the environmental factors of time and place have been controlled in the selection of comparison vehicles it is useful to examine these characteristics.

Place

Most crashes (21/26) occurred at intersections; 10 occurred on main roads in the urban core, 8 on suburban arterial roads and 6 on roads in residential neighbourhoods. All but two roads were described as straight and on the level. The speed limits in effect at these locations reflect their urban nature. Only two crashes occurred on roads where the speed limit was higher than 60 km/hr (37.5 mph). The mean speed of traffic observed when controls were selected was within 20% of the posted speed limit for 17 crashes. For two crashes it was not obtained, but in the remaining 7 crashes it was at least 20% higher than the posted value.

Volume of traffic was observed for each vehicle, excluding only the two in driveways. The distribution was quite skewed, so percentiles have been used to describe this characteristic. The median (50th percentile) was 37 cars in a 15 minute interval, or 1 car every 24 seconds. The interquartile range (25th to 75th percentiles) was 21 to 100 vehicles, or from one car every 45 seconds down to one every 9 seconds. Comparing traffic volumes for target and bullet vehicles using the ratio of traffic volumes (target:bullet), the median and interquartile range of this ratio was 0.61 and 0.30 - 1.60. Thus, for every six vehicles travelling in the direction of the target vehicle, there are, on average, 10 travelling in the direction of the bullet vehicle.

Drivers

Mean age was 38.7 ± 15.2 years for drivers of target vehicles and 33.5 ± 9.3 years for drivers of bullet vehicles. The proportion of male drivers was 75.0% and 81.2% in target and bullet vehicles respectively, and the proportion in which someone reported an injury was 60.0% and 56.0% respectively. Although the prevalence of injury looks high, the injuries were uniformly minor, with no abbreviated injury score (AIS) greater than 1. The AIS score ranges from 0 (no damage) to 6 (not survivable), rated at each of six anatomical domains (e.g., head, abdomen etc.) (O'Keefe & Jurkovich, 2001).

Vehicles

Five crashes had little information on either vehicle, and data on the bullet vehicle was lacking in two additional crashes, leaving 40 vehicles in 21 crashes. The license number was able to produce a VIN and associated vehicle information for 129 of the 161 controls. The vehicle characteristics are given in Table 1. Comparisons of bullet and target vehicles were made, using the paired *t*-test on the 19 crashes where data on both vehicles were available. Except for vehicle age these were not statistically significant ($p > 0.05$). Vehicle age was different, largely due to the restriction among target vehicles to late model vehicles, which did not apply to bullet vehicles. Mean values are very similar but the variability is significantly greater ($p < 0.05$) among the bullet vehicles for all characteristics except number of cylinders and vehicle width.

Table 1. Characteristics of Crash and Control Vehicles by whether Bullet or Target Vehicle

Vehicle characteristics	Bullet Vehicle		Target Vehicle	
	Crash	Control	Crash	Control
Number of vehicles	19	60	21	69
No. of cylinders	5.38±1.59	5.11±1.30	5.00±1.21	5.39±1.70
Engine size (L)	2.84±1.43	2.70±0.92	2.61±0.86	2.85±0.93
Wheelbase (cm)	276.5±39.9	271.7±18.3	270.7±15.6	274.8±23.7
Length (cm)	474.2±60.5	469.6±28.8	473.0±30.3	474.1±35.9
Width (cm)	180.5±13.5	179.7±10.5	180.9±11.6	179.6±9.7
Height (cm)	151.8±23.4	151.6±17.1	144.7±11.7	149.9±14.9
Curb weight (kg)	1491±480	1489±373	1378±210	1477±297
Vehicle age at crash (yr)	5.12±4.69	2.02±1.83	1.52±2.29	2.34±1.70

Comparison of crash and control vehicles begins with the summary of vehicle characteristics given in Table 1. Patterns are difficult to discern; for many measures the value of the mean appears very similar in all four groups. The proportions of vehicles with a variety of safety features are given in Table 2. Front seat airbags are common, having been required equipment on new vehicles in the USA and Canada since 1994; nevertheless, the proportion of vehicles with this feature was lower in bullet vehicles involved in these crashes. Side airbags were much less common in all groups of vehicles, but especially in crashing vehicles. Antilock braking systems (ABS) were very common in all groups except crashing bullet vehicles. Traction control was found in only 25 vehicles overall, and nearly all of these were control vehicles. When vehicles were grouped by model, the largest number of vehicles in all four groups was four-door sedans; the distribution of the remaining vehicles in each group varied, with two-door vehicles (coupes, convertibles etc.) being more common among crashing vehicles and other vehicles (minivans, sport utility vehicles (SUVs) and light trucks) more common among controls.

Tests of significance have not been done on these summary statistics, because such tests should accommodate the matched selection of control data. Conditional logistic regression, however, does this, and the results appear in Table 3 for most of these variables. For direct measures of size and power, odds ratios are all very close to the value 1.0, and $p > 0.30$, providing no evidence that these factors are associated with crash involvement. Two-door vehicles, such as coupes and sports cars, however, have an odds ratio of 3.45 ($p = 0.014$) relative to four-door sedans. Vans, SUVs and other truck-like vehicles used to carry passengers have an $OR = 0.61$, suggesting a reduced rate of crashing for these vehicles. However, it is not statistically significant ($p = 0.16$) with a 95% confidence interval of 0.23 - 1.71.

Table 2. Numbers (and Percentages) of Vehicles by Style and with Selected Safety Features

Safety feature	Bullet Vehicle		Target Vehicle	
	Crash	Control	Crash	Control
Antilock braking system	5/12 (41.7)	38/42 (90.5)	18/20 (90.0)	36/39 (92.3)
Traction control	0/18 (0.0)	10/48 (20.8)	1/18 (5.6)	14/51 (27.5)
Front airbags	13/18 (72.2)	59/59 (100.0)	20/21 (95.2)	67/68 (98.5)
Side airbags	0/18 (0.0)	24/55 (43.6)	4/21 (19.0)	23/61 (37.7)
2-door (coupe etc.)	6/17 (35.3)	7/60 (11.7)	4/21 (19.1)	4/68 (5.9)
4-door sedan	8/17 (47.1)	38/60 (63.3)	15/21 (71.4)	50/68 (73.5)
Van, SUV or light truck	3/17 (17.7)	15/60 (25.0)	2/21 (9.5)	14/68 (20.6)

Table 3. Results of Conditional Logistic Regression: Odds Ratios for Side Impact Crash Involvement

Vehicle Characteristic	Odds ratio (OR)	*p*-value	95% confidence interval
No. of cylinders	0.95	0.73	0.71 - 1.26
Engine size (per litre)	0.93	0.71	0.64 - 1.36
Curb weight (per 100 kg)	0.95	0.35	0.85 - 1.06
Wheel base (per 10 cm)	1.02	0.85	0.86 - 1.20
2-door vs 4-door	3.45	0.014	1.29 - 9.22
Other vs 4-door	0.48	0.16	0.17 - 1.34
Traction control	0.09	0.018	0.01 - 0.66
ABS: bullet vehicles	0.10	0.017	0.02 - 0.66
ABS: target vehicles	1.11	0.96	0.16 - 7.70

ABS and traction control both appear to have a markedly protective effect. The effect of traction control was equally strong for target and bullet vehicles; for ABS, however, the effect existed only for bullet vehicles, not for target vehicles. While the comparison was not quite statistically significant at the 5% level ($p = 0.083$), the size of the difference in odds ratios was substantial. The relationship with front or side airbags has not been considered in these regression analyses. There is no role for these devices in the prevention of the crash itself; they act only once a crash is underway in attempting to reduce the risk of injury to vehicle occupants.

Discussion

This study has demonstrated that some vehicle characteristics are associated with substantial differences in the risk of side impact crashes. ABS appear effective in reducing the risk in bullet (striking) vehicles but not in target (struck) vehicles. It makes intuitive sense that vehicles hit from the side would not be as able to use brakes to avoid a crash as would the striking vehicle. Other studies of ABS have not looked so specifically at side impact crashes, nor been able to differentiate between bullet and target vehicles, and have not demonstrated such a strong effect (Farmer, 2001; Farmer, Lind, Trempel, & Braver, 1997).

Traction control appears to have a protective effect for both bullet and target vehicles. This may be an artifact of control selection, however. A safety feature that is not available or optional in one year may become popular and be standard equipment in later model years of the same vehicle. Controls that were selected a year or more after the crash occurred may be more likely to have safety features in these circumstances than the crashing vehicle they are compared to. Such an effect would influence bullet and target vehicles alike; thus this is not a likely explanation for the effects seen for ABS.

Disparity in vehicle size, particularly the trend to larger passenger vehicles built on truck frames (Gladwell, 2004), has been of concern, particularly related to the risk of injury in the second, smaller, vehicle. These data do not suggest that size, whether expressed as curb weight, engine size or by external dimensions, affects the risk of side impact crashes for either target or bullet vehicle. When these vehicles are classified roughly by type, two-door passenger cars have a higher risk of crashing compared to four-door sedans. Minivans, SUVs and light trucks have a reduced risk of crash involvement, although this study is not large enough to be confident this was not a chance event.

Two-door vehicles are often smaller than other passenger cars; the fact that a comparable effect was not found for other more direct measures of size suggests that other characteristics of two-door vehicles or their drivers may be implicated. The elevated crash risk for two-door vehicles may be due to factors that we were unable to examine, such as vehicle conspicuity, driver actions and other characteristics. For example, vehicles without easy access to the rear seat may be less likely to be used as family vehicles, and driven in different circumstances.

Many studies of vehicle characteristics have used data that is limited to crashes, or to crashes in which someone has been killed. This is helpful when the research question addresses the risk of injury when a crash has occurred, but is less useful in addressing fundamental questions of factors affecting crash occurrence in the first place. This study is unusual because, unlike most of the crash investigations supported by Transport Canada, these have included a selection of control vehicles. As pointed out early in this paper, this is similar to the methods described by Haddon et al. in 1961. The similarities and differences are instructive.

Haddon and his colleagues stopped passing pedestrians on the street, and succeeded in identifying four controls for each person in his series of cases, with one exception. This was the only crash in which two pedestrians had been killed so eight control pedestrians were required. The site was a street in Harlem, and the time was early on a Sunday morning. By the time they had identified six controls, a group was beginning to gather, and the intrepid investigators decided to beat a retreat. This procedure, involving direct contact with control pedestrians, was essential to obtain the information they wanted, including a sample of breath to assess blood alcohol concentration. Clearly they required controls to consent to take part and cooperate with observers.

In Haddon's study, it is not clear how the team decided which pedestrians to stop; one difference that became quite apparent when the data were analysed was that controls were generally younger than the cases. Haddon does not appear to have taken this to be a genuine effect; i.e., that older people on the street are more likely to be run over and killed by passing cars than younger people. Instead he conducted his analyses so as to control for age when looking at other factors such as alcohol. How did this age difference happen? Perhaps team members, who were instructed to match only on whether pedestrians were men or women, found younger people easier to approach; perhaps older people were less likely to agree to being stopped in the street. In the present study, investigators merely observed passing license numbers; no attempt was made to get more information by stopping cars or making direct contact with drivers. By not attempting to contact vehicle owners or drivers in the present study, we may have reduced the risk of selection bias among the control vehicles we did obtain. Nevertheless, we failed to obtain as many controls as planned for 20 crash-involved vehicles. In addition, the license numbers recorded did not give adequate information from the corresponding VIN in 32/161 (19.8%) of control vehicles where a license number had been obtained. This may have been due to errors in transcribing the license number (lighting conditions and the speed and density of traffic made transcription difficult in some circumstances), ambiguities in the data provided by the manufacturer (in a few cases, the same model code was used for several vehicles, or the vehicle description was very non-specific) or clerical errors in recording the VIN itself.

Use of the VIN proved surprisingly difficult. This was due, in large part, to the lack of consistency among manufacturers in the level of detail or format they adopted. Data on external dimensions were available, and engine size and number of cylinders were usually provided. Information pertaining to safety features like seatbelts and airbags was more variable, but usually available. Data on ABS and traction control were more difficult to obtain. In many cases, unless we could establish that these features were either standard equipment or not available, we had to treat the variable as missing. This has led to reduced power and precision for the results of these variables.

In general, the confidence intervals for these estimates of effect are broad, indicating a lack of precise estimation of the real size of the protection involved. While one can be assured that the effects of ABS (in particular) are real, a larger sample size is required to estimate the actual magnitude of effect with any confidence. Haddon's design has proven useful in these circumstances. It has allowed us to control for considerable variety in the environment in which these crashes have occurred. It is not possible to control for specific weather conditions, such as blizzards or thunderstorms, but more stable conditions related to season of the year, time of day and traffic conditions have been addressed. This allows comparisons to concentrate on vehicle characteristics, and to examine the possibility of different effects for each vehicle involved in such crashes.

ACKNOWLEDGEMENTS

I am grateful to Transport Canada for allowing use of crash data; I am indebted to my colleagues Ian Hale, Bhagwant Persaud, John Bou-Younes and Ravi Bhim from Ryerson University in Toronto, Canada for assistance with control data. I wish to thank the Ministry of Transportation in Ontario for providing VIN for control vehicles. This research is part of A01-ACI: Crash investigations: implications for vehicle design, funded by Auto21, The Automobile of the 21st Century, a Network of Centres of Excellence research program. This support is also gratefully acknowledged.

REFERENCES

Chipman, M. L. (2004). Side impact crashes – factors affecting incidence and severity: A review of the literature. *Traffic Injury Prevention, 5*, 67-75.

Farmer, C. M. (2001). New evidence concerning fatal crashes of passenger vehicles before and after adding antilock braking systems. *Accident Analysis & Prevention, 33*, 361-369.

Farmer, C., Lind, A., Trempel, R., & Braver, E. (1997). Fatal crashes of passenger vehicles before and after adding anti-lock braking systems. *Accident Analysis & Prevention, 29*, 745-757.

Gladwell, M. (2004, January 12). Big and Bad: How the SUV ran over automotive safety. *New Yorker_LXXIX*, 28-33.

Haddon, W. Jr. (1972). A logical framework for categorizing highway safety phenomena and activity. *Journal of Trauma, Injury, Infection & Critical Care, 12*, 193-207.

_____ . (1973). Energy damage and the ten countermeasure strategies. *Human Factors, 15*, 355-366. (Reprinted in *Injury Prevention* 1995; 1(1): 40-44.)

Haddon, W. Jr., Valien, P., McCarroll, J. R., & Umberger, C. J. (1961). A controlled investigation of the characteristics of adult pedestrians fatally injured by motor vehicles in Manhattan. *Journal of Chronic Diseases, 14*, 655-678.

NHTSA (National Highway Traffic Administration). (2003). Available online at nhtsa.dot.cov/cars/testing/ncap/Info.html, accessed 12 Oct 2003.

O'Keefe, G., & Jurkovich, G. J. Measurement of injury severity and co-morbidity. In F. P. Rivara, P. Cummings, T. D. Koepsell, D. C. Grossman, & R. V. Maier (Eds.), *Injury Control: A Guide to Research and Program Evaluation* (pp. 32-34). Cambridge, UK: Cambridge University Press.

ORSAR (2001). Table 3.4: Initial impact type by class of collision 2001, In: *Ontario Road Safety Annual Report 2001.* Available online at *http://www.mto.gov.ca/english/safety/orsar01/chp3a_01.htm*, accessed January 11, 2004.

Pruesser, D., Williams, A., Ferguson, S., Ulmer, R., & Weinstein, H. (1998). Fatal crash risk for older drivers at intersections. *Accident Analysis & Prevention, 30*, 151-159.

Selvin, S. (2001). *Epidemiologic Analysis: a Case-oriented Approach* (pp. 124-125). New York, USA: Oxford University Press.

Waller, P. F. (2002). Challenges in motor vehicle safety. *Annual Review of Public Health, 23*, 93-113.

Zhang, J, Fraser, S., Lindsay, J., Clarke, K., & Mao, Y. (1998). Age-specific patterns of factors related to fatal motor vehicle traffic crashes: focus on young and elderly drivers. *Public Health, 112*, 289-295.

In: Contemporary Issues in Road User Behavior and Traffic Safety ISBN:1-59454-268-6
Editors: D. Hennessy and D. Wiesenthal, pp. 275-283 ©2005 Nova Science Publishers, Inc.

INDEX

A

Abtahi, A., 10
abuse, 26, 54, 158, 159, 178, 179, 180
accelerator, 234
accident cause, 106
accident metamorphosis, 7
accounting, 27, 43
accumulation, 160
accuracy, 253, 256, 258, 259, 260, 262
activities, 32, 61, 66, 167, 201, 240, 245
adaptation, 3, 6, 9, 10, 11, 122, 131, 132, 133
adjustment, 209
administration, 60, 133, 148, 153, 156, 164, 165, 166, 182, 212, 217, 224, 243, 264, 266, 267, 274, 275, 283
adolescents, 9
adulthood, 45, 156
advances, 272
advertising, 168
advocacy, 168
age, 16, 17, 19, 24, 34, 42, 45, 51, 55, 82, 85, 86, 87, 88, 89, 90, 91, 93, 94, 95, 96, 99, 100, 104, 107, 113, 114, 115, 116, 117, 119, 122, 132, 153, 155, 161, 177, 235, 239, 263, 264, 265, 269, 271, 273, 274, 276, 278, 279, 281
agent, 120, 210
aggression, xiii, xiv, xv, 14, 15, 19, 20, 23, 24, 25, 33, 34, 35, 36, 37, 39, 40, 41, 42, 43, 44, 45, 46, 51, 58, 62, 63, 64, 66, 67, 68, 69, 70, 71, 72, 73, 75, 76, 77, 78, 83, 84, 110, 156, 178, 179, 242
aging, 272
aging population, 272
aid, 54, 245, 262
air bags, xi
alcohol, xiv, 3, 9, 47, 54, 82, 105, 108, 135, 136, 137, 138, 139, 143, 145, 146, 147, 148, 149, 153, 154, 155, 156, 157, 158, 159, 160, 162, 163, 164, 165, 166, 167, 168, 169, 170, 172, 174, 175, 176, 177, 179, 181, 182, 187, 228, 231, 235, 241, 281
Anger, W.K., 10
anxiety, 51, 53, 57, 63, 162, 201, 206, 209, 210, 211
appreciation, 110
Aschenbrenner, M., 10
Asia, 101
assessment, 20, 44, 49, 50, 53, 54, 56, 111, 121, 137, 156, 161, 163, 165, 196, 202, 204, 208, 211, 243, 274, 276
assessment techniques, 202
association, 27, 43, 92, 93, 94, 160, 163, 277
Assum, T., 10
Australia, 24, 81, 82, 83, 136, 147, 148, 149, 164, 223, 228, 241, 242, 273
authenticity, 74, 180
authority, 18, 72, 107, 261

B

barriers, 3, 136, 259
base rate, 213
behavioral change, 185, 223
behaviour, xiv, xv, 20, 33, 34, 36, 45, 60, 70, 98, 100, 111, 121, 122, 186, 193, 195, 197, 223, 240, 242, 250, 261
benefits, 5, 8, 9, 57, 58, 76, 97, 109, 156, 217, 240, 245, 261, 271
best practices, xi, xii
beverages, 136, 138
bias, 19, 20, 256, 282
Biehl, B., 10
Bjørnskau, T., 10
black, xiv, 72, 86, 133, 145, 178, 188
bonds, 155, 164
boredom, 5
Branas, C.C., 10
Britain, 24

Brown, J.D., 10
business, 4, 138, 139, 142, 168
business cycle, 4

C

California, vii, 7, 9, 11, 23, 27, 29, 31, 32, 35, 148, 157, 165, 182, 274
campaigns, 32, 168, 216
Canada, xi, xii, xiv, xv, 11, 33, 34, 35, 36, 45, 69, 120, 135, 138, 139, 147, 148, 149, 164, 165, 166, 167, 182, 187, 188, 195, 215, 223, 224, 227, 228, 242, 245, 246, 250, 251, 263, 264, 265, 266, 272, 273, 274, 275, 276, 279, 281, 282
candidates, 62, 63
capacity, 115
carelessness, 3
casualty(ies), xi, 5, 104, 137, 145
cell phones, xi, 26, 228, 241, 245
Chandon, J.L., 10
change, 119, 148, 160, 181, 195, 219, 220, 221, 222
channels, 260
chaos, xv
character, 104, 180
Chebat, J.C., 10
childcare, 96
children, 8, 26, 96, 116, 117, 118, 119, 237, 239, 253
China, 101, 106, 107
classes, 102
Clinton, President William, 195, 232
clusters, 204, 208, 211
collaboration, v
collision investigation, xi, 253, 261
collisions, xi, xiv, 7, 14, 18, 133, 139, 146, 193, 215, 216, 228, 229, 230, 231, 237, 239, 240, 241, 242, 243, 246, 251, 252, 253, 254, 257, 258, 260, 261, 263, 264, 265, 266, 267, 269, 270, 271, 272, 274
Columbia, 36
commercial, 43, 85, 95, 245, 277
commitment(s), 180, 195
Commonwealth, the, 241
community(ies), xv, 32, 35, 48, 55, 70, 139, 168, 169, 175, 177, 194, 195, 196
competition, 24, 64, 181
competitive, 156
compliance, 117, 193, 196, 274
computers, 250
conflict, 66, 71, 174, 177, 178, 262
confusion, 4
congress, 47, 224
consent, 66, 72, 73, 156, 174, 261, 281
consumption, 3, 9, 108, 136, 137, 138, 139, 142, 146, 148, 156, 159, 164, 168, 174, 177, 185, 186, 223
content, 8, 28, 32, 43, 111, 162, 237
content analysis, 28, 32, 111
contribution, 122
corporations, 168
costs, xi, 5, 9, 76, 138, 142, 199, 202, 218
crash helmets, 7
credibility, 171, 257
credit, 61, 73
crime, v, 43, 70, 148, 240
cross-fertilization, xiv
cultural, 5, 32, 77, 89, 96, 98, 107, 109, 110, 111
cultural differences, 32, 110
culture, 68, 83, 93, 98, 100, 109, 111, 169
curiosity, v
Czech Republic, 227

D

danger, 3, 24, 26, 66, 171, 187, 228, 240
data set, 88, 113, 140, 146, 218
de Bruin, R.A., 10
death rate, 4, 5, 7, 10, 145, 274
decision-making, 131
deficit, 67
delivery, 193, 239
demand, 43, 111, 233, 234, 237, 239, 242
demographics, 82, 113, 156, 263, 272
Denmark, 216, 223
dependent, 3, 8, 10, 14, 66, 73, 74, 189, 230, 234, 252, 261
Derrig, R.A., 10
deterrence, 120, 168
developing countries, xi
development, xi, xv, 20, 32, 36, 44, 46, 55, 59, 68, 109, 110, 122, 132, 199, 202, 204, 205, 206, 207, 208, 209, 210, 211, 214, 223, 242, 249, 251, 260, 261, 262, 263, 273
Diagnostic and Statistical Manual of Mental Disorders, 202, 212
differentiation, 116
discourse, 171, 181
disease, xi, 8, 230
dispersion, 141, 223
disposable income, 5
distortions, 55
donations, 186
driver, v, xi, xiv, xv, 3, 7, 8, 10, 11, 14, 15, 16, 17, 18, 19, 20, 23, 24, 25, 26, 27, 28, 29, 30, 31, 32, 33, 34, 35, 36, 44, 45, 46, 48, 50, 51, 52, 53, 54, 55, 58, 59, 60, 66, 68, 69, 71, 72, 73, 74, 75, 76,

77, 78, 81, 82, 83, 88, 89, 90, 93, 95, 96, 98, 99, 100, 104, 105, 106, 107, 108, 109, 110, 111, 113, 115, 117, 118, 119, 120, 121, 122, 126, 127, 130, 131, 132, 133, 142, 143, 145, 146, 147, 154, 162, 164, 165, 166, 169, 170, 171, 172, 173, 175, 176, 177, 178, 179, 180, 181, 194, 196, 197, 217, 222, 223, 224, 227, 229, 230, 231, 235, 236, 237, 239, 240, 241, 242, 243, 245, 249, 250, 251, 252, 253, 254, 255, 256, 257, 258, 259, 260, 261, 273, 274, 275, 281
driver behavior, xi, xiv, 14, 28, 73, 82, 98, 99, 110, 120, 217, 223, 249
driving license, 109, 113
driving violations, viii, 79, 81, 82, 84
drugs, 105, 159, 164, 165, 166
DSM-II, 60, 156, 158, 203
DSM-III, 60, 156, 158, 203

E

economic, 4, 5, 10, 55, 101, 135, 146, 147, 149, 176, 178, 199, 202, 241, 274
economic growth, 4
education, 8, 32, 55, 99, 104, 105, 108, 110, 119, 122, 147, 156, 157, 162, 168, 181, 182, 194, 195, 196, 209, 210, 211
efficiency, 32, 108, 187
election, 175
elementary (primary) school, 8
email, 245, 250
empirical, 10
employees, 9, 72
employment, 62, 156
employment status, 156
engineering, xi, xiv, 3, 7, 275
England, 27, 133, 231, 242, 243
English, 106
enthusiasm, vi
environment, xiv, 35, 44, 68, 69, 70, 195, 196
epidemiology, xi, 229
equipment, 62, 83, 223, 232, 237, 247, 277, 279, 281, 282
ergonomics, xi
Europe, 136, 215
evaluation, xv, 11, 13, 34, 39, 44, 137, 138, 149, 181, 182, 194, 196, 212, 213, 215, 216, 217, 220, 222, 224, 239, 262
examinations, 186
expectations, 19, 40, 62, 109, 155, 191

F

facial expression, 75
family, 58, 71, 135, 156, 157, 162, 164, 169, 179, 180, 201, 281
family members, 58, 169, 201
fatalities, xi, 7, 10, 11, 26, 47, 71, 103, 104, 138, 139, 140, 142, 143, 145, 146, 147, 153, 154, 155, 167, 187, 215, 231, 246, 263, 273
feedback, 187, 194, 195, 196
female(s), 15, 20, 32, 39, 45, 62, 64, 74, 75, 76, 77, 81, 84, 85, 86, 87, 88, 90, 105, 107, 114, 115, 120, 157, 158, 207, 211, 236, 264, 265, 266, 271, 272, 274
Finland, 24, 45, 99, 236
firearms, 27
firm, 77, 136, 239
Fitch, H.G., 11
flight, 243
Florida, v, 37, 61, 70
food, 142, 196
Ford, D., 72, 186, 187, 192, 193, 194, 197, 234, 236
Fosser, S., 10
Fox, D.K., 10
France, 227
free, 5, 8, 9, 11, 18, 48, 82, 83, 110, 138, 175, 176, 229, 230, 232, 233, 234, 235, 236, 237, 239, 241, 246, 247, 249
freedom, 95, 174
frequency, 29, 30, 31
friction, 11
friends, 39, 40, 58, 158, 169, 170, 173, 174, 175, 176, 177, 180
funding, 98

G

gasoline, 5
gender, xiii, xv, 16, 17, 24, 34, 42, 51, 55, 67, 73, 74, 76, 77, 82, 85, 88, 90, 93, 94, 95, 98, 107, 113, 119, 205, 207, 211, 274
gender roles, xiii
generation(s), 110, 233, 260, 267
Georgia, 218
Germany, 24, 164
goals, 24, 49, 63, 178, 263
government, 108, 264, 273
grant, 7, 11, 163
Green, Carole, 19, 108, 111
Groeger, J.A., 10
groups, 16, 48, 49, 51, 52, 53, 56, 63, 70, 95, 96, 97, 98, 117, 139, 140, 143, 155, 157, 158, 159, 160,

161, 163, 168, 175, 207, 209, 210, 230, 232, 265, 267, 269, 270, 271, 279
growth, 4, 43, 228
guilty, 178

H

Harano, R.M., 11
harmonization, 148
Haynes, R.S., 11
hazardous substances, 3
health, 3, 4, 8, 9, 10, 11, 35, 56, 58, 136, 147, 149, 153, 162, 169, 182, 185, 186, 195, 275
health care, 186
health services, 149
high school, 72, 155, 157, 185, 195
history, 19, 24, 45, 83, 139, 156, 157, 159, 160, 162, 164, 205, 206, 211, 251
HIV, xi
Hong Kong, viii, 101, 102, 103, 104, 105, 106, 107, 108, 110, 111, 112
Hopkins, B.L., 10
house, 60
household income, 156
Hubert, D.E., 11

I

ideas, v, vi, xiv
Illinois, 78
inattention, 3, 27, 28, 107, 193, 228, 233
incentive(s), 8, 9, 11, 156, 194
income, 5, 122, 156, 157, 162
Indian, 239
indicators, viii, 162, 163, 165, 199, 202, 204, 205, 206, 211, 219, 221
industry, 39, 250, 272
information, viii, 82, 111, 125, 126, 157, 282
information processing, 235, 239, 256
infrastructure, 60, 101, 241
input, 245, 249, 260
inspectors, 137
instability, 163
institutions, 169
instruments, 52, 99
insurance, 5, 9, 11, 60, 167, 177, 181, 227, 261, 274
integrity, 200
interest, v, xiii, 3, 66, 72, 169, 201, 230, 234, 246, 247, 249, 252, 261
internet, 20, 37, 168
interpreting, 155, 219, 222, 230, 259
intervention, viii, 8, 49, 50, 54, 55, 56, 59, 76, 108, 138, 140, 143, 147, 163, 165, 167, 168, 169, 171, 173, 175, 176, 178, 180, 181, 195, 208, 209, 210, 211, 212
Ireland, 223
Israel, 24
issues, iv, xiv, 29, 32, 72, 111, 159, 163, 167, 196, 254, 272
Italy, xv, 24, 34

J

Japan, 106
jobs, 231
Jones, Bill, 13, 20, 125, 132, 155, 165, 168, 181
jurisdiction, 6, 96, 145
justice, 157, 165

K

Knudson, M.M., 10

L

labeling, 69
labor, 218
lane changing, 105
language, 101, 106, 109, 239, 262
Late, 118
Lave, L.B., 11
Lave, T.R., 11
law enforcement, 58, 108, 122
laws, 7, 10, 100, 101, 105, 120, 156, 168, 261, 272
learning, v, xiii, 58, 109, 230
legal, iv, 101, 108, 139, 156, 178, 187, 235, 261
legislation, 7, 83, 108, 110, 169, 195, 227, 239
liberalization, 135
licenses, 109
limitation, 41, 194
listening, 116, 117, 118, 119, 232, 234, 235, 246
literacy, 168
literature, 35, 39, 47, 48, 49, 50, 62, 66, 68, 71, 75, 106, 109, 136, 204, 216, 228, 239, 256, 270, 283
Liu, L.L., 10
long distance, 62, 63

M

male(s), 15, 17, 20, 32, 39, 42, 45, 64, 72, 74, 76, 81, 84, 85, 86, 87, 88, 90, 95, 96, 100, 105, 107, 114,

115, 120, 121, 155, 207, 228, 231, 236, 241, 264, 274, 278
management, 6, 11, 33, 99, 195, 196, 210, 211, 215, 223, 224, 249
marital status, 99, 156
market, 262
marketing, 8, 122, 168
May, A.D, 11
mean, 17, 49, 50, 52, 53, 64, 75, 88, 217, 219, 220, 221, 222, 248, 278, 279
measurement, xiv, 69, 110, 169, 208, 218
measures, 4, 8, 10, 37, 39, 40, 41, 42, 43, 50, 51, 52, 54, 55, 57, 63, 66, 67, 81, 107, 109, 113, 136, 138, 143, 153, 155, 156, 160, 161, 162, 163, 202, 204, 209, 211, 216, 219, 223, 233, 237, 246, 251, 262, 279, 281
media, 5, 6, 43, 47, 75, 99, 168, 182, 228
media messages, 168
mentorship, v
methodology, xi, 67, 238, 243
mining, 10
minority, 78
minors, 168
Missouri, 47, 165
mobile phone, 82, 83, 84, 85, 95, 110, 240, 241, 242
model, 24, 66
modelling, 100
momentum, 189
money, 158
monitoring, 56, 194, 247
monopoly, 138
Moore, Dennis, 258, 262, 274
mortality rates, 146
mother tongue, 239
motorcyclists, 7, 26, 106, 109, 112
movement, 107, 112, 189, 259

N

Napier, V., 11
negligence, 106, 107, 110
Netherlands, xi, 10, 24, 148, 227, 274
networks, 107
New South Wales, 81, 136, 148
New Zealand, 164, 212, 235
newspapers, 34
normative, 99, 195
North America, 24, 25, 26, 44, 45, 63, 216, 217, 227, 263, 264

O

objectives, 217, 264
observation, 218
Oregon, 7
Organization for Economic Cooperation and Development (OECD), xi, 11
outline, 271
output, 245, 259
ownership, 105, 174, 227, 232, 260

P

parachute jumping, 7
parents, 8, 72
participation, 195, 246
partnership, 113, 121
peers, 5
Pennsylvania, 231, 237
percentile, 219, 220, 221, 222
perceptions, xiii, 8, 14, 20, 32, 35, 77, 132, 137, 138, 162, 236, 246
performance, 10, 14, 34, 68, 73, 82, 83, 111, 113, 127, 167, 170, 172, 175, 194, 218, 222, 223, 224, 232, 233, 234, 235, 236, 237, 238, 239, 241, 242, 245, 250, 260, 261, 272, 273
personality, 13, 19, 34, 43, 44, 45, 51, 58, 104, 105, 111, 113, 155, 164
Perth, Australia, 136, 147, 148, 228
physical injury, xiv, 207
Pine, A.R.C, 11
police, 26, 32, 51, 52, 53, 58, 66, 71, 97, 106, 107, 114, 115, 119, 120, 137, 139, 168, 177, 194, 228, 231, 232, 254, 257, 259, 261, 276
policy(ies), viii, xii, 3, 33, 101, 110, 122, 135, 136, 139, 146, 147, 148, 149, 186, 195, 223, 228, 240, 241, 242, 261, 274
policy makers, xii
politics, 168
poor, 10, 107, 176, 201, 209, 256
population, 4, 6, 7, 10, 48, 49, 50, 51, 53, 54, 57, 58, 76, 97, 98, 101, 102, 107, 108, 109, 122, 135, 136, 153, 154, 155, 164, 215, 263, 264, 265, 267, 270, 271, 272, 273, 274
post-traumatic stress disorder, 107, 112, 214
power, xiii, 34, 45, 136, 137, 143, 145, 169, 173, 204, 229, 253, 259, 279, 282
pre-planning, 173
prices, 138
primary, 8, 28, 56, 126, 131, 232, 234, 238, 247
priorities, 180
privatization, 135

problem solving, 25, 209, 210
production, 4, 110, 179
productivity, v, 9, 199
profanity, 180
prosperity, 5
public buses, 106, 108, 111
public health, 4, 58, 136, 147, 153, 275
public policy, 186, 228, 240
punishment, 56, 107, 120, 168, 194

Q

questioning, 170
quizzes, 37, 43

R

radio, 82, 108, 231, 232, 234
rain, 105, 116, 117, 119
ratings, 14, 16, 17, 62, 206, 234
recovery, 70, 207, 209
regulation(s), 44, 135, 137, 138, 145, 168, 224, 227, 228, 242, 251, 261, 264, 265, 272
relevance, 111, 163, 175
reliability, 16, 24, 27, 28, 40
religion, 25
research, v, xi, xiv, xv, 14, 19, 24, 27, 28, 32, 33, 34, 43, 48, 53, 55, 58, 60, 62, 63, 66, 68, 69, 70, 71, 72, 73, 74, 75, 76, 77, 78, 82, 83, 95, 98, 106, 107, 109, 110, 111, 113, 119, 126, 130, 132, 137, 146, 149, 154, 155, 156, 163, 164, 166, 170, 171, 181, 186, 187, 188, 192, 193, 194, 196, 207, 208, 210, 211, 215, 216, 217, 218, 220, 222, 223, 228, 229, 230, 232, 233, 235, 236, 237, 238, 239, 240, 242, 253, 257, 258, 260, 261, 262, 263, 264, 265, 266, 267, 271, 272, 275, 276, 281, 282
resource allocation, 243
resources, 60, 147
responding, 58, 110, 157, 234, 246
responsibility, iv, 5, 15, 16, 18, 169, 174, 176, 180, 181, 205, 207, 213
restructuring, 55, 209, 210, 211
retail, 138, 147, 168, 181
rewards, 8, 9
Rice, J., 156, 164
ripcord, 7
risk(s), xiv, 3, 4, 5, 6, 7, 8, 9, 10, 11, 14, 18, 20, 24, 26, 45, 47, 96, 97, 100, 104, 105, 107, 108, 111, 112, 120, 121, 122, 123, 132, 154, 155, 157, 162, 164, 167, 169, 170, 171, 172, 173, 174, 175, 180, 181, 202, 205, 206, 208, 211, 216, 220, 222, 228, 229, 230, 231, 232, 236, 239, 240, 245, 249, 262, 263, 264, 273, 275, 276, 277, 280, 281, 282, 283
risk perception, 11
road layout, 104, 105
road safety, xi
Rothengatter, T., 10

S

safety, iv, v, xi, xiv, 3, 4, 6, 7, 8, 9, 10, 11, 20, 26, 32, 34, 48, 56, 58, 60, 92, 95, 97, 98, 102, 105, 106, 107, 108, 110, 111, 112, 120, 123, 127, 128, 129, 131, 133, 136, 137, 146, 147, 148, 153, 160, 164, 165, 166, 167, 169, 171, 172, 175, 176, 180, 181, 187, 193, 194, 195, 196, 197, 215, 216, 217, 218, 219, 220, 221, 222, 223, 224, 228, 229, 233, 234, 236, 240, 241, 242, 243, 249, 250, 257, 260, 261, 262, 263, 264, 265, 272, 273, 274, 277, 279, 281, 282, 283
Sagberg, F., 10
sales, 138, 142
sample, 49, 50, 57, 155
SARS, v
saving, 174, 178
science, 72
seatbelts, 7, 107, 108, 282
second language (ESL), 239
secondary, 214, 238
security, 175, 180
Segui-Gomez, M., 10
self, 13, 14, 19, 20, 25, 34, 40, 43, 44, 45, 48, 51, 54, 56, 57, 59, 82, 99, 107, 109, 122, 126, 131, 133, 156, 157, 161, 163, 165, 168, 169, 170, 175, 181, 186, 196, 200, 210, 212
self-esteem, 157, 161
sensitivity, 204, 211
serious injuries, xi, 269, 270
services, iv, 137, 149, 276
severity, 7, 25, 47, 48, 52, 56, 58, 106, 107, 112, 154, 155, 156, 158, 163, 205, 206, 211, 216, 217, 222, 251, 254, 255, 258, 260, 261, 262, 263, 264, 266, 267, 269, 270, 271, 272, 273, 274, 283
signage, 104
simulation, 232, 234
Singapore, 106
skills, vi, 14, 20, 54, 55, 111, 121, 125, 171, 209, 210
skydiving, 7
Slade, J., 11
social institutions, 169
social order, 107
social resources, 33
social stress, 69
social support, 208, 213

socio-economic status, 55
speeding, viii
sport, 7
St. Louis, 47, 165
stability, 162, 171, 262
staff, 110, 156
Standard deviation, 222
standards, 215, 263, 264, 271, 272, 273, 276
statistic(s), 17, 52, 53, 57, 102, 111, 138, 147, 149, 165, 263, 274
storms, 105
strategies, xii, 33, 35, 51, 54, 58, 164, 166, 168, 173, 180, 181, 185, 194, 195, 196, 209, 213, 215, 223, 224, 251, 283
study, xiii, 7, 8, 9, 10, 11, 13, 14, 15, 18, 19, 23, 28, 29, 32, 33, 34, 35, 39, 40, 43, 45, 46, 48, 54, 55, 56, 58, 60, 63, 69, 71, 72, 73, 75, 78, 81, 82, 83, 84, 86, 94, 95, 96, 98, 99, 100, 106, 107, 108, 110, 111, 120, 122, 136, 137, 138, 139, 146, 155, 163, 170, 178, 189, 193, 199, 202, 204, 207, 208, 209, 210, 212, 213, 214, 215, 216, 217, 218, 219, 220, 222, 223, 230, 231, 236, 240, 241, 246, 247, 249, 253, 256, 257, 258, 261, 264, 265, 266, 267, 269, 270, 272, 274, 275, 276, 280, 281
supervision, 109
surveillance, 249
survey, 55, 122, 138, 141, 147, 156, 166, 243
sustainable, 121
Sweden, xi, 6, 163, 216, 224, 236
Sydney, 34

T

targets, xv, 35
taxis, 102, 108
technology, xi, 63, 194, 196, 243, 246, 250, 259, 260, 261, 262, 273, 274
telephone, 26, 27, 28, 81, 100, 110, 111, 196, 227, 228, 229, 230, 231, 232, 233, 234, 235, 236, 237, 238, 239, 240, 241, 243
television, 43
terrorism, v
Texas, 199
therapy, 60, 164, 212, 213
threats, 179
threshold, 130, 132, 161, 162, 163, 264, 274
threshold level, 162, 163
time frame, 24, 32
trade, 97, 149
trading, 136, 146, 147, 148
traffic accidents, xv, 11, 37, 59, 99, 104, 105, 107, 112, 115, 121, 122, 123, 148, 162, 196, 213, 243
traffic density, 104, 105, 234
training, 7, 10, 32, 55, 121, 131, 168, 181, 232, 243
transit bus operators, 9
transition(s), 83, 142, 161, 216
translation, 109
transport, xii, 10, 11, 45, 60, 83, 99, 100, 101, 102, 103, 104, 106, 108, 110, 111, 112, 113, 117, 121, 122, 123, 139, 147, 167, 182, 224, 242, 245, 246, 249, 250, 251, 253, 264, 265, 266, 272, 274, 276, 281, 282
transportation, 168, 169, 171, 261, 276
Trimpop, R.M., 10

U

unemployment, 4, 162
United Kingdom (UK), xi, 20, 23, 34, 99, 100, 101, 102, 103, 104, 106, 108, 111, 113, 119, 120, 121, 122, 227, 283
United Nations, xi
United States (US), xi, 4, 7, 24, 26, 46, 71, 122, 135, 139, 166, 167, 169, 187, 199, 202, 211, 215, 227, 228, 257, 262, 264, 272, 275
universities, 276

V

Vaaje, T., 11
values, 90, 93, 94, 109, 111, 128, 129, 140, 168, 181, 217, 219, 221, 222, 252, 269, 271, 277, 279
variable(s), viii, 94, 117, 185
vehicle, iv, xi, xiii, xiv, 3, 10, 15, 16, 19, 26, 27, 28, 35, 42, 47, 49, 51, 56, 66, 67, 72, 73, 74, 75, 82, 83, 84, 85, 86, 87, 88, 89, 90, 91, 92, 93, 94, 95, 96, 98, 100, 102, 104, 105, 106, 107, 108, 110, 112, 113, 114, 121, 122, 125, 126, 127, 129, 130, 132, 137, 138, 139, 142, 143, 145, 146, 158, 165, 171, 172, 173, 174, 175, 189, 193, 196, 199, 212, 213, 214, 215, 216, 217, 219, 223, 224, 227, 228, 230, 231, 232, 233, 234, 235, 236, 239, 240, 242, 243, 245, 246, 249, 250, 251, 252, 253, 254, 255, 256, 257, 258, 259, 260, 261, 262, 263, 264, 265, 266, 267, 271, 272, 273, 274, 276, 277, 278, 279, 280, 281, 282, 283
violence, xiii, xv, 16, 25, 27, 34, 35, 43, 69, 70, 76, 78, 147, 171, 178, 179, 180
vision, xi
vulnerability, 34, 106

W

Wales, 81, 136, 148, 231, 243
Warner, K.E., 11

wealth, xiii, 260
Western countries, 108
white, 63, 67, 70, 186, 196
wholesale, 136
Wilde, G.J.S., 10
women, 42, 62, 72, 76, 77, 78, 122, 195, 236, 246, 263, 281
work, v, vi, viii, xiii, 10, 39, 41, 54, 61, 62, 64, 65, 69, 72, 96, 99, 105, 111, 122, 132, 153, 172, 176, 208, 215, 216, 217, 218, 219, 220, 221, 222, 223, 224, 241, 252, 276
workers, 223